Corporate Strategy in Construction

Understanding today's theory and practice

Steven McCabe

Birmingham City Business School
Birmingham City University
Perry Barr
Birmingham

WILEY-BLACKWELL

A John Wiley & Sons, Ltd., Publication

This edition first published 2010
© 2010 Steven McCabe

Wiley-Blackwell is an imprint of John Wiley & Sons, formed by the merger of Wiley's global Scientific, Technical and Medical business with Blackwell Publishing.

Registered office
John Wiley & Sons Ltd, The Atrium, Southern Gate, Chichester, West Sussex, PO19 8SQ, United Kingdom

Editorial offices
9600 Garsington Road, Oxford, OX4 2DQ, United Kingdom
350 Main Street, Malden, MA 02148-5020, USA

For details of our global editorial offices, for customer services and for information about how to apply for permission to reuse the copyright material in this book please see our website at www.wiley.com/wiley-blackwell.

The right of the author to be identified as the author of this work has been asserted in accordance with the UK Copyright, Designs and Patents Act 1988.

Library of Congress Cataloging-in-Publication Data
McCabe, Steven.
 Corporate strategy in construction : understanding today's theory and practice / Steven McCabe. – 1st ed.
  p. cm.
 Includes bibliographical references and index.
 ISBN 978-1-4051-5912-8 (pbk. : alk. paper)
 1. Construction industry–Management. 2. Strategic planning. I. Title.
 HD9715.A2.M3474 2010
 624.068′4–dc22
                                        2009029834

A catalogue record for this book is available from the British Library.

Set in 9.5 on 10.5 pt Times by Toppan Best-set Premedia Limited
Printed and bound in Malaysia by Vivar Printing Sdn Bhd

1 2010

Contents

Foreword

The phrase 'May you live in interesting times' is often thought to be Chinese, but the reality is that no-one seems to know its origin. Nonetheless, it is a useful phrase to use at this time and, given the uncertainty and doubt that surrounds the origins of the current global financial crisis, doubly apposite.

A long period of certainty has come to and end, a period when it seemed that we understood what drove financial markets and where the risks of global collapse of the magnitude seen before in history could be avoided – the emphasis was on exploring ever newer ways of exploiting markets and, in the process, many hard lessons that had been learnt over the years were forgotten or ignored. As we discovered only too harshly, this turned out to be an expensive mistake. Businesses have been lost, livelihoods blighted, and the future filled with fear and uncertainty.

We are now entering a new era, when these old certainties have been shown to be illusory, and we need to remind ourselves that the lessons of the past, built up over a long period of time, as a result of much thinking and analysis, still hold true.

This book will, I am sure, make a significant contribution to helping organisations work out the best way forward, by providing a structured way of developing their corporate strategy. Steven McCabe rightly suggests that this process is like an organisation setting out on a journey, and I do hope that many organisations use the insights contained in this book to help on this journey. The fortunes and hopes of everyone involved depend on this, and I congratulate Steven McCabe on his vision and endeavour in producing this book.

Stephen Brown
Head of Research, RICS

Preface

'Collapse in confidence in banks' Title of webpage by BBC Business Correspondent Robert Peston on Monday 19th January 2009

When this book was originally commissioned in late 2006 the world had a different feel about it. There was a sense of confidence in the financial system upon which all major economies tend to be based. In particular, those who engaged in property development were enjoying gains that were in most cases impressive and, it must be stated, in some instances spectacular. Property development cannot take place without the sustenance of finance. Investment banks and financiers were acknowledged to be the key to success. They lent to the developers who used the real estate they procured to build whatever would make the highest returns; residential and commercial being particularly popular. As some believed, those in the financial system who created new and innovative methods of raising money, and passing on risk through what became known as 'derivatives' were 'Masters of the Universe'.

Over two years later these 'masters' are now presiding over a system in crisis. As Robert Peston's comments makes clear, banks no longer enjoy the confidence or status they once had. Equally, property, once seen as a 'rock-solid' investment no longer gleams so brightly. Quite the contrary, property throughout the world has declined very rapidly and in some American cities where houses were sold to the so called "NINJAs" of the sub-prime market (those with No Income, No Job and no Assets), is now pretty much worthless. As newscasts stress each day the speed at which the financial system has collapsed – with its attendant consequences for jobs and future investment – is unprecedented. This is not strictly true. Those who have knowledge of Japan point to its so called 'lost decade' in the 1990s when it experienced conditions very similar to those that currently afflict all economies; most particularly America, the UK and the Eurozone. Some commentators ask, 'How did we allow this to happen, and why did no one warn us?'

What is important, of course, is that it tells us much about the apparent collective willingness to believe that markets can only go up. More crucially, what does it tell us about the ability of those who take key decisions about the future; those whose task includes the need to consider strategic options for their organizations? The havoc that the current financial conditions are wreaking on all major economies suggests that those who were "masters of the universe" a couple of years ago are now seen as, at best, having 'feet of clay', at worst, being villains who plundered the market. Strategies that were once successful for property developers and financiers are now seen as being flawed. However, like all previous crises that have afflicted humanity from time-to-time, civilization will continue. Sadly, some organisations may cease to exist; at the time of writing many retailers and automotive producers are being severely strained by lack of sales. However, a great many will continue. Indeed, those that ensure society functions effectively, such as building and maintaining the infrastructure, and are therefore considered crucial to our daily well-being, will not be allowed to disappear. As Chapter One will explain, construction covers a wide gamut of activities and certainly includes the building and maintenance of roads, hospitals, schools, and a great many other administrative functions that we depend upon. All those engaged in providing the skill and expertise to carry out construction, large contractors, jobbing builders, small subcontractors, suppliers are vital. When the current economic crisis abates, it is important that there is a sufficient skill base to ensure that growth can be achieved.

This book describes the reasons why every construction organisation must have a strategy. Similar to taking a trip to a new area, an organization without a strategy will be likely to get lost along the way (assuming it even knows what it wants to achieve). Therefore, having some idea of what the 'journey' entails and making adequate preparations to get to the 'intended destination' will always be a worthwhile exercise. As explanation of strategic theory describes, even though the eventual destination may not be the one that was originally intended, the ability to cope with change and respond appropriately will ensure survival. Indeed, given the current problems that face many organisations, survival may be the only short-term strategy. Long-term development and growth may have to wait.

Crucially, therefore, understanding the basis of strategic theory is important to practitioners and students alike. Appreciating the basis of concepts associated with strategy and how they may be applied is essential. This book will assist you in understanding the context in which construction takes place. It provides you with a detailed description of how the major concepts may apply in order to pursue strategic objectives. A central theme of this book is the importance of people in creating successful organisations. So called 'World-Class' organizations have consistently demonstrated the causal link between highly dedicated employees – through training, education and continual encouragement – and achievement of strategic goals in terms of productivity, sales growth, customer satisfaction and return on investment. The case study material contained in Chapter Twelve reinforces this message.

About the author

Dr Steven McCabe is Director of Research Degree Programmes in the Business School at Birmingham City University and lectures in aspects of general management and strategic and human resource management on postgraduate courses across the faculty. He also teaches management to postgraduate students in the school of Property Construction and Planning in the faculty of Technology, Engineering and the Environment at Birmingham City University. Prior to his academic career, Steven was employed by organisations working in both the private and public sector of construction. His research focuses on how the effort of people is a vital part of the desire to achieve organisational excellence through improvement of processes in construction. As a result, Steven has published widely on quality management, the implementation of TQM, benchmarking, process improvement and operational techniques based upon 'lean' principles. He has also carried out research into the socio-historical development of class relations in British construction.

Acknowledgements

There are a great many people who have assisted in the preparation and writing of this book. Without their continued support and encouragement it would not have been possible. Firstly, I would like to thank Madeleine Metcalfe (Senior Commissioning Editor at Wiley-Blackwell) who saw the potential for a new book on strategy. I would also like to thank Lucy Alexander and Tristan Lee (editorial assistants at Wiley-Blackwell) for their continual advice and guidance which has been most appreciated. I'd like to compliment Les Glazier's diligent editing of this text which has improved the end product. Any book that describes strategy but does not contain empirical descriptions of what actually happens will always be viewed as less than complete. Therefore I wish to particularly extend my gratitude to the contributors who describe their experiences of strategy in Chapter Twelve.

There are others without whose continuous support and encouragement, particularly during the bad days, would mean this book could never have come to fruition. I want to acknowledge the inspiration provided by those academics whose work has demonstrated the importance of people in construction. Most especially, I wish to thank my wife Gráine and dear sons James and John for their support and perseverance; without whom I could not achieve.

Dedication

As those who read my contribution to *People and Culture in Construction – A reader* edited by Dainty et al. (2007) will know, I dedicated the chapter that described the development of industrial relations in the industry to my father James who died in 2004. Following on from that it is important to recognise the efforts of all construction workers whose dedication enables the industry to make almost anything possible. This book is in memory to all who have endured pain and suffering to as a consequence of their employment in construction. Most especially I wish to particularly remember construction trade unionist Desmond "Dessie" Warren who died in 2004 aged 66 after suffering Parkinson's disease which many argue was caused by the "medication" he received whilst imprisoned for his involvement in the national building strike of 1972. I sincerely hope that his vehement argument against exploitative practices in construction employers will continue to be heeded, particularly during difficult economic times when there will undoubtedly be a temptation to revert to the 'bad old ways'. As those familiar with Dessie's case will know, his stance caused him to pay an extremely high price.

Chapter 1
Introduction

'It's tough making predictions especially about the future', ex-baseball player and sometime pundit, Lawrence Peter 'Yogi' Berra

1.1 Trying to predict the future – a task fraught with risk

The quote above is amusing but when judged on another level it makes perfect sense! We all know what it feels like to try and make plans only for something unexpected to crop up. Sometimes whatever it is that occurs is trivial and causes minor inconvenience. Sometimes the magnitude of what happens is so great as to cause everything we planned to become either impossible or utterly pointless. Whilst we may be affected on a personal level, it may be that we are inconvenienced on a collective level. As this book explains, strategic management is about the organisation and, as a consequence, about all of those who are involved (employed or giving their services voluntarily): those who may provide products or service and the consumers. The real dilemma for those who have responsibility for the organisation (the custodians) is in making the right decision on the basis of the information they have available. There are, it seems, no magical ingredients[1].

The politician, Harold Macmillan, was once asked what it was that caused those in power to be unable to deliver promises made in election manifestos. His reply, after a somewhat long dinner, was pithy but instructive: 'Events, dear boy, events'. This quotation is frequently used by commentators to summarise the dilemma that all of us face in making plans for the future. If we make assumptions about what we think may happen, we should acknowledge the potential for things to change. At the time of writing this particular part of the book there have been two anniversaries, the original events of which have had an immense impact on the world through both politics and the economic system.

In the first event, the attacks of September 11th 2001 were immediately obvious in terms of their significance and are rightly perceived as a defining moment of history. Like a generation before who had witnessed the assassination of President John F. Kennedy, you may be asked where you were or what you were doing, when you first heard news of the awful events at the World Trade Centre. Beyond the shock that this event caused in terms of realignment of relationships and the beginning of the so-called 'War on terror', we have witnessed an increase in security (especially at airports) and the impact of uncertainties in economics which affects us all. For a short period, many seriously wondered whether people would be prepared to work in high-rise buildings again. Those involved in the design and construction of such beings were certainly

---

[1] The French general, Napoleon, was reputed to judge the promotion of officers by considering, amongst other things, their luck.

required to explain why those trapped in the burning towers could not escape more easily and how any deficiencies might be avoided in the future. Sadly, it seems, the ingenuity (and, it seems, evil intent) of terrorists such as those who perpetrated this attack makes the construction of absolutely safe buildings almost impossible.

The second event is one that may not have had the immediacy and visual impact of the September 11th attacks, but its significance is being felt by all who have a vested interest in the continued well-being of the economic system. This event, the beginning of the so-called 'credit crunch' in America in late summer 2007, has created an apparent 'tidal effect'. As well as making everyone much more aware of the intricacies and complexity of the banking and financial system, including the peculiar and exotic lexicon employed by practitioners, it has caused panic and fear by investors. Most especially, a financial system that almost everyone believed was perfectly safe (including well-known banks) has been shown to be reliant on decisions and strategies that were based on extremely risky assumptions (at best) and downright greed and hubris by individuals who believed that their actions, as well as making them wealthy, were unquestionable.

The effect of the credit crunch has been to undermine the state of property worldwide (which has traditionally been a very good long-term investment) and simultaneously increase the cost of borrowing and reduce the amount of credit available. House prices in America have rapidly declined (now being experienced elsewhere, notably Dubia) and the confidence and the belief that increasing value in real estate gave to everyone has disappeared. People are finding that their ability to spend on the basis of credit has been stymied. Reduced income because of the effects of the credit crunch either directly or indirectly means that people consume less. Governments are finding that their spending plans will be negatively affected. Firstly, there are likely to be by reduced tax receipts if people spend less and, potentially, if unemployment increases. The latter means that the financial burden required to pay benefits increases. Added to that, there have been examples in the UK and America[2] of the need to provide vast amounts of financial assistance to failing financial institutions. This is because the risk of allowing what is commonly referred to as the 'toxicity' of so-called 'sub-prime' lending could potentially create a 'contagion' that would mean that the entire system that it facilitates by homeowning would collapse. Because this is a vista far too awful to contemplate for governments, they are willing to provide such assistance – regardless of any proclaimed beliefs in the market.

The consequences of the events that have stemmed from the credit crunch have gone beyond the effects of our capability to consume or repay the mortgage or on any government's ability to manage the economy. Those organisations that can claim to be unaffected are extremely lucky; albeit there are always those who gain in any crisis. Those firms in industries that were reliant on never-ending consumption are now experiencing extremely tough trading conditions. More especially, any that assumed property values were unidirectional (upwards) have probably found that long-term plans are now erroneous. For an industry like construction the effects of the credit crunch have been particularly severe on some organisations and firms. Managers taking strategic decisions in property investment and development or speculative housebuilding will be likely to have found to their cost, like Harold Macmillan, that events really can cause unanticipated problems. When one consults the list of activities contained in the Standard Industrial Classification (see below), it is not hard to see why any downturn in the economy has an immediate effect through reduction in spending on construction (in terms of labour, materials, plant and associated expertise and professional services)[3].

As the next section begins to consider, what can strategy really do to assist people? Strategic theory, if it is assumed to be about certainty about the future, is bound to fail. Those whose task

---

[2]Freddie Mac and Fannie Mae in America and Northern Rock in the UK.
[3]The terms 'accelerator' and 'multiplier' are sometimes used. These describe the way in which money can be invested in particular parts of the economy to create activity (which will often require construction) and therefore accelerate economic growth. Such growth (through employment and production of materials) in turn leads to further demand which ensures that the effect of the original investment is increased (multiplied).

is to make strategic decisions are often paid large amounts of money to make confident predictions on which a great deal rests – in terms of investment in resources and productive capability. As some may ask, can strategic management provide tools and techniques that demonstrate the logic of decisions taken. If it cannot, what does it provide?

1.2   A journey towards strategy: art or science?

Strategic management has many definitions which causes something of a dilemma for those who suggest that there is one best way. However, some commentators recognise that 'managing strategy' requires a combination of both the 'hard-headed' skills of carrying out calculations as to likely outputs from certain inputs, as well as intuitive skills which rely on things like hunch and 'good judgement'. So, for example, one definition I came across advised the need for a combination of art and science. Putting these together, it was explained, would allow those involved in strategic management to be involved in the formulation, implementation and evaluation of cross-functional decisions which would allow an organisation to achieve its objectives.

Wickham makes the point that strategy is always going to be difficult because, as he states, 'organisations are complex … and made up of people with minds, wills and interests of their own' (2000, p.1). Therefore, managers who need to make strategic decisions must appreciate that there will be effects which impact on all those people involved in the processes which must be implemented in pursuance of whatever objectives have been decided. As Wickham believes, if managers have only limited ability to control the 'human aspect' within their organisation, they can do even less about the external environmental aspect. The environment is, he contends, 'constantly shifting' and will provide the influences that create opportunities and threats that may potentially enable the organisation to, ideally, prosper or, if the wrong decisions are made, to decline. As history has demonstrated all too frequently, success today provides no guarantee of survival tomorrow.

The desire of those who provide definitions to include achievement of objectives is understandable. The emphasis of strategy is on making sure that the organisation (see below for a consideration of what this word means) is collectively working towards whatever preconceived ends have been decided. However, in setting organisational objectives we must be aware that there are many competing forces that must be reconciled before potential objectives can be agreed. As will be explained, organisations usually have both a technical and a social side. There may be an assumption that because the former, i.e. the technology and equipment or plant, can be programmed to perform, then why should not the latter also be considered in a similar way?

However, if you consider how the world works on a day-to-day basis, we know that sometimes things don't always work out as we expected. Chance and, frequently, events intervene to change the way that the intended outcomes are produced. Sometimes, such serendipity allows consequences which are, despite not being intended, beneficial. Sometimes, of course, the consequences are not beneficial and cause alteration to plans which, in turn, are disruptive and potentially costly. Whatever happens, as sentient beings, we acknowledge that this is not a bad thing. If life were as straightforward and as linear as some suggest it should be, it would be a lot more dull and predictable than it often is. This is the same for organisations. If organisational strategy were simple and success guaranteed, then it would make no sense to do other than follow the 'standard formula'. Why would you deviate from it?

On the basis of what has already been stated, strategic objectives are problematic in that there are no guarantees in terms of attainment. But this does not mean that they should not be set at all. At the very least, they should act as beacons which will, like guidance for a ship, provide a sense of direction. Whilst the captain of a ship will, if at all possible, attempt to go in a straight line, they will alter their course to take account of potential dangers or any obstructions. The objective is to reach the final destination safely. It is perhaps worth acknowledging that for many managers in organisations the difficulty is in selecting the 'right' beacon. And continuing the analogy of the ship, sometimes a radical alteration in course is required. As will be explained, in dynamic and turbulent markets, the 'art' of organisational strategy is being able to have sufficient resources that are capable of responding to events and influences that no one had even envisaged.

1.3   Strategy, a problem of expectation?

In his seminal book, *What is strategy and does it matter?*, published in 1993, Richard Whittington acknowledges that there is no shortage of publications with the words 'strategy' and 'management' in their title. As he explains, they tend be 'filled with charts, lists and nostrums, promising the reader the fundamentals of corporate strategy[4]' (1993, p.1). There is a problem with such texts, Whittington suggests, because, given that they tend to present similar 'matrices, [and] the same authorities', how can they live up to the promise of giving the answer to how to 'do' strategy? Given, at the time that Whittington wrote his book, that such books tended to sell at approximately £25, he is led to conclude:

> There is a basic implausibility about these books. If the secrets of corporate strategy could be acquired for £25, then we would not pay our top managers so much

Such sentiment is hard not to agree with. Undoubtedly, every author believes (or would like us to believe) that they have discovered the secret of strategy. Accordingly, to be able to demystify such secrets is no mean feat and would prove popular among those managers whose task includes strategic management. Equally, the academic community would be equally keen to know what the magical ingredients of successful strategy are! The publication of *In Search of Excellence* by Tom Peters and Robert Waterman in 1982 is a case in point of what happens to those who believe that it is possible to derive all the answers from one text. Peters and Waterman argued that, on the basis of research carried out into sixty-five so-called 'excellent companies', they had identified eight attributes which were crucial to their success. Word spread among the management community, especially at senior level, that all that was required for success was to read this book and ensure that you had implemented the eight attributes. A significant footnote to this book is that a number of these excellent companies went bankrupt. Clearly, it seems, Peters and Waterman, even though their book has become a legendary best-seller (some six million copies), had not 'cracked strategy'.

The author's own experience is that the most enjoyable books have a definite appreciation of what the potential readership believe they will want and, as a consequence, aim to provide material that matches perceived expectation(s). Academic books are no different. They attempt to distil informative and useful theory that has proven application in the empirical world. Students who read such texts are expected to be able to articulate such theories (certainly in work that is submitted for assessment) and, it is hoped, in the workplaces that they attain employment in. Strategic management textbooks, including this one, should not depart from this principle and, therefore, the attempt here is not to pretend that there are magical formulas which will provide guaranteed success.

A logical conclusion is that there are no absolutes as far as achieving corporate success is concerned. In so far as the practice of implementing strategy is concerned, much would seem to

---

[4]Corporate strategy is an expression that implies business-oriented organisations.

depend on who takes the key decisions in any organisation. These people are usually, though not exclusively, senior managers[5]. An interesting observation made to me many years ago was that senior managers are frequently guided by nothing other than their instincts and that such instincts are rarely (if ever) based on reading of texts intended to present strategic theory.

Completing the writing of this book in 2009 it seems that the world in which organisations – and, more especially, managers – have to operate is less certain now than in any time in history. At first glance this contention may seem somewhat facile. However, there is little doubt that the difficulty of making key decisions concerning resources which an organisation should deploy in pursuit of certain goals is harder now than ever. For those of us who live in the developed world, we normally have a surfeit of choice in everything we consume. Even though we may not have enough money to buy all that we want (or are persuaded that we must have), there is usually constant competition for our custom. For those organisations that can give us the best bargains or value, provide us with the widest choice or, increasingly, it seems, offer exactly what we want (customise), there is the hope of gaining what is known as 'competitive advantage' (see below).

The word 'organisation', according to Sims *et al.* is a 'notoriously difficult concept to define precisely' (1993, p.277). However, they make the point that an organisation is more than the technology and systems that are used to produce, distribute and deliver products and services to potential customers (this is a word that will be analysed subsequently). Most especially, organisations must involve people for the purposes of setting some sort of direction. Those who make such decisions, at whatever level in the organisation, will expect to be judged on the outcomes. Accordingly, those who make decisions that result in success can expect plaudits and, usually, financial reward. For those whose decisions result in failure, whilst the consequence is not always guaranteed to end in unemployment, they should expect others to question their judgement and ability in the future. Managers in construction organisations are treated no differently and, as analysis of the historical development of the industry suggests, can expect little sympathy if they make 'poor' strategic decisions.

1.4   The dilemmas of a formal definition of strategy

Such dilemmas inevitably require more precise consideration of what it is that managers are actually doing when they involve themselves in strategy. Like many management concepts, whilst there is broad agreement about the generalities of what strategy is, many definitions have particular nuances that suggest personal differences in interpretation. Many definitions have their provenance in what is often referred to as the 'classical' view of strategy (see below), which assumes that decisions can be made on the basis of rationality and that desired outcomes can be attained as a result. The word strategy has its origins in Greece at around 500 BC when it was used in a military context as a means to identify best how to organise soldiers. At a similar time in China, the need to understand the political and economic context in which battles take place was identified as a key to success by Sun Tzu in his book, *The Art of War.*

As some conclude, the idea that running a business is similar to war – in that there are two opposing forces willing to engage in conflict (and that there can be only one victor) – is still viewed by some as the basis of 'effective' strategy. Accordingly, the definition provided in the online edition of the Oxford English Dictionary draws inspiration from the historical military applications:

---

[5]Whittington believes that the faith that is placed in senior managers to make the 'right' strategic choices is the reason they are paid so much.

> In (theoretical) circumstances of competition or conflict, as in the theory
> of games, decision theory, business administration, etc., a plan for success-
> ful action based on the rationality and interdependence of the moves of the
> opposing participants (Oxford English Dictionary Online, 2008)

Ironically the end of the Second World War and the development of more formal approaches
to business meant that the emphasis of strategy moved from how to win battles and wars to the
more widely accepted view that it is about how to organise resources to effectively compete in
business[6]. The 1960s saw the publication of a number of seminal books (Chandler, 1962; Sloan,
1963; Ansoff, 1965a) which helped to consolidate the belief that managers, especially those at
executive level, were expected to dedicate themselves to the pursuit of efficiency and effectiveness. Alfred D. Chandler, an academic who researched and wrote about strategy was among
many influential commentators who used General Motors (GM). His definition was that it is 'the
determination of the basic, long-term goals and objectives of an enterprise' (1962, p.13). Alfred
Sloan, who was a former President, published his biography, *My Years with General Motors*, in
which he was absolutely specific about what strategy is:

> ... the strategic aim of a business is to earn a return on capital, and if in
> any particular case the return is not satisfactory, the deficiency should be
> corrected or the activity abandoned (1963, p.49)

Given that Sloan was writing about his time as President of General Motors which, at that time,
was a successful car-making corporation employing over half-a-million people, his views were
hugely influential. He, like Chandler, made clear his belief that strategic management required
careful consideration of the externalities and opportunities of the marketplace. Successful strategy
was based on exploiting those opportunities which offered the greatest potential market.
This task, it was stressed, was not easy and, significantly, because it required absolute dedication
by those managers it concerned, meant that they should not be encumbered by what was seen
to be the day-to-day (and therefore relatively straightforward) issues of operational management.
GM's success, Chandler contended, had been due to the ability of executives to be 'removed
[...] from the more routine operational activities,' and, moreover, this gave them 'the
time, information, and even psychological commitment for long-term planning and appraisal'
(1962, p.309).

Four perspectives for considering strategy[7]   The belief that strategy can be managed by senior
managers and based on the rigid application of top-down, rational decisions has become dominant. As advocates of this view assert, it is the best (only) way to achieve success. So, once senior
managers have taken key decisions concerning strategy, all that remains is for other managers to
implement them. Because of the dominance that this perspective has had since being proposed
by Sloan, it is known as classical. As Whittington explains:

> Flattered by the image of Olympian detachment, lured by the promise of
> technique-driven success, managers are seduced into the Classical fold
> (1993, p.17)

The classical approach is one that is based on a desire to maximise **outcomes** (whether
measured in profit or otherwise) by the deliberate application of predetermined **processes**. Given

---

[6]The word 'business', it is accepted, is usually considered to be profit-centric.
[7]These are more fully explained in Chapter 2.

the ontological basis on which the classical perspective is based, i.e. rationalism, many have argued that other perspectives are required to account for the imperfections of the empirical world in which all organisations operate (implement strategy). As in general studies of management, there is an increasing awareness that because strategy is implemented by, and has effects upon, social beings (people) the assumption of rationality is suspect. Therefore, three other perspectives have been proffered as alternatives: evolutionary, processual and systemic. Like the classical approach, each may be considered by the two key dimensions of outcomes and processes.

As the word implies, the **evolutionary** perspective is based on a belief that only organisations that are able to quickly adapt to the vagaries of the market will survive and prosper (much like Darwin's belief in the evolution of species). Whilst they will still seek to maximise outcomes, their use of processes will not be predetermined. Rather, processes will be developed during the course of practice; they will 'emerge'.

The **processual** perspective is one that acknowledges markets, people and organisations as being far from the perfect entities that proponents of the classical perspective believe is the case. Therefore, the processual perspective accepts that strategy is a matter of accommodation of the multiplicity of outcomes needed to satisfy both those involved and the opportunities that present themselves in the markets. Accordingly, outcomes are 'pluralistic' and processes, similar to the evolutionary perspective, are 'emergent'.

Finally, the **systemic** perspective is one that recognises that culture and powers are influenced by the extant characteristics of the social systems in which the organisation operates and the relevant 'markets' it serves. In terms of outcomes it is similar to the processual perspective in that they are pluralistic. However, in terms of processes, the systemic perspective fully embraces the belief that a deliberate approach is appropriate.

Following on from this brief analysis, it can be discerned that there are a multiplicity of approaches to understanding what strategy involves. In subsequent chapters further consideration will be made of the range and importance of definitions. However, as the title of this book makes clear, the context in which analysis of strategy takes place is construction. Therefore, it is useful to consider the essential characteristics of construction and whether they provide a particular (or peculiar) climate in which strategic management must take place.

1.5   The context of construction – a truly unique industry?

One of the challenges in teaching students enrolled on construction courses is to explain how strategy can be understood in terms of an industry that, to a very large degree, is reliant on what happens in the national economy and which, in turn, is affected by global events. Accordingly, if the state of the economy is good, then the sense of well-being and optimism will result in a higher demand for goods, services and the general infrastructure – all of which will stimulate demand in construction. As some might argue, traditionally, strategy, as far as construction organisations are concerned, is simply a matter of being able to 'read' the ongoing trends in the economy. To do this successfully requires either unique insights (something that can be notoriously difficult), or to be 'fleet of foot' in being able to shift resources or, as is often the case, relying on the subcontracting system to fulfil short-term needs.

Many commentators, most notably those who wrote the 1998 report, *Rethinking Construction* (Construction Task Force), assert that the industry's apparent unwillingness to invest time and money into long-term development and improvement of processes used in production has been one of the reasons why it has failed to match the expectations that some suggest are possible. Their use of other industries with which to make comparison is intended to present a stark contrast between those that are dynamic and highly innovative and construction which they consider to be, in general terms, patently not. Such analysis is too simplistic and, as this book will explain, consideration of strategy is vital in every organisation regardless of its size.

The fact is that every member of society relies on the output of construction. The industry is responsible for creating what is referred to as 'the built environment'. That is: everything we see around us that did not occur naturally. Construction, therefore, has a connection with every aspect of our daily lives. It includes every building that we use for residential purposes, education, industry and leisure. Construction was involved in creating the traffic infrastructure for all roads, motorways, railways (and their precursor, canals), airports, harbours and their attendant structures. It is fundamental in terms of the infrastructure that creates the sense of well-being we enjoy as a civilised society: clean water, sewage disposal, and the multitude of utility services that allow us to live comfortably all year round regardless of external temperature or environmental conditions (gas, electricity and telecommunications).

However, this does not tell us exactly what construction consists of. In order to fully appreciate the extent of what construction actually conists of and its impact on creating the built environment we all rely on and 'enjoy', it is necessary to consult the Standard Industrial Classification of Economic Activities (which is published by the government through the Office for National Statistics). This provides a fully comprehensive categorisation (under Section f) of every activity that is considered to be construction. Indeed, it is deemed to be:

> … general construction and specialised construction activities for buildings and civil engineering works. It includes new work, repair, additions and alterations, the erection of prefabricated buildings or structures on the site and also construction of a temporary nature. General construction is the construction of entire dwellings, office buildings, stores and other public and utility buildings, farm buildings etc., or the construction of civil engineering works such as motorways, streets, bridges, tunnels, railways, airfields, harbours and other water projects, irrigation systems, sewerage systems, industrial facilities, pipelines and electric lines, sports facilities etc. This work can be carried out on own account or on a fee or contract basis. Portions of the work and sometimes even the whole practical work can be subcontracted out. A unit that carries the overall responsibility for a construction project is classified here. Also included is the repair of buildings and civil engineering works. This section includes the complete construction of buildings (division 41), the complete construction of civil engineering works (division 42), as well as specialised construction activities, if carried out only as a part of the construction process (division 43). The renting of construction equipment with operator is classified with the specific construction activity carried out with this equipment. This section also includes the development of building projects for buildings or civil engineering works by bringing together financial, technical and physical means to realise the construction projects for later sale. If these activities are carried out not for later sale of the construction projects, but for their operation (e.g. renting of space in these buildings, manufacturing activities in these plants), the unit would not be classified here, but according to its operational activity, i.e. real estate, manufacturing etc. (Office for National Statistics, 2007, p.173)

As this definition describes, there are three main types of construction activity (referred to as divisions): 'Construction of buildings' (41), 'Civil engineering' (42) and 'Specialised construction activities' (43). Each division is further subdivided into groups and classes. As the following list demonstrates, it is within these groups and classes that the true complexity and wide variety of tasks can be fully appreciated:

41 Construction of buildings
41.1 Development of building projects
41.10 Development of building projects
41.2 Construction of residential and non-residential buildings
41.20 Construction of residential and non-residential buildings
41.20/1 Construction of commercial buildings
41.20/2 Construction of domestic buildings
42 Civil engineering
42.1 Construction of roads and railways
42.11 Construction of roads and motorways
42.12 Construction of railways and underground railways
42.13 Construction of bridges and tunnels
42.2 Construction of utility projects
42.21 Construction of utility projects for fluids
42.22 Construction of utility projects for electricity and telecommunications
42.9 Construction of other civil engineering projects
42.91 Construction of water projects
42.99 Construction of other civil engineering projects
43 Specialised construction activities
43.1 Demolition and site preparation
43.11 Demolition
43.12 Site preparation
43.13 Test drilling and boring
43.2 Electrical, plumbing and other construction installation activities
43.21 Electrical installation
43.22 Plumbing, heat and air-conditioning installation
43.29 Other construction installation
43.3 Building completion and finishing
43.31 Plastering
43.32 Joinery installation
43.33 Floor and wall covering
43.34 Painting and glazing
43.34/1 Painting
43.34/2 Glazing
43.39 Other building completion and finishing
43.9 Other specialised construction activities
43.91 Roofing activities
43.99 Other specialised construction activities
43.99/1 Scaffold erection
43.99/9 Specialised construction activities (other than scaffold erection)

Every group and its classes are fully explained in the guidance document that is provided by the Office for National Statistics. As will become obvious, from the list above and consultation of the associated guidance notes, construction is crucial to every aspect of our existence. Literally nothing we do on a daily basis will not involve some contact with construction. Moreover, as the standard industrial classification makes clear, construction, as well as being a set of activities that ensures a fully functioning society can be maintained, has an economic importance that cannot be overestimated. It is a major employer. Its contribution to the economy is crucial. The most up-to-date output figures produced by the Office for National Statistics show that the value of construction is in excess of £80 billion per year (some 6% of Gross Domestic Product) and that for the second quarter of 2008 the contribution of various segments of construction has amounted to a total provisional output of over £21 billion:

- Housing new work: £3,027 million;
- Non-housing new work: £8,358 million;
- Housing repair and maintenance: £4,972 million;
- Non-housing repair and maintenance: £4,655 million.

Current statistics show that construction employs over 2.2 million people (the vast majority[8] being male). Trying to gain an idea of how many firms there are in construction is notoriously difficult because of the very high number of organisations employing very few people. This, together with the transient nature of the industry means that accurate figures are hard to come by. By way of providing some indication of the number of 'enterprises' that operate in construction, Sommerville and McCarney presented a study in 2003 that suggested that whilst there were almost 700,000, some 571,455 employed no one at all (consisting of the self-employed) and that of the remainder that do have employees (120,345), 85,445 employ four or fewer and 113,755 employ fewer than 20. As will become important in terms of considering how strategy is practically applied, this study showed that there were only 855 that employed 100 or more people; and only 140 employing 500 or more. Construction, therefore, is an industrial sector that is dominated by micro enterprises.

The difficulty is in considering how construction really operates and how particular influences exist, or are managed (manipulated) to create the conditions in which 'activity' is required. As Hillebrandt (1984) argues, it is an industry which has characteristics that make it different and, most especially, the desire of the person(s) who wish to procure products or services is frequently determined by the prevailing economic conditions. She believes that direct comparison with other sectors, although less straightforward than some suggest, is possible:

> The construction process is long and involves a large number of separate organisations in design, costing, pricing and production of each product often with a client, private or corporate, who has never or rarely before been involved in a major construction operation and for whom the expenditure is among the largest [they have] ever made. Not only is the new building or works normally designed by a separate organisation from that which will construct it, but since it must usually be priced before it is produced, there are great uncertainties for the contractor in that pricing process. Furthermore, each product tends to be large in relation to the size of the contractor undertaking its construction so that the risks for that firm are considerable. The physical process of production is often a messy one carried out substantially in the open and subjected to the vagaries of the weather. Conditions of work are generally inferior to those in factories. (1984, pp.2–3)

Therefore, the simple assumption that applying strategic principles in construction that have worked elsewhere would seem to have problems. Nonetheless, as many acknowledge, the characteristics of construction have caused the industry to be viewed by outsiders in a highly negative way. Almost a quarter of a century since Hillebrandt's observation (and over ten years since the recommendations contained in *Rethinking Construction*), it is worth remembering that many organisations have striven to ensure that the industry is less prone to the sort of conflict and labour problems that used to be commonplace in the 1970s and 1980s. As well as this, no matter what the general public thinks of what the construction industry achieves (and the way it does it), it

---

[8]Whilst there is some minor variation from quarter to quarter, male employment in construction is usually some two million representing over 90%.

is vital to us. Its output is the consequence of deliberate implementation of processes that, usually, frequently involve many individuals, organisations and agencies, all of which will have particular strategies.

The history of construction adequately demonstrates that those individuals (or groups), who had the foresight to consider the future demand of clients, and who assembled the right mix of resources involving equipment and plant, were able to achieve remarkable success in terms of completed projects. Anyone who has marvelled at the ingenuity of the work of the great engineers such as Telford and Brunel should realise that their vision required a supply of men (as workers in construction were[9]), equipment and materials. Compared to contemporary construction sites, the work that was carried out in, for example, the building of the canals, railways and the huge Victorian municipal schemes would probably seem chaotic and, certainly, dangerous. However, all of these works required a degree of organisation on a day-to-day basis that enabled work to take place. Whilst it is unlikely that those carrying out such organisation would have described their efforts as being strategic, they would have acknowledged the importance of planning and procuring men and materials in a way that was timely and cost-effective. For the small contractors who emerged at this time, longer-term considerations were not a high priority. In effect, it might be said that their approach to strategy was one that evolved. Those construction firms which evolved at the same time as the Industrial Revolution provided the production capabilities to build the rapidly developing towns and cities (and the accompanying infrastructure). A short history of the development and emerging importance of firms in construction, and the strategies they pursued, is presented in Chapter 3.

1.6   Developing an understanding of who 'consumes' construction

One of the central assumptions of strategy is that having developed an analysis of the general business environment and the particular markets for a product or service (see Chapter 4), it is possible then to allocate resources in accordance with demand patterns (see Chapter 6). For products and services that are considered to be stable (e.g. food and fuel), demand is consistent, based on the existing population who normally make such purchases. However, there can a dramatic (albeit temporary) shift in demand when, for example, there is a perceived crisis: such as the potential for exceptionally bad weather when people will over-buy in order to stock up on both food and fuel. When the crisis is over, though, demand normally returns to a consistent level. Services, such as provision of transport (buses, trains), are similar in that there are consistent levels of use. Interestingly, though, demand for certain modes of travel, like particular food purchases, can be altered by the provision of alternative choice – especially if it is based on price. Air travel has shown tremendous growth in the last decade following the expansion of budget airlines. Nonetheless, once the demand patterns are established, further growth has a tendency to be consistent (assuming the costs do not shift, such as by the introduction of taxes or if fuel prices increase).

Crucially, it is recognised that items like food or fuel are things we buy on a very regular basis. Construction is not something we purchase in the regular way that we tend to purchase foods or fuel, certainly as individuals. Even though some clients (organisations representing a wide spectrum of activities) may procure facilities that are used to carry on their business or provide services (especially in the public sector), they will normally do so on an intermittent basis which is determined by need (such as expansion) or a requirement to update, both of which will be

---

[9]It was not uncommon for women (and children) to accompany men who worked on the great civil engineering schemes. The conditions they endured, living in the makeshift huts and shanty towns, being nothing more than hovels, were unsanitary and prejudicial to health.

dependent on finance[10]. A more likely situation involving regularity would be for a client to contractually arrange for repairs and maintenance or a 'rolling contract' of refurbishments to modernise their existing premises (which Marks and Spencer is undergoing at the time of writing). The problem for those who wish to consider strategy in construction, especially from the perspective of a manager, is how demand patterns can be similar to those for staple items or services.

There is an inherent difficulty in trying to predict the needs and demand patterns of consumers of construction. As Boyd and Chinyio (2006) suggest, there is a very wide diversity of types of construction client. Using work carried out by Mbachu (2002), they provide a model which presents such categorisation by 'form and use':

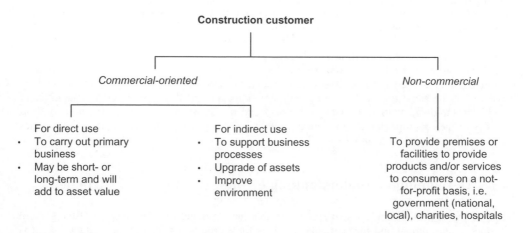

Figure 1.1 Categorisation of construction customer (adapted from Mbachu, 2002).

As this diagram shows, the main division is between those clients that are profit-orientated (private sector) and those whose objective is not to make profit (usually in the public sector). Within each of these divisions there are numerous sub-divisions that create particular uses for the final products that will be used (consumed) by particular individuals or groups. From a strategic perspective, it is in understanding how the decision-making process is carried out to satisfy the needs that particular consumers of construction have. As Boyd and Chinyio acknowledge, because the process of arriving at their final decision is both complex and frequently shrouded in mystery ('either consciously or unconsciously'), it is extremely difficult to develop an accurate understanding (2006, p.11). Nonetheless, for those making judgements about the magnitude of existing or potential markets, it is important to recognise that even though completed buildings are the culmination of the construction process, this will be a long process involving the input of many organisations who will need to supply both products and services. Importantly, their ability to be able to supply will be based on their own assumptions about the potentiality of markets and their choices concerning requisite resources. Failure to correctly interpret indications about the way that particular parts of the construction market will behave (what consumers will need and demand) will create problems of either over- or under-supply which will cause consequential problems for the industry in general.

---

[10]During periods of rapid expansion some clients can require construction on a very regular basis (as food chain McDonalds did during the 1980s and 1990s).

1.7   The structure of this book

Whilst this book is not formally divided into parts, it can be considered as consisting of the following elements:

- An historical overview of both the evolution of strategic theory (Chapter 2) and a socio-historical analysis of the British construction industry (Chapter 3).
- A number of chapters that deal with the background theory for strategic management and which explain concepts that are crucial to understanding how organisations identify the influences in the environment in which they operate (Chapter 4), how they can appreciate the needs of potential consumers and customers (Chapter 5), how an appropriate resource-base can be developed and maintained (Chapter 6) but that the most essential one is their people (Chapter 7).
- The next two chapters are deliberately intended to describe the importance of constant evolution in order to be prepared for the future. Chapter 8 is concerned with how any organisation must be aware of the way in which knowledge, innovation and technology can be used as the basis for strategic development. Chapter 9 describes the fact that organisational development and change have become a constant. In particular, the 'quality revolution' has demonstrated the importance of organisational culture and embracing changes that deliver increased customer value (a fundamental of the 'Egan agenda').
- Chapters 10 and 11 respectively explain what is involved in considering the development of strategic options and how any option chosen will involve careful management, constant review and adaptation to deal with issues and dilemmas that arise.
- The final chapter is an empirical presentation of a number of contributions that have been provided by managers of construction companies 'doing strategy'. As such, they provide their reflections and observations on what this requires their organisations to do and, most especially, cope with the problems that have been caused by the economic downturn resulting from what is now referred to as the 'credit crunch'.

The full list of chapter headings is as follows:

Chapter 2 – Strategic management theory: its origins, development and relevance to contemporary organisations

Chapter 3 – A short socio-historical analysis of the development of the British construction Industry

Chapter 4 – Understanding the environment – markets and competition

Chapter 5 – Strategy and its connection with consumers and customers – the arbiters of success

Chapter 6 – Developing and maintaining organisational resources – the basis for delivering strategy

Chapter 7 – 'Organisation[al] matters' – a strategic perspective of the importance of how to manage people

Chapter 8 – Knowledge, innovation and technology – the 'keys' to the future

Chapter 9 – Change – the only constant in strategy

Chapter 10 – Considering the development of strategic options

Chapter 11 – Implementing the strategy – issues, dilemmas and delivery of strategic outcomes

Chapter 12 – Turning theory into practice – some empirical examples of strategy in construction organisations

Chapter 2
Strategic management theory: its origins, development and relevance to contemporary organisations

> 'The underlying principles of strategy are enduring, regardless of technology or the pace of change', seminal strategy theorist, Michael Porter

2.1 Objectives of this chapter

This chapter considers the historical origins of strategy. Its historical origins are examined in order to appreciate why concepts have been developed which provide the basis of a subject that is now considered worthy of study. Even though strategic management is a relatively new subject, and may certainly not be considered 'classical' (as mathematics, Latin or medicine might be), its popularity increased rapidly in the second half of the twentieth century. The reason, it seems, is that managers taking key decisions concerning the future of their organisations became more aware of the complexities of coping with a rapidly changing environment. Trying to appreciate the influences that create patterns of demand and consumption became crucial in trying to better understand the choices they had in terms of selecting the most appropriate path, i.e. the one most likely to ensure successful attainment of objectives.

   The second part of the chapter deals with the various concepts that underpin the subject of strategy. These concepts will be examined in order that their relevance to those who take strategic decisions on behalf of their organisations can be understood. Finally, the comparison will be made between what is known as **intended** (also known as prescriptive or deliberate) and **emergent** approaches to strategy. As will be explained, whilst some believe these may be considered as contrasting choices that may, according to the beliefs of the organisation (or certainly the managers responsible), be selected as the most appropriate basis for developing and implementing strategy, others think they can coexist. Ultimately, however, as has been stressed previously in this book, strategic theory should not be considered as providing guaranteed outcomes. Success (or failure) is hugely dependent on a combination of experience, judgement, intuition and, as history appears to demonstrate, being able to make the right 'call' – to be able to guess correctly.

2.2 Early origins

There is a strong military background to the study of strategic management and terms such as objectives, mission, strengths, and weaknesses were formulated to address problems on the battlefield. The word 'strategy' has its origins in the Greek expression *strategos* which referred to a military general, and which was a combination of *stratos* (army) and *ago* (to lead). Even though society has always undoubtedly been interested in ways of organising commerce and trade, the main priority of ancient civilisations such as the Greeks or Romans was both attack and defence. The former was about acquisition of territory and resources (including people who, because they could be used as slaves, were seen as a commodity). Because those who conquered were able to increase their wealth and power, others would covet what they had and, in turn, consider taking it by force. Therefore, defence was required, it can be assumed, to deter such attack and, if it did occur, to resist.

Whether the objective is to attack or defend, men would have to be mobilised to create armies. These men had to be trained, provided with weapons and, when required to engage in activities, to be organised. Those men who were able to organise their men effectively became highly regarded and, usually, were promoted. Indeed, any cursory analysis of military history will demonstrate that those men who led their armies most effectively achieved legendary status. The skills and attributes that these leaders seemed to possess, i.e. courage and confidence in their decision-making – especially when in apparent crisis or under pressure – are believed by many to be still relevant for any manager whose job includes matters of strategy.

The crossover of strategy (*strategos*) into civilian usage occurred in 500 bc when Athens was ruled by the Kleisthenes using a military council of ten tribal units. These units were known as *strategia* and the leader of each used the title *strategoi*. Bracker (1980) makes the point that Socrates was supposed to have explained to Nichomachides that being a general in the army is not dissimilar to being a businessman in that both require the ability to plan and use resources. As the Romans began to dominate Greece, the *strategia* became the individual state ruler. Whilst, in effect, the word came to be accepted as meaning leadership and governance, it still assumed that those in control would have armies at their command. As we all know, what the Romans achieved across Europe was truly impressive. Notably, the vision of rulers in using armies to assist in constructing the infrastructure of roads and planning of habitation (water and sewers) was vital to their success[1].

During the period of dominance by the Romans, the need to organise the empire required clear thinking about objectives (and how to ensure sufficient resources – labour and materials) and necessitated planning. The reasons why the Romans lost their empire may be argued; perhaps, more careful long-term consideration might have avoided such loss. Nonetheless, what cannot be denied is that, by using the techniques developed by military engineers, they were able to create the basis of towns' development and transport infrastructure that exists in many parts of this country. The Middle Ages were a time of war in Europe which saw borders expand and contract depending on who ruled and their inclination to gain territory. Strategy based upon conflict was the way to achieve such gain and you should note the names of the Prussian commanders and theorists, Clausewitz and von Moltke.

Whilst strategy may have been less important after the fall of the Roman Empire (Whittington, 1993), its derivations were used. For example, in the Middle Ages the word *stratagem* was used to mean any attempt to second-guess and beat an enemy. During the Napoleonic wars in the eighteenth century, the belief was that a strategy was required to win battles. Strategy was, as a consequence, something largely used by military leaders to consider best how to win battles and use their resources. Indeed, the early development of strategy as a formalised approach to business and management, certainly in America in the aftermath of the Second World War, was based upon using the experience of leaders who had developed their skills in war.

It is recognised that the fundamental difference between military and business strategy is that the latter is formulated, implemented, and evaluated with an assumption of **competition**, whereas military strategy is based on a belief that **conflict** (or at least its threat) will be the consequence.

---

[1] The term, 'civil engineer', it should be noted, was developed to distinguish them from their military counterparts who would have existed in such armies.

Nonetheless, military conflict and business competition are so similar that many strategic-management techniques apply equally to both. Business strategists have access to valuable insights that military thinkers have refined over time. Superior strategy formulation and implementation can overcome an opponent's superiority in numbers and resources.

2.3   The rise and rise of strategy as a corporate tool

2.3.1   Early years: the origins of theory

Strategic management is a relatively new subject – most certainly when compared to the classics. As the introductory chapter explained, strategic theory is a subject for which there is an abundance of definitions but no precise agreement on what it actually is or means to those who 'practise' it. Part of the apparent problem of definition, it seems, is that theory has largely followed in the wake of empirical evidence of what works. As will be described, in the aftermath of the Industrial Revolution managers had to cope with the expectations placed upon them as best they could; there were no 'established' theories from which they could draw inspiration and potential guidance. However, those who used the tenets of what was known as scientific management (and early management theory) found that there was tremendous corporate advantage in terms of reducing cost and production time. Because many of those who have adopted various approaches to strategic management are American, there is a tradition for the subject to be viewed as having emanated from there. Whilst this is not strictly true, it would be unfair to underestimate the influence that America has had in developing the early origins of strategy.

The Industrial Revolution created the need for bigger organisations to cope with considering more than just on a day-to-day basis. If you wanted to operate a factory, you needed to hire labour, have a power supply (coal, water) and procure the raw materials. This required those managers involved to consider markets and how they could ensure adequate supply to match demand. Because there was little or no prior experience to draw upon, managers coped as best they could and any decisions taken were based upon hunch (intuition) as to what would suffice in the short term. In small local markets this proved adequate. As students of management discover, the beginnings of theory were provided by writers such as Frederick Winslow Taylor and Henri Fayol who suggested, on the basis of their experience, that there were 'rules' which provided better ways of managing. Such theories are categorised as being central to what is known as 'scientific management'. Taylor's theories, published in his book, *The Principles of Scientific Management,* in 1911, were seminal and became the model for how to organise production, frequently being referred to eponymously as 'Taylorism'. Even though scientific management is criticised by those who argue that it is detrimental to people working in the production system and, as a consequence, leads to a reduction in innovation and quality of the end product, it is still considered by some to be the best way to organise resources.

The development of scientific management coincided with tremendous growth in the demand of society for goods produced. This was an era when things changed very quickly and for those with sufficient ability to assemble the capabilities to satisfy such demand, potential new markets could be exploited. As an American industrialist demonstrated by his utter devotion to the tenets of scientific management, the benefit of a strategy based upon mass production which slashed unit costs of production was the ability to become dominant which, of course, enabled huge profit to be made. The industrialist in question, Henry Ford, created an organisation that became synonymous with using what has become known as 'mass-production' techniques which are based on increasing efficiency and reducing unit costs. By adopting this principle, Ford demonstrated that mass production could stimulate mass consumption. So successful was Ford's application of the principles of scientific management that the word 'Fordism' is considered to be interchangeable with 'Taylorism':

- Standardisation of parts used (referred to as 'armoury practice' in reference to the way that small arms were produced in government factories);
- Flows process in which cars were produced along 'lines'[2] by components being attached by workers using automated tools to ensure speed;
- Division of labour into workers who carried out simple and highly repetitive tasks.

Ford's production system enabled dramatic gains in efficiency and reduction in cost which translated into sales:

> In 1913–14, first for sub-assemblies and then for putting together the car itself, Ford's new assembly lines made extraordinary gains in efficiency for a remarkably small monetary investment. Between October and December 1913, the first assembly line for chassis reduced the time required to put a car together from twelve-and-a-half hours to two hours and forty minutes. Model-T production jumped from 13,840 in 1909 to 585,388 in 1916; at the same time the price dropped from $950 to $360. (Pursell, 1994, p.98)

The effects of Ford's approach to strategy based upon mass production set the benchmark that many manufacturers continue to follow. Even though many, especially the Japanese in the aftermath of the Second World War, sought to learn the principles of mass production, the belief that slavish devotion to cost reduction is always positive has been questioned. As Ford was to discover in the 1920s, the reliance on one product (the Model-T) meant that customers had no choice. Variety, it was believed, was counter-productive and, in the 1920s, when General Motors, led by Alfred Sloan, were able to offer precisely this, sales of the Model-T declined. However, neither Ford nor Sloan saw any strategic advantage in looking to workers on the production line to offer anything other than an ability to work as fast as possible (and certainly as quickly as the assembly line ran).

Ford and Sloan managed their organisations on the basis of what **they** thought worked. Even if they had sought to consider other theories they would have been disappointed: none existed. In effect, what they did was the only strategic theory that existed. Consequently, if alteration was felt to be necessary to achieve their objective, it was done. In Ford's case change and management was by diktat (often accompanied by the threat of the sack for non-compliance). Sloan, at least, conferred with senior and middle managers before implementing any change. However, the overriding objective that they sought was to maximise profit.

In the 1930s and 1940s America and Europe were affected by depression and, of course, war. In times of privation and scarcity of resources (and money), organisational managers believed their main role to be one of ensuring survival. The war, in particular, created circumstances in which people were expected to contribute to the war effort (most especially in America). Not for the first time, the military context provided the climate for strategic thinking which was primarily dedicated to ensuring sufficient armaments to win against both the Germans and their allies, and the Japanese. The aftermath of the war saw the emergence of America and the Soviet Union as the dominant political and military influences in the world. Economically, however, America was utterly dominant.

---

[2] It is said that Ford took his inspiration for production lines from the great Midwestern slaughter house and that, instead of the animals being taken apart as they moved on a continuous chain, cars were assembled.

2.3.2   Post-war strategy: a bright new future?

America's ability to produce and organise efficiently during the war meant that those responsible for the successful outcome, both from a production and military aspect, were in strong demand. If this is what could be achieved during conflict, it was reasoned, then applying the same principles would enable organisations to enjoy similar benefits in peacetime. Strategic management became something that all managers were encouraged to consider as a way to make their organisations operate more efficiently and, it was anticipated, profitably. Coincidentally, there were those who, although not managers in practice, believed that they could provide explanations (theories) for what strategy involves. Such theories were based on the desire to logically consider options and to systematically plan the use of resources (Ansoff, 1968). As Johnson *et al.* suggest, as well as being 'highly influential', managers were advised that they must be able to analyse every aspect of the organisation for which they have responsibility:

> ... managers can and should understand all they possibly can about their organisational world; and that by so doing they can make optimal decisions about the organisation's future. (2005, p.20)

The period between the end of the Second World War and the 1970s was one of comparative stability. As wealth increased, especially after reconstruction in the 1950s, markets for goods increased rapidly. Society changed in a way that meant producers could assume increasing consumption and, as a result, gear their production accordingly. Strategic theory largely concurred with such beliefs and suggested that strategy was merely based upon the need to continue to plan and use techniques such as operational research to ensure effective application and use of resources. Change, even though it might not be an intentional objective, could be dealt with by careful planning. Indeed, organisations that were clever enough could capitalise on new opportunities, and certainly perform better than those that simply reacted. Such an approach is known as 'prescriptive' (Lynch, 2006, p.36) and is described in greater detail below.

Those who manage strategy like stability and optimism. Therefore, the 1960s were, in large part in America and Europe, ideal. The destruction and privations of the 1939–45 war and its immediate aftermath were becoming a memory. Unemployment and the problems of lack of income for even the poorest were to be consigned to history. Consumption and tastes changed to reflect such optimism and all things seemed possible, including, as a result of a strategic vision by President Kennedy that, by the end of the decade, a man would walk on the moon. Construction benefited from a desire to recreate the infrastructure, the standard of housing, the way we worked, shopped and enjoyed leisure time. Unfortunately, in the desire to modernise, it is now recognised that, all too frequently, more was lost (in terms of heritage and historic structures) than was gained[3]. As the next decade would show all too clearly, long-term goals based upon a vision of a 'perfect' world were to be fleeting.

2.3.3   The 1970s and beyond: change as a way of life

If the 1960s are now seen as a decade of hope, the 1970s are seen as being a decade during which, certainly in Europe, unforeseen economic and political change (and conflict) caused society to reconsider both its values and priorities. Rapid price increases in oil caused by conflict

---

[3]Construction has benefited from the demolition and reconstruction of the 'blight' of 1960s building.

in the Middle East caused turmoil in markets and meant that producers had to factor in far higher costs than anticipated. Industrial unrest became common (including a national building strike in 1972). Plans that strategic thinkers had put in place had to be radically altered to cope with the prevailing circumstances. As commentators on strategy asserted, the belief that assumptions could be made on the basis of certainty and long-term stability was flawed. Appreciation of practice in general management had demonstrated that theory could not be assumed to have universality. Many commentators and researchers on strategic management explicitly acknowledged the problems of trying to manage organisations in a world that constantly altered (sometimes radically). For example, Quinn (1980), Stacey (1992) and Mintzberg (1994) argue that the world was a far more complex place than it had been. As Johnson *et al.* explain, the difficulties of dealing with a world in which the old ways no longer seemed appropriate had to be acknowledged:

> Its complexity and uncertainty meant that it was impossible to analyse everything up front and predict the future, and that the search for optimal decisions was futile (2005, p.21)

Therefore, a key belief of the classical approach to strategy (see Chapter 1), that it was possible to predict the future with any accuracy, was accepted as extremely difficult – even when times are reasonably stable. Unforeseen events intervene. More especially, who has the skill to make 'accurate' predictions, no matter how extensive their experience and knowledge? Therefore, the belief that strategy can be a deliberate and rational process which, once commenced, will not need to change, is mistaken. Rather than continuing with such fallacious assumptions, critics believe, there is a better way.

Prescriptive strategy, until then the predominant theory, was challenged. Instead, critics argued, another approach to strategy was needed: one that accepted both that change was likely and could be adapted to. Such a strategy, what is frequently referred to as being 'emergent', was proposed. In effect, it was based on the belief that an organisation should be capable of responding to exigencies that prevailed depending on particular influences that might have an impact upon success. Another trend that became apparent was that western organisations could no longer rely on customers to engage in patriotic buying. As such organisations began to discover, as their sales began to decline, customers were buying goods manufactured abroad: in the case of cars, especially from Japan.

When western manufacturers started to investigate why people would switch allegiance from 'home-produced' goods some salient facts emerged. In order to gain access to the market, overseas producers were prepared to sell at lower cost. Whilst this might be sufficient to attract some to consider alternatives, if the goods were not comparable in terms of usage over their lifetime, customers would return. However, what manufactures had recognised, but failed to deal with, was that their own goods frequently failed too often. The Japanese, in particular, had also recognised the propensity of failure and consciously devised their production strategy to ensure that not only were their goods cheaper to produce (by reducing waste), but, significantly, that they were also able to perform far better in terms of failures or breakdowns. This was the beginning of what was seen to be a 'quality revolution' that caused organisations, regardless of size, to examine their strategy for dealing with production quality. Crucially, organisations that assumed a reliable customer base found that they now had to try much harder to ensure repeat business and loyalty.

Japanese manufacturers demonstrated that it was possible to use mass- production techniques in a way that reduced unit cost and, therefore, potential selling price (which if all other things are equal provides an advantage). Most significantly from an operational point of view, they showed that the traditional belief in standardisation of all aspects of production, including the most vital component of organisations, workers, was counter-productive. As even Ford had discovered when he instituted his production system, workers treated in a way that considered them to be unthinking tended to lead them to boredom and frustration. At one point this led to a strike which Ford dealt with by raising wages, but not by dealing with the long-standing issues of motivation and sense of well-being. The Japanese, on the other hand, viewed people as an integral component of production. People (workers), regardless of what they did, were capable of making a valuable contribution towards long-term improvement, both in terms of the end products and the processes used to assemble them. This belief may have seemed fanciful to western organisations. However, as many manufacturers discovered, the quality level the Japanese produced was to become a potent weapon of competitive advantage. Car manufacturers, for example, found that it was not only more difficult to compete, but that even their very survival was threatened[4].

2.4   So what are the main concepts of strategy – back to the problems of definition?

In Chapter 1, the difficulties inherent in providing a formal definition of what strategy involves were considered. Trying to define something that can be so widely varied, especially depending on the particular circumstances in which it is applied, does mean that there is merit in exploring its core concepts. What can be said with certainty is that managers regularly make strategic decisions. We can reasonably assume that the way they make such decisions is guided by some concepts that may be defined as ideas or principles that are used to inform action (and make sense of the situation)[5]. As the previous sections explain, the objective for those managers concerned with the long-term future of any organisation is to make choices about what they believe is the most effective (or efficient) way to both procure and deploy resources.

According to Huff *et al.*, there are a number of 'characteristics' that are exhibited in strategy which an organisation has implemented and found to be 'effective' (2009, p.6). These characteristics are summarised as follows:

1. Communication of 'a compelling purpose or vision' to everyone within the organisation and all those outside who would be expected to understand what it 'stands for';
2. That there is a connectedness between the organisation's strengths (which are internal) with the 'environmental opportunities';
3. 'Generate more resources than in uses';
4. Coordination of activities in a way that is sensible and dedicated to achieving success;
5. Able to respond to 'new conditions'.

They define strategy as being something which:

> Defines a desired objective and communicates what will be done, by whom,
> how, for whom, and why the output is valuable (Huff *et al.*, 2009, p.6)

In order to assist in understanding the importance of this definition, they pose a number of questions which are shown in the following diagram:

---

[4]Companies that in the past had been major names in motor manufacturing have, because of their inability to compete, disappeared, including the last British mass producer, Rover Cars.
[5]It's worth stating that it has long been recognised that what managers believe they are doing is imprecise and frequently is hard to reconcile with what actually happens.

Figure 2.1 The key questions for strategic decision-makers (adapted from Huff *et al.*, 2009, p.6).

Their belief is that by carefully considering all of these questions in a coordinated way the organisation will be able to ensure that the five characteristics shown above are more likely to be attained.

Wickham believes that there are four 'concerns' to the management of strategy (2000, pp.1–2):

1. Consideration of the organisation as a whole so that the parts are integrated cohesively;
2. Ensuring that what the organisation seeks to achieve is appropriate to the environment in which it operates;
3. Acquisition of the resources that are necessary to 'create value and prosper', i.e. to succeed in attainment of objectives;
4. The ability to compete against other organisations which offer comparable products and/or services or, alternatively, offer a substitute which potential customers perceive to be better.

In making choices (and dealing with these concerns), those managers charged with the task of 'thinking strategically' must recognise the difficulties and dilemmas that exist within the context in which the organisation operates. As such, they must be aware of the fact that what they do is going to be long-term. Additionally, there is likely to be complexity in the decision-making process; most especially if there is a variety of activities and outcomes. Consistent with Wickham's four concerns, decisions should acknowledge the dynamic nature of the environment and markets. Attempting to second-guess potential change is not easy and, as examples of corporate failure attest, may potentially result in failure.

Because an organisation is essentially a collection of people cooperating (or not!) in order to achieve particular outcomes, it is a social entity. Strategy based upon agreement of outcomes (either on a voluntary or forced basis) by all members of the organisation would appear to be the main objective. Consequently, strategy then becomes a matter of ensuring that targets consistent with strategy are achieved. Strategy on this basis, therefore, is reliant upon effective and efficient operations management. However, to consider strategy as simply being a matter of operations is, according to some, mistaken:

> An operational manager is most often required to deal with the problems of operational control such as the efficient production of goods, the management of the salesforce, the monitoring of financial performance or the design of some new system that will improve the level of customer service. These are all very important tasks, but they are essentially concerned with effectively managing resources already deployed, often in a limited part of the organisation within the context of an existing strategy. Operational control is what managers are involved in for most of their time. It is vital to the success of strategy, but it is not the same as strategic management. (Johnson *et al.*, 2005, p.15)

Consequently, according to Johnson *et al.*, strategy is about being able to understand the multiplicity and complexity of 'ambiguous and non-routine situations' all of which have an influence on the whole of the organisation. As they advise, being able to 'conceive of the whole rather than the parts of the situation', strategic managers must be able to see the longer-term implications of all factors that will influence strategy. Accordingly, strategic management is concerned with strategic position, strategic choices and being able to translate strategy into action. Each of these is influenced by other variables (see below).

2.4.1   Strategic position

Johnson *et al.* define strategic position as being 'concerned with the impact on strategy of the external environment, an organisation's strategic capability (resources and competences) and the expectations and influence of stakeholders' (2005, p.17). It therefore considers the way that an organisation fits into its **external environment** which, of course, is constituted by so many variables that provide new opportunities or, alternatively, threats. Whether an organisation can realistically deal with every aspect of the environment is one of the difficulties of strategy. This will be dealt with in a subsequent chapter. **Strategic capability** is concerned with what skills and resources the organisation possesses and which will provide either strength or weakness in the marketplace when compared to other competitors. Whilst an organisation may be unable to exert as much influence over the external environment as it might like, it may alter its capabilities (by training or recruitment) to make it more able to cope with potential changes in the former which are anticipated.

The final influence on strategic position, **expectations and purpose**, is concerned with the desire to fully appreciate exactly what the organisation seeks to achieve, and how the various stakeholders influence outcomes. In doing this, managers need to be aware that different people have different values. As such, they will work towards attainment of things which they believe are congruent with what **they** believe is important. The problem arises for managers who seek to achieve unity of purpose among groups that have different expectations. As such, part of their role will be to negotiate agreement which, at least, attempts to gain some broad consensus. If they do not, the organisation's strategic position may be undermined and, potentially, it is less likely to achieve its intended objectives[6].

---

[6]Many suggest that organisations are similar to 'living organisms' and, therefore, strategy should have resonance with their apparent ability to cope with change by constant adaptation.

2.4.2   Strategic choices

Strategic choices are concerned with the way that the organisation (or at least its 'custodians'), develops an understanding of what is required for it to continue to function at all levels (both business and corporate) and the options that are available to pursue a particular direction. Therefore, at **business level**, the organisation will need to compare what it does and what it possesses (capability) in comparison to others against which it is in competition. This is very much considered with respect to what is required in the short to medium term. At **corporate level**, however, strategy choices must be made with a much longer perspective. There will be a greater exploration of how existing markets may alter. There is consideration of whether potential new markets may open which are consistent with current or future capabilities (opportunity) or, alternatively, whether markets may change in such a way as to be inconsistent with current capabilities (and be difficult to procure). The latter is, obviously, a threat which would need to be carefully managed in order to avoid the fate that has befallen some organisations in the past.

2.4.3   Strategy into action

This aspect of strategy considers how strategic position and choices can be effectively translated into 'actions' that are likely to be successful. The importance of 'action' cannot be overstated. As we know from personal experience, good intentions are fine. However, if we don't do what we set out to achieve because of poor organisation we do a disservice to ourselves and anyone else we disappoint. The description of the development of quality, service and excellence of the last twenty-five years has demonstrated the criticality of reputation. If you make promises to be better than others with respect to the build and performance of your product, anything less will be perceived by customers as not being 'value'[7]. What Japanese producers so amply demonstrated was the need to ensure that all parts of the production process are dedicated to producing the highest possible quality (and that this is a never-ending quest for excellence).

Consequently, ensuring effective action requires that every part of the organisation is considered (technology, people, communication, management) and that they are linked to create a fully integrated whole. Construction has suffered from the belief by customers, whether they are large corporate clients or individuals having work carried out on their home, that it will be unlikely to deliver either on time or budget (and, all too frequently, both). Even worse, the finished quality may be poor. Commentators who have analysed why this should be the case discover that the organisation of trades and workers is a great deal more fragmented than would be found in other industries. The argument that 'this is the way it happens in construction ...' is increasingly being challenged, especially by those who have experienced superior quality in other purchases. Increasingly, there are calls for manufacturing to use its expertise with respect to cost, quality and delivery in construction. Perhaps the experience of British car manufacturing may be instructive to organisations in construction. Those who do not adapt to changing expectations and the innovative methods of production and management may be beaten by others who offer alternatives that are demonstrated to be superior.

2.5   Mintzberg's 'Five Ps' conceptualisation of strategy

Henry Mintzberg, the seminal management commentator and theorist, asserts that strategy should be thought of as being based upon five concepts which begin with the letter P (1987):

[7] It is debatable whether under-promising the attributes will mean that you can always exceed expectations.

1. Plan;
2. Ploy;
3. Position;
4. Perspective;
5. Pattern.

These concepts are not inconsistent with models proposed by others (such as Johnson *et al.*). Nonetheless, his explanation is that the combination of all of these will provide an effective strategy that aims to deliver the intended outcomes. So, the plan is the way that the intentions are pursued. Ploy, he explains, is about ensuring that there are options which can be used to attempt to beat any potential competitors or circumstances which will undermine the outcomes. Position is important in that it either provides differentiation from others or, alternatively, makes distinctiveness or uniqueness explicit. Perspective is important, Mintzberg believes, because whilst the intention of managers may be to assume that every decision they make can be entirely based on rationalism and objective criteria, much of what is considered is based on intuition. Where there are 'signals' available, we make sense of them but as we will know from experience, draw on own personal perspectives and experiences in order to arrive at a particular conclusion – frequently referred to as cognition. An organisation in which there are many decision-makers involved will have to try to develop a perspective that unites them in their stance and belief in what constitutes the best strategy. Finally, pattern is the consistency by which strategy is directed towards particular outcomes. Those organisations that have developed long-term consistency are usually secure in their belief about the quality of what they offer and the likelihood of consumers to continue their loyalty. Whilst this is undoubtedly good, it is important not to become complacent. Equally, organisations that are inconsistent in their patterns will potentially suffer the consequences of engaging in erratic behaviour even though they may justify this by the volatility of the external influences.

2.6   The hierarchy of strategy

In any organisation there are likely to be levels of strategy that are dependent on the place within the organisational hierarchy to which they are applied (see diagram below). Whilst such differentiation is sensible in that it means that those who are responsible for that level of strategy can concentrate on making appropriate decisions, it is important to be aware of interconnectedness. A decision taken at the highest level will have impact at operational level which, in turn, may mean that it is more difficult to meet corporate targets. Additionally, such decisions will be made with respect to relevant timeframes.

As this diagram shows, at the highest level there is what is known as network and groups in which organisations may collaborate in the development of strategy that is mutually beneficial to them and, possibly, their customers. For instance, where there is a strong technological case for the development of a unified and coordinated approach by designers and suppliers of components in an industry, this would be useful. The potential danger of doing this is that organisations collaborating in this way might be accused of engaging in a cartel which can be viewed as restrictive to free trade.

For an organisation the highest level is **corporate strategy** which is concerned with considering the overall direction and purpose of the organisation. As such, the organisation will pursue goals which are consistent with all of those who have a vested interest. In profit-making organisations these are likely to be shareholders. In the case of non-profit-making organisations the desires of the wider community will be important. Those who are involved, certainly including some senior managers, must be aware of the **context** of the environment in which the organisation exists. In a competitive environment they will consider overall markets for their product(s) and/or service(s). Markets will be affected by influences such as economic and social trends. Where organisations

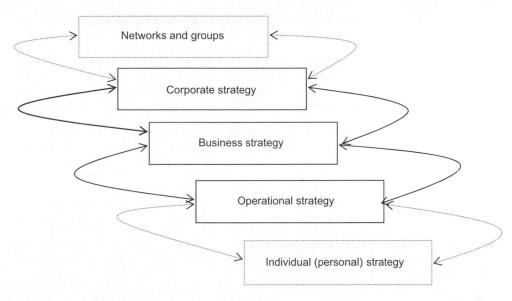

Figure 2.2   A conceptualisation of the hierarchy of strategy (adapted from Huff *et al.*, 2009, p.15).

operate in non-competitive environments, such as government bodies or charities, there will still be an onus on decision-makers to be cognisant of the limitations of available resources and of their dependence on finance. Decisions taken at this level are likely to be broad in scope and concerned with a sufficiently long period of time to potentially make major shifts in strategy, i.e. many years. Decisions are aimed at ensuring that the organisation has a sufficient spread of resources to provide every department (sub-division) with what will be required in order to achieve the expectations set out in the overall corporate plan.

The next level at which strategy is determined is **business strategy**. Because of the decisions taken at the level above, the broad parameters of what must be considered will be defined. However, this is not to suggest that those making such decisions have no freedom (it is highly advisable that they will have been consulted before corporate decisions are taken). Business-level strategy is concerned with understanding and considering particular markets or areas of operation. Most especially, strategy at this level is intended to identify how the product(s) and/or service(s) can be provided in a way that ensures that they are perceived as being better than those offered by competitors, to gain a so-called 'competitive advantage' (see below). For those organisations where profit is not the motive, strategy at this level will be concerned with ensuring that what is offered is the best that is possible with the resources available. In order to achieve this aim, much will depend on the ability of those involved in production and/or delivery (the third level). Any inconsistency will be quickly discovered and, as the experience of organisations that ignored this connection discovered, result in both a loss of reputation and enable competitors to gain advantage. This, in turn, will potentially undermine the corporate strategy. Whilst this level of strategy has a shorter timeframe than corporate strategy, it will still be likely to be many months (and potentially years).

Operational strategy is primarily concerned with the ability of the organisation to deliver on a day-to-day basis. This will involve all of those with responsibility in developing strategy (or a series of strategies) that ensure that there are adequate resources to deliver the product(s) and/or service(s) in a way that is consistent with the intentions set out in higher-level strategies. Therefore, if you intend to supply cheaper than others, operational strategy is dedicated to achieving precisely that objective (so-called 'price differentiation'). Equally, if the promise to consumers is either speed of delivery or end quality, then the operational strategy must achieve that. As explained already, as many customers have discovered, some organisations (especially some Japanese car and electronic manufacturers) can achieve what seemed impossible: to deliver high quality at lower cost and within rapid times compared to competitors. The argument that you cannot match what others do because you do not have sufficiently skilled labour or advanced technology is unlikely to be met with sympathy (no matter how loyal customers have been in the past)!

Finally, as the diagram which is based on the consideration of Huff *et al.* of 'levels of strategizing' (2009, p.15) shows, there is **individual (personal) strategy** which strictly goes outside of organisational concerns. However, as subsequent chapters will explain, the organisation is a social entity which should be collective in both action and thought. Achieving the former may certainly be possible by the use of rules, backed up by the threat of punishment. Such an organisation will be seen as highly bureaucratic and the culture, which will be described in more detail in a subsequent chapter, will be one that is unlikely to instil in people a belief that they should be fully committed (or passionate). It is much better that the organisation is one in which people feel that the organisation's purpose is one that they support, and that it has been developed with their cooperation. Engaging people in the development and implementation of strategy is a recurring theme of this book.

2.7   Terms associated with strategy

In understanding strategy there are a number of terms that are common to the subject. These are:

- Mission;
- Vision;
- Goal;
- Objective;
- Capability;
- Systems;
- Models;
- Control;
- Strengths;
- Weaknesses;
- Opportunities;
- Threats;
- Values;
- Initiatives.

Many of these will be obvious. However, it should be made clear that, as in much of what constitutes management, meaning can be interpreted depending on those using the words. With this in mind it is important to appreciate that the context of the organisation is important. Large organisations will be likely to have a much more formal approach to strategy than those which are smaller in size. Additionally, as explained in Chapter 1, the context of construction is different from that of other sectors (such as retailing or manufacturing), or the public sector. However, construction is required to provide premises and facilities which will allow others to carry out their daily functions. This means understanding the strategic intent of organisations which procure work and, implicitly, the importance that they place upon particular aspects of strategy.

2.8   Attaining success: the importance of analysis, development and implementation

Those managers responsible for corporate strategy must be aware of the overall intentions of the organisation 'in the broadest possible terms' so as to be able to match these with the most likely scenario that will unfold. Predicting likely scenarios, of course, is the most difficult part of strategic management; in an increasingly uncertain world nothing can be guaranteed! However, as subsequent chapters will explain, there are two very different approaches that can be employed to attempt to carry out **analysis**. One, planned, assumes that it is possible to make choices based on a rational analysis of the future which will give a logical conclusion as to what is the best outcome. The other, emergent, suggests that because of the volatile (and increasingly unpredictable) nature of markets and society predictions are unlikely to be accurate. There is no definite view as to which is the best way to carry out analysis. As many organisations now believe, they must prepare a range of scenarios and maintain sufficient flexibility to deal with each of them (as well as the ability to react to unforeseen events)[8].

What is important is that analysis is carried out in order to ascertain the resources that will be required to be able to cope and, ideally, to be better. This means that resources should be **developed**. Consideration and development of resources will certainly involve people who will be employed (directly and indirectly). They should possess the sort of skills and knowledge that will allow the organisation to cope with whatever changes are likely to occur. Closely linked to people is technology. The world we work in today is radically different to that which our parents worked in. Technological change affects almost every aspect of working lives and there is no reason to believe that change will not continue. Indeed, it would be foolhardy of managers to believe that their consideration of strategy should ignore potential technological change. Additionally, those organisations which manage their relationships with others in order to become intimate and symbiotic can find that they enjoy a competitive advantage – usually referred to as 'supply-chain management'. In effect, even though they may not directly own or control the resources of others (such as suppliers and subcontractors), the relationship becomes one that is seamless and coordinated. Achieving such developed (and some would say 'mature') relationships takes time and effort.

Having carried out analysis and considered what development of resources is needed to cope with potential scenarios, it is necessary to make decisions as to implementation. This becomes the key test. If the organisation is successful, the implementation of strategy is appropriate. It follows that future strategy is likely to be implemented similarly (at least if the scenarios remain consistent). However, if the organisation does not enjoy the anticipated success that it was believed would occur, this does not mean that decisions were necessarily incorrect, simply that events may have altered. Obviously, at this point it is probably advisable that the strategy being implemented is reviewed with urgency and, if required, altered to be consistent with the prevailing circumstances.

2.9   The three essentials of strategic decision-making

Acknowledging the uncertainty of strategic decision-making is the dilemma that those responsible must accept. In so doing, they should recognise that what they are engaged in is a **process** of attempting to meld the influences of what is going on in the environment (**context**) with the most appropriate match of skills and resources available (**content**). Pettigrew and Whipp (1993)

---

[8] As such, analysis is considered as a continuum with 'planned' at one extreme and 'emergent' at the other.

propose a model which incorporates all three of the so-called 'essential dimensions' of strategy which, they believe are subject to continuous change (see below).

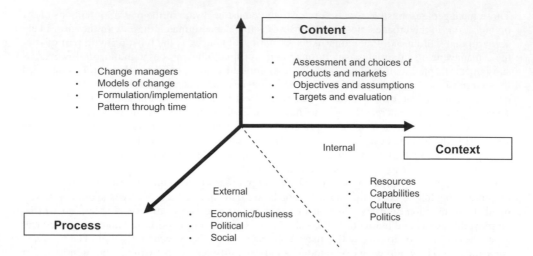

Figure 2.3 'Understanding strategic change: three essential dimensions' (from Pettigrew and Whipp, 1993a, p.26).

Pettigrew and Whipp advise that the essential dimensions should be considered as being influenced by aspects such as 'hierarchical, organizational and economic circumstances from which they emerge' (1993, p.26). Indeed, as they state, the process part of strategy is crucial in that it should not be thought of as linear, but as being 'continuous, iterative and uncertain' (1993, p.27). Process, therefore, may be considered as playing a vital role in appreciating the theory of corporate strategy. Lynch contends that, because of the 'breadth and complexity of the subject', process may be used as the way to consider the two main approaches to understanding the theoretical basis of 'corporate strategy development' (2006, p.16). These are 'prescriptive' and 'emergent' and are considered in detail in a subsequent section of this chapter. As the definitions which are provided by Lynch suggest (2006, p.17) [the use of italics are his], the major difference is in terms of the way that development and implementation phases are differentiated from each other (in the former, analysis, development and implementation are sequential and discrete, whereas in the latter, they are much more closely interrelated):

- A prescriptive corporate strategy is one whose *objective* has been defined in advance and whose *main elements* have been developed before the strategy commences.
- Emergent corporate strategy is a strategy whose final *objective* is unclear and whose *elements* are developed during the course of its life, as the strategy proceeds.

Lynch provides a diagrammatic representation of each of these in which there are three common elements in both:

1. Strategic analysis – involving the environment, resources and the vision, mission and objectives of the organisation.
2. Strategic development – which requires consideration of options and the selection of the one (or more) which is most appropriate and timely as the 'route forward'. In carrying out

consideration of strategy, it is necessary to choose an organisational structure and a 'style' of managing.
3.   Strategy implementation.

The main difference is that, in using a prescriptive approach, the three elements are distinct. In adopting an emergent approach, however, whilst analysis is distinct from the other two, they (development and implementation) are interdependent. Accordingly, prescriptive is an approach that assumes deliberate predetermined choices with the intention to create particular actions. Emergent strategy is an approach that emphasises the use of trial, experimentation and learning from experience. Whilst these two approaches may be viewed as being exclusive of one another, some (most notably Henry Mintzberg, 1987) argue that it is entirely possible for both to coexist without problem. Indeed, Lynch uses the analogy of the human brain in which both sides must work together for the person to 'function properly' (2006, p.16).

2.10   Lynch's test of 'good' strategy

The difficulty, of course, is in selecting the strategy that is most likely to provide success. Whilst there can never be guarantees of what will provide success, there are various tests that can be used by organisations in considering strategy. Such tests, according to Lynch, may be considered to be either based on what works in practice ('application-related') or based on logical analysis and rational method ('academic rigour'). So, the former takes into consideration the following questions:

●   Will the strategy deliver what is known as 'value-added' in so far as what is implemented (use of processes) will deliver higher value from the inputs than they had previously? Every organisation will be assumed to want to be engaged in improving its 'value-adding' position so that it is able to serve the needs of customers or consumers.
●   Is it consistent? The strategy the organisation wishes to implement should be consistent with the resources available and the environment in which it operates.
●   Will it enable the organisation to develop (or sustain) **competitive advantage** (an expression that is embedded within strategic management theory) over its competitors?

The academic-rigour tests are intentionally less practice-orientated than those which are application-related and can include analysis of the following:

●   Originality;
●   Purpose;
●   Logical consistency;
●   Risk and resources;
●   Flexibility.

Each of these should be considered in the context of what actually goes on and, it is anticipated, works in terms of moving the organisation towards its key objective(s). Therefore, the first of these – originality – considers whether the strategy is able to achieve something radical and innovative. Doing the same thing may be sensible (particularly if you cannot think of how to improve the product and/or service). The trouble is: customers are encouraged to expect constant change which, everyone assumes, makes the experience better. Undoubtedly much of our contemporary experience validates this assumption. But as Lynch warns, a danger lurks within the desire to be original in that there is a temptation to use it as 'just another excuse for wild and illogical ideas' (2006, p.20).

Purpose refers to the desire to ensure that whatever strategy is adopted there is a congruence between what is proposed and the key intentions of the organisation (as contained in the mission). Purpose, it is important to understand, comes from those who have most influence. Therefore, leaders' aspirations play a vital part in proposing, developing, shaping and implementing purpose through strategy.

Logical consistency is intended to provide a test which asks whether the strategy being considered is the most sensible choice based on the information (facts) that has been used as the basis for decision-making. The dilemma that managers taking such decisions face is how much they trust the information that they receive, particularly if it is based on signals that come from customer behaviour or market conditions.

As the fourth test implies, there are two considerations. Firstly how much risk is involved in implementing the strategy? This is allied to the 'resource-base' that currently exists. If the organisation wishes to implement a strategy based on capability and expertise that already exists (or can easily be adapted), there is unlikely to be great difficulty. The risk will be determined by how much the strategy requires the organisation to shift from traditional markets and established customers or whether it is intended to consolidate them. Judging risk and making the right decision carries great responsibility. If successful, those who made the 'right call' can expect to be held in high esteem and to be well rewarded. Lack of success (failure) may not result in outright criticism[9]. However, their credibility may be called into question in the future.

The final test, flexibility, is one that is extremely apposite in the environment in which this book is being written. Specifically, how much does the strategy require the organisation to commit its expertise and resources in a particular direction? Increasingly a key characteristic of an organisation is to be able to be nimble in switching resources and adapting its tactics to suit the prevailing environmental or market conditions. Organisations are expected to behave in a way that is analogous to ships. Whilst they set a path that may seem logical at the outset they should alter course if there are conditions that put the craft at serious risk. No captain would deliberately sail their ship into rocks. Likewise, those managers given the task of deciding on a strategy will be assumed to want to implement it regardless of any unanticipated conditions that will, like rocks, cause the organisation to be scuppered.

2.11   What will really work? An exploration of major theoretical perspectives

In many major texts dedicated to strategy, there is general agreement that there are two main theoretical approaches to strategy. The first, as described above, is that strategy can be rational and deliberate; it is 'intended' (see Johnson *et al.* (2005) or 'prescriptive' (see Lynch (2006)). The second, also described above, owing to the inherent uncertainties of the world in which organisations operate, is that strategy must be developed in accordance with the forces or 'influences' of the prevailing environment. As explained above, strategy which is implemented in this way ebbs and flows or, alternatively (depending on your preferred metaphor), it waxes and wanes. Such an approach is usually referred to as being 'emergent'[10].

Whilst thinking in two apparent extremes is useful, and applied accordingly, it is accepted that the two can be operated in accord with each other (see below). Prior to considering how the organisational strategy can be implemented in a way that incorporates both approaches, it is necessary to examine each in a more detailed way that elaborates on what each consists of.

---

[9]There is a view that too many very senior managers expect high financial reward for success but suffer disproportionately less than lower-paid employees as a consequence of failure. The ramifications of the financial crisis of 2008 would appear to suggest that this tradition continues.

[10]There are exceptions but such distinctions are considered unimportant.

Importantly, each generic approach consists of theories that are based on particular assumptions.

2.11.1   *Deliberate strategy and its theoretical models*

This type of strategy assumes that there can be a logical and structured approach to the planning and delivery of intended outcomes (Ansoff, 1965b, Andrews, 1971). Lynch provides a step-by-step approach to how a deliberate (prescriptive) approach will be applied (2006, p.39).This has been adapted to include the following essential elements:

1. Developing the organisation's key aims and objectives – its mission.
2. Carrying out an analysis of the environment in which the organisation operates which will consider all of the key influences and determinants on demand (including consumer choices, trends and market behaviour).
3. Reconsidering step 1 in the light of step 2 to ensure that the mission is appropriate.
4. Developing a number of strategic options that might be applied to attain the key objectives.
5. Using an agreed and 'trusted' methodology to select the option that is considered **most** likely to attain strategic success (actual attainment of objectives).
6. Implementing the option chosen in step 5.
7. Monitoring the results to provide feedback and historical data which can inform the strategic decision-making process in the future.

Grant (2003) provides a strategic planning cycle that diagrammatically summarises all of these steps:

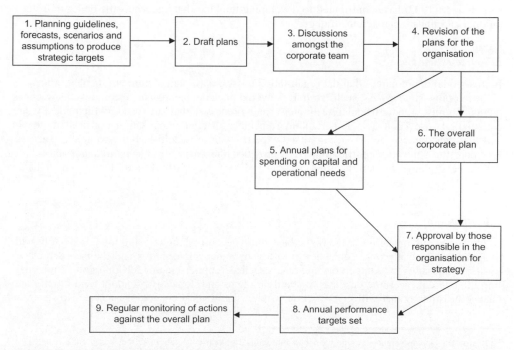

Figure 2.4   Diagrammatic representation of a typical planning cycle (based on Grant, R. 2003, p.499).

As Johnson *et al.* maintain, adopting a deliberate approach to strategic planning will assist the organisation in the following ways (2005, pp.569–70):

- It will provide a formalised 'structured' way of thinking about the future.
- It will encourage questioning as to options (and should allow options to be challenged).
- It will consider periods that are appropriate (usually long-term) over which the strategy will be put into operation; whilst 2–5 years is considered normal some organisations attempt to consider the really long-term (measured in decades)[11].
- It will enable managers to attempt to create coordination between departments and throughout the hierarchy that may exist.
- It will provide a basis for communicating a consistent message.
- It will allow measurement against targets (and milestones for interim assessment).
- The existence of a plan is believed to be better than having nothing; if people can be convinced, then it gives them something to work towards.

According to Lynch, there are four main 'areas' of deliberate strategy:

1. Industry- and environment-based theories;
2. Resource-based theories;
3. Game-based theories;
4. Cooperation- and network-based theories.

Industry- and environment-based theories

In this model the key objective is ensuring that every decision is based on the desire to engage in maximising profit (surplus) by using tactics that ensure competition is the overriding principle. As such, the works of Chandler (1962), Sloan (1963), Ansoff (1965a) and, more latterly, Wheelan and Hunger (1992) have contributed to developing thinking that underpins the belief in using a rational and linear approach to strategy.

Resource-based theories

In this model the resources that the organisation possesses (or can obtain) are the key influences in developing strategy. As such, resources would include the physical components such as plants, premises and particular technology and processes that are used (which ideally are patented). However, it also important to remember that an organisation consists of people and that their knowledge and accumulated expertise can provide the ability to deliver success. As such, the following contributors have provided the basis for this particular approach to deliberate strategy: Wernerfelt (1984), Nonaka (1991), Kay (1993), and Hamel and Prahalad (1994).

Game-based theories

In this deliberate model for strategy the basis of decision-making is using models (ideally with assigned values that provide objectivity) to test hypothetical scenarios and their consequences for the organisation, its competitors and others such as suppliers and/or subcontractors. In modelling the consequences, the desire is to avoid problems and, logically, to adopt (and modify and adapt) the strategy that will give the organisation the greatest chance of success. However, as

---

[11] I heard of a Japanese organisation in the 1990s that had a 100-year corporate plan; that takes real foresight!

experience in all games tell us, sometimes luck plays its part (especially where chance, as in cards, is involved). The work of Dixit and Nalebuff (1991) has been influential in developing this model of strategy.

Cooperation- and network-based theories

This model is based on recent empirical evidence that shows that organisations that are wiling to cooperate with others and develop relationships that are mutually beneficial enjoy no less success than those engaged in outright competitiveness. Whilst it is logical to develop cooperation with those in the supply chain (suppliers, subcontractors and distributors), the cooperative network may also incorporate competitors. Because of the huge costs involved, major motor manufacturers are happy to cooperate in research and development into new technology and innovations that can be shared. This reduces the burden and allows every customer to enjoy the benefits, safety being a prime example. The work of Child and Faulkener (1998), Inkpen (2001) and Eisenhardt (2002) has been seminal in developing this particular approach to deliberate strategy.

2.11.2   *Emergent strategy and its theoretical models*

Because of the criticisms of a deliberate approach to strategy (see particularly Mintzberg in his seminal 1994 *Harvard Business School Review* article and Stacey in his book, *Strategic Management and Organisational Dynamics,* 1993), the assumptions of using a deliberate approach were challenged and an alternative approach was proposed. This alternative should, critics of the deliberate approach maintained, be one that eschewed the belief that there could be certainty about the future. It should recognise that strategy is often developed in response to the changes that occur in the environment and may be negotiated by the various parties (individuals and coalitions) involved. Crucially, the notion that the process is logical and sequential and the outcomes entirely predictable should be rejected as being flawed in the extreme.

  What has become known as the emergent approach developed as a direct result of assumptions that characterise the deliberate approach being rejected. This is an approach in which the final objective is fuzzy (a guesstimate) and can be altered to suit the way in which the environment changes. As such, strategy is never entirely certain but constantly adaptive; strategy **emerges** rather than being decided *a priori.* Strategic implementation is something that becomes almost a by-product. As well as requiring an entirely different sort of organisation (in terms of those who lead, those who must cope with the inherent uncertainty and the embedded culture that exists), it has an impact on what goes on (as Lynch states, the effects are 'profound', 2006, p.44). The following consequences are likely:

● Strategy is likely to always remain at a nebulous level and, therefore, issues that would be resolved at the outset under the deliberate approach must remain vague.
● Resources must remain fluid and flexible to respond to needs as they emerge.
● The process of strategic development is one that occurs after implementation; it is a reflective exercise which may not accurately recall actual events (people have a tendency to idealise situations).
● There will be a need to do things and see what happens. This means that there is a need to more readily tolerate failure if things do not work out.
● The organisation's culture should be one that is framed by a desire to learn from doing rather than being instructed (people will be willing to engage in practices without formal instruction or protocol).

- There is maturity in terms of communication and exchange of information and data between members.
- Whilst explicit knowledge remains important, greater importance is attached to the tacit aspect (knowing who knows what is more important than finding the correct procedure).
- Managers must develop more attuned sensitivity in their relationships.
- Feedback and personal development assume much higher significance.

In essence, whilst it is difficult to summarise the emergent approach to strategy in a simple diagram, it is one that emphasises the desire to engage in constant dialogue and compromise. The results may not always work but important lessons may be learnt in the process.

The emergent approach is not without critics – most especially advocates of the deliberate approach. Their concerns are summarised as follows:

1. No organisation can afford to have resources and expertise on 'standby' awaiting opportunities as they emerge.
2. Managers find the culture of being adaptive and experimental stressful and alien. The culture of management tends to be based on a belief that they should make decisions and instruct others to implement them. 'Muddling through' is hard to cope with.
3. Uncertainty creates complexity and confusion.
4. Not all organisational members may be prepared to do the same thing when a solution emerges; it may suit some but not all (creating coalition and alliances needs time and dedicated effort by supporters of a particular approach; situations may not allow this).

Criticisms notwithstanding, there are, according to Lynch, four main areas (theories) that are associated with the emergent approach to strategy. These are:

1. Survival-based theories;
2. Uncertainty-based theories;
3. Human-resource-based theories;
4. Innovation- and knowledge-based theories.

Survival-based theories

This is resonant with the Darwinian principle of the survival of the fittest. Therefore, the key to survival in markets is to be able to react quickly to events and changes. On the other hand, those who cannot adapt sufficiently quickly will be found out by the market and suffer. One way to do this, according to Henderson (1989), is to be differentiated from competitors in the hope that customers will want what others cannot provide. Others argue that in competitive markets the main way to survive is to be the fittest (to operate with maximum efficiency). If an organisation has sufficient resources, it might provide a range of products or services which it will launch either simultaneously (or over a period of time) with the purpose of allowing the market to decide what is most popular in terms of sales.

Uncertainty-based theories

This is closely aligned to the belief in chaos theory: that something apparently insignificant can cause a major disruption as the effects ripple throughout systems and, in particular, markets (Gleick, 1988). Accordingly, being able to react quickly to events is essential. This requires good intelligence in order to constantly identify risk, and flexibility in terms of the way that operations and resources are used.

Human-resource-based theories

This approach to emergent strategy explicitly recognises the importance of people in developing solutions that can be effectively implemented and will enjoy support. There is a belief that people will, if allowed (and encouraged to do so), constantly adapt their approaches and the processes they use in order to arrive at solutions. This assumption leads to what Quinn (1980) describes as being 'logical incrementalism' and which is based on our apparent desire to evolve. Recent strategic developments that have been based on the experiences of Japanese approaches to management have emphasised the importance of people and their contribution to creating what is known as a 'learning organisation' (Senge, 1990). This requires that organisations which implement human-resource-based theories exhibit the following:

- People can cope but need confidence based on training and education;
- They are willing to engage in **learning** about new situations;
- They should be treated in a way that **values** their contribution to creating solutions (they must be informed and made to feel 'included').

Innovation- and knowledge-based theories

As students of organisational development will discover, especially in the case of those organisations that have been in existence for a considerable period of time, willingness to develop and implement new ideas is crucial. Whilst consumers may like traditional brands, they also want to feel that they are getting the latest developments (in terms of technology or benefits from the latest research). No one can ignore the profound effects that the Internet has had on organisations, certainly in terms of communicating ideas or connectedness with other organisations. Accordingly, writers like Kay (1993), and Nonaka and Takeuchi (1995) have identified the importance to organisations to explore ways to innovate and use their knowledge to create added value or novelty as a way of developing strategy.

In order to stimulate an environment in which members can be encouraged to reconsider practices and routines in a different way, to think about radical solutions to existing problems, and certainly to develop products and services that are radical to potential markets, a conducive climate is required. There should be a culture in which every person sees this as the norm – rather than as a task done by others in specialised departments. This is something that is entirely consistent with what Japanese manufacturers did in becoming 'world class'. Strategically, becoming what is known as a **learning organisation** is shown to have significant advantage.

2.12   The relevance of strategic theory and some considerations of Whittington's generic perspective

In many of the theories described above, there is an inherent assumption that influences outside of the organisation are the only determinant on strategy. However, strategy may be influenced by what goes on within the organisation and, as with us as individuals, history can play its part. The importance of the historical development of an organisation in terms of how it develops its purpose and, more especially, its strategy was first identified by Edith Penrose in 1959 in her book, *The Theory of the Growth of the Firm*. This work was developed by Alfred Chandler in 1962 in his book, *Strategy and Structure*. As Lynch believes (based on the work of Teece *et al.*, 1997), the impact of history on an organisation's strategy requires consideration of three 'areas':

1. **Processes** – which considers the way in which an organisation has developed its structure, relationships, leadership, technology, and all other assets (all of which may be either fixed or intellectual).
2. **Position** – which is comparative to others operating in the same sector; most especially, what advantage does it possess?
3. **Paths** – in which analysis is carried out in terms of key decisions (by whom?) and the way that aspects of organisational culture has shaped opinions and perceptions.

In the previous chapter Richard Whittington's four strategic perspectives were briefly described. These four perspectives – classical, evolutionary, processual and systemic – provide different generic views as to how organisations can 'orientate' strategy to respond to forces that will determine survival (in a very competitive environment) or, ultimately, perhaps, success. In the first perspective (classical), the emphasis is entirely congruent with the belief that strategy is deliberately planned and based on rational assumptions that maximisation of profit is the key objective. The systemic perspective, whilst also being deliberate, rejects profit as the key motive. Rather, in this perspective, pluralism is the objective in that there is a multiplicity of outcomes which will be mediated and negotiated among the main participants. The other two perspectives, evolutionary and processual, are both emergent. In the former, the objective is maximisation and in the latter, the objective is pluralism.

These perspectives provide an overall framework in which the models described above can be implemented. The dilemma for those taking the decision is, of course, knowing how to judge the environment correctly in order to select the right approach. Thus, the classical perspective will be appropriate for those managers in organisations where the emphasis is on a highly rational, planned approach to strategy. Because of its importance, developing strategy is a task that can be done only by senior management. Managers of lesser seniority are responsible for its implementation (successful or otherwise). It is also associated with the concept of what is known as 'rational economic man' which assumes that the main objective of all individuals, and, by virtue, organisations, is profit-maximisation. This assumption draws inspiration from the eighteenth-century economist, Adam Smith (author of *The Wealth of Nations* in 1776), and requires 'prudence' based on reason and self-command. The former is concerned with being able to contemplate the consequences of actions, and the latter to forgo short-term gain if the long-term will produce more return. Obviously, those managers using approaches or models that are consistent with the classical perspective should be very confident in their ability to predict the future and to know how their decisions can ensure maximum reward on investment (and effort). Given the uncertainties that are inherent in construction, it is worth asking whether the classical perspective still has validity.

In the systemic perspective, whilst the emphasis is on the use of rational plans to determine the future course of the organisation, the behaviour of individuals involved in the creation of strategy reflects the 'particular sociological contexts' in which the organisation operates. As Whittington explains:

> In the systemic view, the norms that guide strategy derive not so much
> from the cognitive bounds of the human psyche as the cultural rules of the
> local society. (1993, p.28)

As he elaborates, the internal aspects of strategy take into consideration 'social groups, interests and resources of the surrounding context'. This 'surrounding context' includes, amongst other things, he advises, 'class and professions, nations and states, families and genders' (1993). Whittington concludes that, whatever many believe about the apparent homogenisation of strategy which world economics and globalisation seem to have brought, 'the peculiarities of history and society still matter' (1993, p.29). As Chapter 3 will describe, the historical development, that has created the particular factors that intrinsically form the context that surrounds the British construction industry, creates powerful influences that determine the strategies that organisations involved tend to adopt.

Evolutionary strategy is based on the concept that survival is achieved by being fittest and nimblest in order to respond to the vagaries of the particular environment and/or markets that an organisation operates in. As Whittington remarks, evolutionary strategy requires that managers 'select markets, rather than being selected by markets' (1993, p.19). Moreover, he recommends, because evolutionary strategy is competitive, 'coexistence [becomes] impossible if organisations make their living in an identical way' (1993). The import for managers is that the most effective way to survive is to differentiate. Importantly, the ability to adapt to changing circumstances is crucial. Quoting from Willamson, he stresses the need for managers to concentrate on efficiency by dealing with their transaction costs of organising and coordinating:

> … a strategizing effort will rarely prevail if a program is burdened by significant cost excesses in production, distribution or organization, [nothing will work if the organisation] is seriously flawed in first-order economising respects (1991, p.75)

Evolution, according to Whittington, is therefore about trying to be fleet of foot and trying many small (and cost-effective) initiatives. Success is based on building on those that work and dumping those that don't. Crucially, though, evolutionary strategy is never-ending; what works now will alter in the future (and sooner than expected!). Construction organisations, perhaps, have always had a tradition of being flexible and, therefore, evolutionary. The problem, as will be elaborated on in Chapter 3, is that the constant use of cost-effective measures to reduce cost in order to ensure survival tends to be destructive. Weeding out the weakest takes its toll on those individuals affected. Reducing cost by lowering wage rates and avoiding expenditure of aspects of human resource development (training and education) stores up longer-term consequences which cause everyone (clients and participants alike) to suffer.

The last of the four perspectives, processual, rejects the assumptions of rationality that characterise the classical and systemic approaches. It also eschews the belief that maximisation of profits is essential and fundamental to the evolutionary perspective. Instead, advocates of the processual perspective believe, strategy is based on what will work under the existing circumstances; the key objective is to satisfy. Based on an understanding of the difficulties and dilemmas involved in human cognition (Mintzberg, 1978) and appreciation of the 'micro-politics' that exist in any organisational setting (Pettigrew, 1973), the processual perspective has been advanced.

Accordingly, processual strategy is based on an acceptance that humans will only be able to deal with a limited amount of information and data; all of which is interpreted in ways that are dependent on the individual's personal experiences and upbringing. Cyert and March advanced the proposition that people are 'boundedly rational' (1963). They also drew attention to the reality of organisations as being social entities made of groups who may form coalitions that suit particular personal or joint objectives. Organisations, therefore, are almost entirely not likely to be the singular homogenous entities that are postulated as being in existence by the other three perspectives. However, because there is a need to get some strategy in place, anything will be better than nothing. Using an analogy of someone crafting clay to make something useful (or attractive), strategy requires those involved in key decision-making to be intimately involved in the process and to be prepared to 'work it' (like clay) into something that is useful. As Mintzberg contends, 'crafting strategy is a continuous and adaptive process, with formation and implementation inextricably linked' (1978, p.25). Processual strategy also recognises the need to use logical incrementalism which, according to Whittington, 'is committed to experimentation and learning' (1993, p.26).

For managers adopting a processual view, there is a need to accept that the best that they can do is to cope with uncertainty by constantly adapting; strategy will therefore emerge. This requires those involved, managers and workers alike, to be able to deal with systems and approaches that are less rigid and imposed than might be relevant in, for example, the classical perspective. Strategy is developed on the basis of 'core competences' that the organisation possesses based, in large part, on the expertise and commitment of its people. Whittington postulates key characteristics that describe the processual perspective:

1. Strategy become a convenient 'construction' to explain what has been done **post-**implementation;
2. Plans may exist simply to provide managers (and others) with reassurance that something has been done;
3. 'Strategy is not just about choosing markets and then policing performance, but about carefully cultivating internal competences' (1993, p.27);
4. Goals are vague and 'slippery'.

The impact for managers of using processual thinking, therefore, is to emphasise their inability to make grand plans. Being able to get agreement on strategy will need them to negotiate their objectives. In so doing, however, they need to recognise that accommodation may be worthwhile to keep all participants in 'broad' harmony. As a strategic approach that might apply in construction, managers must adopt and adapt their use of tools and techniques to suit the needs of their clients.

2.13   Conclusion

Kurt Lewin contended that there is nothing so practical as a good theory. Theory: a word that can have a multiplicity of definitions. One way of considering theory may be that it is intended to provide explanation for the behaviour of phenomena. Crucially, the most important aspect of a phenomenon is the context (or environment) in which it occurs. Strategic theory is hugely dependent on the particular circumstances in which it is applied. Whatever the general concepts that provide the basis of strategic thinking and management (see subsequent chapters), their use in creating success (or otherwise) will depend on the combination of factors that might be thought to be hierarchical. To start with, strategy is used by organisations which, of course, range from the micro to the very large. In turn every organisation is usually thought of as being one entity among many others. This may be a market or the circumstances that create a need for its products or services. This, in turn, creates the overall environment in which each organisation must compete for scarce resources such as land, labour or capital; even governments must do this.

At the time of writing, the UK government has used a financial 'instrument' known as 'quantitative easing' to increase the overall supply of money. Advocates have argued that it has been used to increase spending and, in turn, increase confidence. Critics, on the other hand, suggest that its use will merely undermine confidence in the British economy even further. Some who are particularly apocalyptic cite the precedent of Germany in the 1920s when economic crisis led to the destruction wrought by Hitler. However, one tactic, which was used by the Nazis and has been employed on a regular basis since, is to invest in the infrastructure which, as well as creating employment, improves the economic capability of the country. As the next chapter explains, there is a long tradition of construction being used as an 'accelerator'. The important thing is that organisations that collectively make up the industry are capable of responding. But as the next chapter makes especially clear, there is an equally long tradition of treating its resources (particularly people) in a way that is cavalier, exploitative and considers them to be capable of being 'turned on and off' like water from a tap. Whilst some may argue that this is a theory in that it explains behaviour, it is one that has not served the industry nor, equally importantly, its customers well.

Chapter 3

A short socio-historical analysis of the development of the British construction industry

'Those who cannot remember the past are condemned to repeat it.'
George Santayana, philosopher

3.1 Objectives of this chapter

In writing this chapter the objective is to provide an historical overview of the influences that have created what is accepted to be the modern construction industry. In so doing, this chapter should be judged as a backdrop of the major periods and 'events' that have contributed to creating the circumstances that have led to the way that the industry has operated in the past, currently conducts itself and, it is suggested by some commentators and critics, should alter in the future so as to attain the desired levels of output, standards of work and levels of behaviour to others, such as those who supply products and/or labour (including professional services).

There is a belief expressed by many that construction, because of its peculiarities and arrangements, is unlike any other industry. Such people assert that construction is **unique** and, as a consequence, application of principles that work perfectly well elsewhere, such as, perhaps, a management theory like strategic management, should not be applied 'blindly' (or certainly not without a deeper appreciation of the context). Indeed, as they would typically argue, to make comparisons is to ignore the particular factors and history that have created a set of processes and contractual relationships, especially in Britain, that are unlike anything found elsewhere.

On the other hand, there are those who eschew the argument of uniqueness. For them, the argument of uniqueness is nonsense. Rather, they believe, many of the criticisms that are levelled at the industry are simply a consequence of a widespread refusal by those involved in all aspects of development, design and production (to name only three parts of the overall process) to both learn from past mistakes and to emulate what other industrial sectors have achieved by **radical change**. Only by doing so, they explain, can the industry escape the historical traditions and relationships that continue to exist.

Exploring the origins of the current institutional and contractual arrangements that typically exist in construction provides a useful 'backdrop' to consider the validity and relevance of strategic management theory that is described in later chapters. As will be suggested, historical traditions that have evolved in the industry, particularly since the Industrial Revolution, created the system known as contracting. Whilst it might be argued that this system has served the interests of some of the key 'players' involved in the process (notably clients), it has also been responsible for others such as contractors and those who supply them to adopt strategies that are deliberately intended to reduce cost.

As critics of the way that contracting has been used argue, it displays the worst aspects of capitalism. Even though reducing cost by improving efficiency may be entirely legitimate, doing so by the use of questionable and immoral practices such as exploitation of labour, or, to use the vernacular, 'screwing the suppliers' (such as by delaying payment or using their power when work is scarce), is certainly not. In contemporary terms, the construction industry in the latter part of the first decade of the twenty-first century has enjoyed a sustained period of both stability and growth. This has allowed organisations to invest in improvements in aspects of the processes used and the development of alternative forms of relationships that appear to provide mutual benefit to all concerned. In an increasingly uncertain world, especially in the aftermath of the financial consequences that have stemmed from the 'credit crunch', there may be a temptation to return to the adversarial relationships and conflict that characterised British construction in the period from the 1970s onwards, when economic conditions were resonant with those being experienced in 2008.

As the penultimate section of this chapter speculates, construction might usefully learn that there are alternatives to strategies merely intended to produce survival (even at the cost of others). Rather, it will be explained, organisations in other sectors have found that there are more fruitful and mutually beneficial strategies that can be pursued. These 'enlightened strategies' are explicitly based on continuous improvement of product and/or service coupled with development of its capability through key resources such as its people.

3.2   Early origins – the beginning of civilisation

Anyone who has observed ancient neolithic monuments such as Stonehenge in Salisbury or Newgrange in Meath (Republic of Ireland) will be likely to be impressed by the ingenuity and dedication that must have been involved in creating a 'structure' which, whatever the purpose, is still in existence many thousands of years later. That there were likely to be only rudimentary tools and that the materials were lifted into place using primitive ropes (despite weighing many tons) makes the feat remarkable. Moreover, the realisation that the stones used were often transported tremendous distances, having been quarried using simple tools, tells us much about the dedication and expertise that those involved must have had.

The effort that went into creating the monuments described must have been considered worthwhile. Whether the same skill and dedication went into creating the dwellings that would have been required is unlikely. However, it created a set of skills that were adapted and developed as civilisation began to develop into the small collections of homes that were the basis of prototype hamlets, villages and towns – some of which still exist but many others of which were abandoned at some point and are now forgotten. Those involved in the construction of such collections of dwellings would have looked for certain necessities, some of which are still crucial today. Food would have been relatively easy to obtain as land was plentiful. Water would likely to have been drawn from a river or stream. But as those who live too close to water, even today, will attest, in times of heavy rain flooding can be a problem. Having sufficient water to satisfy day-to-day needs was important but it was necessary to find ways to get it to dwellings that might be some distance away from the source. The Romans who occupied Britain in 55BC brought with them novel methods and solutions to the development of the towns and cities in which they garrisoned their troops that showed how construction could provide a supply of clean water, sanitation and building that ensured rapid expansion.

3.3   The Roman influence

Roman construction technology and methods are legendary. Using discipline and military organisation to muster men and materials, Roman influence was able to stretch across Europe and go as far as Scotland (Hadrian's Wall being built to provide a barrier to those considered unruly) and Wales. Importantly, they brought systems of building that created the buildings and infrastructure that are still visible in some towns and cities to this day. The development of skills such as stonemasonry and carpentry would be essential in achieving standards of comfort and hygiene that would have been unimaginable to the indigenous occupants of Britain. What was achieved required careful planning and the use of resources that any strategic manager would recognise as being essential to achievement of objectives. Without a detailed analysis of what the Romans did create in Britain, it is important to acknowledge that they laid down the early basis of roads, sewers and supplies of clean water that, in some cases, still exist to this day; see Salway, 1981 and de la Bédoyère, 2001.

3.4   Medieval organisation – the emergence of the guild system

The Romans demonstrated what could be achieved by the use of what were freely available materials – stone and timber – used by craftsmen to construct buildings that were larger in terms of their size and height than anything attempted before. The ingenuity and skills that were brought to this country were developed even after the Romans departed. They were used to construct buildings that still exist in many of the ancient towns and cities: cathedrals and castles.

Anyone who has visited a medieval cathedral or castle cannot but be struck by the sense that such vast buildings were the result of men working with hand tools. But the efforts of these men needed coordination and management. The client, whether the church, the crown or nobility, would procure the skills of those capable of completing the work needed. Whilst it is undoubtedly the case that many of those employed might have merely provided the huge amount of manual input needed, the most highly prized workers were those who possessed knowledge and proven ability to carry out skilled work. In order to ensure that their skill was recognised and that their pay remained high, these workers formed groups to cooperate in maintaining high standards.

These groups, known as **guilds**, developed in towns throughout the country in the early period of the second millennium; although they are believed to have been in existence previously, formal records do not appear before the early Middle Ages. These guilds enjoyed greatest strength and influence in the period between the fourteenth and sixteenth centuries.

Guilds provided mutual support for local craftsmen who could take advantage of being a member. Because standards of workmanship were established – and regulated by the use of fines and exclusion – there was an expectation that using a member of a guild provided a guarantee of work that would be of a high quality. This allowed guilds to set rates of pay for the services of a member. In order to ensure that there was not an over-supply of those with requisite skills, the number of new entrants into a particular guild was strictly controlled. Guilds developed the apprenticeship system which required that someone wishing to become a member must serve a minimum period of training to learn from their masters. During this period, frequently as long as seven years, there would be no payment to the apprentice. Indeed, it was common for payment to be made to the master for the privilege of being able to progress to become a member of the guild[1].

Guilds, according to Clarke, 'epitomise[d] the system of production for use' (1981, p.53). As well as binding members to the regulations that had been decided upon, they also enabled control over the material used – a vital aspect of the process creating the output of a trade:

> Thus masons worked on stone, carpenters and joiners on wood, plumbers
> on lead, and smiths on iron. And originally the trades were responsible

---

[1] And to cover the costs of lodgings and food provided.

both for the production of these particular components of the building process and their assembly on site. (Clarke, 1981)

Elaborating on this point, Clarke contends that organising in this way meant that craftsmen were able to dictate exactly which materials they wished to use; any alteration was their decision and certainly could not be imposed by clients (or at least those carrying out design on their behalf). It allowed them to extend their influence throughout the process from the point at which their services were procured until completion:

> The usual procedure was for a client or those in charge of building opera-
> tions to make a separate contract with each craft under prices fixed through
> the regulated rates, and on a piece-work basis. (Clarke, 1981, p.54)

Importantly, so far as the organisation of construction work was concerned, materials became a part of the process in which profit could be maximised if there was greater control over the choice of materials. Hence, there was the evolution of merchants within the guilds who sought to concentrate on material supply. These merchants (often referred to as **burgesses**) were able to exert control over crafts in two ways: certainly by their ability to purchase and supply the requisite materials; but also by their influence over trade in the local town and the fact that it was they who decided on the procurement of construction work. Even greater influence was created by the state which introduced charters of incorporation which conferred most power to the richest masters. Accordingly the craftsmen were organised into those who were either sufficiently wealthy and powerful to become livery companies (known as 'mercers') or those who carried out work, the yeomanry (small masters and travelling craft-workers, so-called 'journeymen').

3.5   The malign influence of merchants

According to Clarke, the division between merchants and master craftsmen was important in that it meant smaller concerns (those not regarded as being 'master') could obtain materials which significantly reduced the ability of the larger concerns who had enjoyed 'monopoly and regulation of the market' (1981, p.54). As she observes, 'evidence of deteriorating conditions for building workforce during the period of merchant control is numerous throughout the sixteenth and early seventeenth century' (Clarke, 1981, p.55). Not for the first time, the temptation to gain financial advantage through exploitation of the workforce occurred – a culture that has unfortunately continued to characterise the industry to the present (and one that has been used a short-term strategy by many contractors and sub-contractors).

Merchants, unsurprisingly, were keen to maintain their control. They formed what were known as livery companies with the specific intention of increasing both their power and wealth, especially through the acquisition of land and property. To undermine the potential for craftsmen to resist the control of merchants, additional apprentices were recruited by the latter. The effect was to decrease wages and undermine the established system of training that the traditional guilds had developed. This had many effects, the most notable of which were that apprentices became low-paid labourers; journeymen worked for lower daily rates; and smaller masters were increasingly confined to yeomanry.

In the seventeenth century it became easier for workers to gain entry into construction. This was particularly demonstrated by the granting of dispensation to workers in the aftermath of the Great Fire of London of 1666. Thus the influence of traditional guilds was further undermined. The alliance of yeomanry, based on trades that had created the 'craft corporations', which had collectively maintained the interests of members, was under threat. Forced to look for other ways to defend the interests of particular trades, these craft corporations became trade societies which considered wages, quality of workmanship and ensured that the apprenticeship scheme still operated. Those who still carried out work, the journeymen, found that this was their best way of attempting to organise themselves and ensure that their interests were properly considered. As communications between towns and cities improved, there developed a system known as 'tramping'. This enabled craftsmen to move to different areas of the country where work was more plentiful, but in a way that was regulated in order to maintain rates of pay that were adequate for them (and of course for those craft-workers who were indigenous). However, the increasingly transient nature of craftsmen in construction would prove to be to the advantage of those who wished to introduce a new form of managing the construction process: one that relied on being able to mobilise resources as effectively (and cheaply) as possible. The effects of this system, some positive but many negative, still characterise the construction industry to the present day. As many commentators assert, the strategic desire by managers of construction companies has all too often been dedicated to short-term gain rather than long-term improvement in the process.

3.6   The end of artisans and the emergence of alternative arrangements

The Industrial Revolution had a profound effect on the way that many aspects of British society and commerce operated. The migration of people from an agrarian existence to the factories of the rapidly developing towns and cities ensured that Britain changed in a way that could never have been anticipated. Wealth creation undoubtedly made many industrialists extremely rich. For those who worked in the early factories, conditions of work were frequently poor. Equally so, the accommodation they and their families had to endure was often appalling. Those with the skills and ability to marshal resources for building found that they were in great demand. Prior to the Industrial Revolution, construction was largely low-rise housing or agricultural buildings, many of which could be put up with relative ease and without the need for craftsmen. Larger buildings, such as for religious, royalty or defensive purposes, required the input of skilled workers. Consequently they were usually carried out over long periods of time: in some cases, over many years (and decades).

Industrialists, intent on making profit from the opportunities that were possible for those with sufficient capital to build a factory and purchase equipment, were not prepared to wait. Equally, there was the corresponding need for accommodation (and the associated need to provide sanitation and water) to provide housing for workers. This required large numbers of construction workers. The existing guild system could not provide sufficient numbers. Moreover, the existing rates that would be charged were seen as prohibitive to clients. As such, there was increasing pressure to abandon the guild system that protected the skills and pay rates of the journeymen.

An essential requirement for the factories was power which initially was provided by water. However, once coal was found to be more reliable and effective, it became the fuel of choice. The problem was that coal seams were often great distances from the towns and cities in which rapid industrialisation was taking place. Construction solved this problem by the building of canals which needed large numbers of workers whose main skill was to be able to dig what can only be regarded as tremendous amounts of soil (and rock) by hand. Individuals who could provide the requisite men for carrying out such work found that they were in great demand as the canal system rapidly developed to meet the need for coal which, it can be noted, increased exponentially.

Like the demand from factory owners and other clients who saw the opportunity in creating the urban environment through schemes such as speculative housebuilding, circumstances existed in which those who could summon sufficient resources to respond to the newly emerging markets could make profit. For example, Clarke explains how speculative developers were responsible for building the housing that was needed by the new urban population. With a resonance that is currently all too familiar, she details how credit, that was made more widely available to encourage developers and builders, and which 'enormously extended the production of houses', created a 'disturbance in the circulation mechanism [...] causing a financial crisis' (Clarke, 1981, p.62). Hermione Hobhouse (1971) provides a wonderfully evocative and exhaustive account of the origin and development of the 'master builder' Thomas Cubitt who was responsible for much of the more stylish and upmarket housing of the nineteenth century.

Such tremendous growth in demand from industry and commerce and also the rapid urbanisation caused resentment of the guild system which was seen as being self-interested and to stand in the way of progress. Accordingly, the emergence of organisations that provided an alternative system of procurement and production was welcomed. Importantly, these organisations promised to offer clients an 'all round' service based upon a tender that was priced in total, rather than on a retrospective 'measure and value' basis. Better still, certainly as far as the client was concerned, such tenders had to be competitive because submitting as low a price as possible was essential to getting the work. Clarke notes that, 'From about 1811 what we would recognise as 'general contractors' emerged to become the preferred method of 'letting and organising building work' (Clarke, 1981, p.64). The long-term consequence was that artisans and journeymen became more likely to be dependent on contractors for employment. As well as this, they were less able to set the rates at which they would work.

Contractors, like many of the industrialists they worked for, were primarily interested in profit. They had every incentive to reduce their costs by estimating the likely total cost of the work but reducing the price they paid for materials and labour. In the case of the former they would need to negotiate with the merchants who still held great power. Dealing with the guild representing journeymen was, though, something they believed that could be avoided by using workers outside of such a system. As Clarke notes:

> In subsuming the artisan in this way into the process of production, contractors represented an attempt to eradicate feudal restrictions on the sale of labour by altering the conditions of employment. Previously a master tradesman agreed to furnish labour at recognised time and piece rates, or journeymen undertook day work through a subcontract. Now rather than wages being regulated through the corporation or trade societies, individual wage bargaining was introduced. (1981, p.65)

The belief that change was required was reinforced by government legislation that outlawed the practice of journeymen combining to protect wage rates. Contractors, therefore, felt that their desire to reduce the costs they needed to pay labour for carrying out particular aspects of work was entirely legitimate. As Cowley (2001) and Coleman (1965) describe with respect to railway building in particular, some contractors used methods of price reduction that not only were immoral but also were corrupt. For example, Coleman describes the insidious practice of contractors and subcontractors effectively buying work on which they would make profit by the use of a system called 'truck'. The operation of truck relied on having a captive market for food and beer which workers, commonly known as 'navvies'[2], were, as they were itinerant in that they lived where the work was being carried out, which was often far from towns. Contractors and subcontractors went into partnership with local suppliers of food and beer to supply workers who were allowed a form of credit (truck) in which they would pay for what they received at the month's end. The profit was made from the fact that the produce was overpriced and frequently not fit for consumption. As a consequence, these progenitors of dynamic capitalism in construc-

---

[2] A shortening of the word navigator which derived from the fact that when work was carried out to create the canal system, workers were known as 'inland navigators'. When canal building ended, effectively usurped by the railways, the workers continued to be known as navvies.

tion demonstrated what could be achieved against seemingly insurmountable odds. The speed at which the British railway system was created can be viewed as incredible; especially in comparison to the length of time it now takes (see Wolmar, 2007). However, the cost in terms of lives lost and bodies maimed was a fact that tended to be ignored – certainly until the death toll[3] of the building of the first[4] Woodhead tunnel near Manchester between 1839 and 1845 was brought to the attention of the public in the *Manchester Guardian* in a leading article on March 7th 1846:

> The contractors being exposed to fierce competition, are tempted to adopt the cheapest method of working, without any close reference to the danger to which men will be exposed […] Life is now recklessly sacrificed; needless misery is inflicted; innocent women and children are unnecessarily rendered widows and orphans; and such evils must not be allowed to continue, even though it should be profitable. The rights of property are very sacred, and must not be uselessly encroached upon; but life is holier still.

Whilst general contracting did not treat its workers as badly as organisations in the railways, there was a general belief that, if they could get away with it, exploitation was 'fair game'. The desire to make profit was the driver behind changes in both the process and the production technique. As a result, contractors, and more especially subcontractors, were responsible for driving down costs of labour and materials, challenging traditional methods and long-established accepted practices. As the next section explains, for the next one- hundred-and-fifty years, the industry became one in which there were ongoing struggles between those who wanted to use the contracting system to their advantage and those who believed that such change was most definitely against their interests.

3.7   Contracting and its long-term effects on people, processes and production

Mass production, a core element of the factory system that emerged as a result of the Industrial Revolution, created new and cheaper goods. This enabled the development of greater (mass) consumption which, in turn, caused the need for further production. The guild system was viewed as being too antiquated and protectionist to allow the newly emerging organisations to respond to the demands for the built environment that would enable industrialists to develop the factories and associated housing, transport network and infrastructure. Contracting may have been questioned for its morality and practices but, as its supporters argued, how else could construction have risen to the challenges posed by the newly emerged markets?

Contracting had demonstrated that dynamic entities which possessed a 'can do' attitude could achieve almost anything. However, in order to work most effectively, and especially because of the pressure from clients, outdated conventions and traditions had to be replaced by new methods of operation and alternative skills developed by workers. As such, there was inevitably conflict between those who wished to maintain the exalted position of particular trades and those who argued that resistance to change slowed progress. The fact that many trades were typically involved in any building meant that those who wished to challenge established orders could

---

[3]Because records were not kept, it is only estimated that over thirty men were killed.

[4]There was a second tunnel dug between 1847 and 1852 and even though there were significantly fewer fatalities as a direct result of working, 28 people died as a result of cholera in the shanty towns that were a common feature of such works and which were frequently occupied by women and children who accompanied the men working on railways.

effectively divide and rule. Contractors, wherever they could, sought to undermine the privileged position that the traditional trades had enjoyed since the formation of guilds.

One of the major advantages that early contractors enjoyed in their attempts to create change was a major influx of cheap labour. This assisted them in undermining the ability of artisans to protect rates of pay and status (Postgate, 1923). Notably, there was a move towards larger concerns as being the preferred organisations to which clients turned to, in order to carry out many of the building schemes that had been generated by the rapid development of urbanisation and need for transport and sanitation infrastructure (Cooney, 1955). Additionally, as Ball (1988) explains, the use of contracting in construction also had the effect of creating what he describes as a 'rigid formalisation of building professions' that were separated from the part of the process in which production took place. So, for example, architects and surveyors had become a major feature of the industry but rarely were directly involved in the way that work was carried out. All of these organisations would, of course, pursue strategies that suited their ability to fulfil the needs of the markets that existed at a particular time.

So far as production was concerned, contractors wished to continue to dilute the practices that stood in the way of progress. Ultimately, this meant confronting the remnants of the guild system. Many of those workers who would have been part of the traditional system now saw that protection and solidarity could be achieved through membership of trade unions. This development was, of course, not welcomed by contractors who believed that their main objective was in responding to the opportunities that the newly developing markets presented. However, the fact that any building traditionally involved a number of trades meant that there was no single union. Ball makes the point that the fact that each union looked after only its own members' interests meant that, by dealing with each one separately, changes were allowed to be implemented – regardless of concerns.

Thus, the mid-nineteenth century saw the move towards payment based on hours worked. In summer when days were longer, workers could earn more than in winter when working hours were reduced and the levels of employment reduced. In effect, there was a move towards what is usually referred to as 'casualisation'. Postgate notes that from 1860 onwards there was the development of what he terms 'labour aristocracy' whereby those whose skills were most in demand could benefit from better terms and conditions compared to those whose skills were believed to be less important. Workers who were semi-skilled found themselves in a more vulnerable position. Any attempt to organise into a union was actively resisted by the unions of more skilled workers. Mainly, though, contractors continued activities deliberately intended to reduce cost. Where technology or alteration in processes could be used to develop new methods of working that achieved this objective, they were warmly embraced. As the next section describes, there have been various attempts to redress the shift in power in construction from workers to contractors; most of these have been in vain.

3.8   The rise and fall of trade unionism in construction

Some may argue that trade unionism has little place in a book dedicated to strategic management. However, the role of trade unions in British construction in challenging the hegemony of the worst excesses of contractors provides a useful narrative as to why the importance of people was identified in *Rethinking Construction* as being a crucial element of overall improvement in the industry's capability in satisfying its customers. As this report and the subsequent recommendations by the 'Respect for People' working group make clear, if organisations ignore the need to consider its 'human resource', they are less likely to achieve their strategic objectives than they might otherwise be capable of.

Whilst the decline of the guilds may have been welcomed by contractors, they did not want skilled workers to organise themselves into interest groups dedicated to protecting the interests of members. However, this is precisely what happened. As Ball explains, workers grouped around particular trades felt that there was an 'ideology of mutual help and solidarity' which allowed them to organise with the specific purpose of recreating what the guild system had done: regulation of conditions and wages to the advantage of members (Ball, 1988, p.63). Any gains by one union were often passed on to others who it was felt should be treated fairly. Unfortunately, though, contractors did not always take such an equitable view and conflict frequently became inevitable. The most pressing need for unions was the belief that they should do all that they could to resist the imposition of hourly rates of pay which was accompanied by the abandonment of a standard day. This, according to Price (1980), resulted in workers being expected to work long hours during the summer. The additional pay that could be earned in summer would be expected to make up for loss of earnings in winter when the day would be shorter or workers laid off (especially when the weather made effective working extremely difficult).

Whatever concerns that craft unions had, and no matter how vigorously they resisted change, by the late nineteenth century the contracting system and its consequences had become the norm. Unions recognised that their best interests were served by compliance rather than by conflict. However, they soon discovered that employers were more than willing to use methods and technology that reduced the reliance on skilled workers which, of course, lessened the influence of unions. There was the introduction of what became known as 'labour only subcontracting' (LOSC) whereby work would be contracted to firms to effectively hire the workers and be responsible for their payment and welfare.

Unions recognised that LOSC undermined their ability to defend their members' conditions and wage rates. The inability of unions to organise to stop such development frequently led to disillusionment – which reduced membership – and led to local disputes. Ironically, employers looked to national and regional official unions to control members' behaviour 'on the ground'. In an attempt to increase relative strength, unions considered amalgamations, something that had previously been inconceivable. Consequently many loose alliances evolved which eventually resulted in the formation of the National Federation of Building Trades and Operatives in 1918[5].

The process of urbanisation continued until the war in 1914 when the mobilisation of men and resources for the military campaign effectively stopped construction activity. Whilst skilled construction workers enjoyed a brief renaissance in their relative importance after the war, this proved to be short-lived. Technological developments such as the increased use of reinforced concrete and greater mechanisation of the process allowed work to be carried out more quickly and, significantly, enabled the use of semi-skilled and unskilled workers (Kingsford, 1973). This trend continued until 1939 when, once again, construction activity was halted for all but essential and military work. Unlike the First World War, the Second had involved bombing of cities and towns which caused great numbers of civilian casualties. Such bombing resulted in significant damage to all parts of the urban landscape and infrastructure. For almost thirty years construction activity was to benefit from, initially, post-war reconstruction and, in the 1960s, modernisation of all aspects of the built environment.

Construction firms that might have been small family concerns in 1945 were able to expand rapidly to cope with the vast increase in demand in almost every aspect of activity; housing and road-building were especially significant. Whilst much work was initiated by private sector clients, the role of the public sector as a client was very influential, accounting for 51% of new work carried out in 1969[6]. Workers were required in large numbers: those with skills but also those who would be prepared to carry out tasks that were less skilled. Unions attempted to organise in order to ensure that they were able to increase membership and, therefore, influence. This task was made difficult by the fact that for those workers willing to work long hours (often seven

[5]Being replaced in 1971 by the foundation of Union of Construction and Allied Trades and Technicians (UCATT).
[6]This percentage has declined ever since and has only been reversed in recent years (if PFI investment can be loosely assumed to be public sector work).

days a week), their earning power was often very high. The practice of payment by 'lump' was widely accepted. Lump meant that workers received their wages without any reduction for tax or national insurance which, as self-employed concerns, they were responsible for[7].

Construction unions, arguing for greater regulation of wages and agreed rates of pay and conditions, which would be less advantageous to workers being paid by the lump system, found that their message was frequently ignored by, most certainly, employers, but also by those whose interests they claimed to be protecting. The argument that once activity declined workers would experience rapid reduction in wages and loss of opportunity to work made little or no impact. Construction unions (most notably UCATT[8]) believed a stand should be made to restore their traditional rights. As they argued, continued diminution of worker importance would allow contractors unbridled ability to exploit a valuable resource. Strategy used by construction organisations frequently involved cutting costs wherever possible. As advocates argued, this was essential to merely survive. Unless you could offer a lower price than your competitors, you were likely to lose business. The overall consequence was reductions in wage rates, training and investment in development of future skills' base. However, as the next section describes, whilst unions attempted to resist this trend – particularly through a national strike – their efforts were doomed to failure. It was to be over a quarter of a century before any attempt was made to redress the shift of power that occurred in favour of employers. In the meantime, the industry suffered from a process of decline (caused by reduced demand) which caused critics to question whether it could respond to the increasingly demanding expectations of its clients.

3.9   The national building strike and its long-term consequences

The national building strike of 1972 might seem to be merely a footnote in the historical development of construction. Indeed, given the climate of industrial unrest which characterised the 1970s, its occurrence might be considered unsurprising. What does make it significant, if not in terms of its significance in the mind of the public (certainly when compared to the miners' and car-workers' disputes), was the legacy of bitterness between unions and employers. This legacy created a pervading sense of mistrust and hostility that undermined any efforts to implement improvement similar to that which occurred in manufacturing; especially as a result of Japanese influences such as production management techniques.

The lead-up to the national building strike of 1972 was characterised by increasingly fraught relationships between unions (UCATT in alliance with TGWU and NUGMW[9] representing construction workers in the public sector), who argued on behalf of an increasingly small membership, and employers, who asserted their right to pay whatever they believed was the market rate at the time for particular skills. The fact that a report by Phelps Brown in 1968, which had been commissioned by the Labour government, had found in favour of the use of subcontracting and labour-only as being entirely legitimate added to the bitterness that preceded the strike. In the early 1970s workload was still comparatively healthy. However, there were indications that the general economic climate was likely to decline. The consequence of such a decline, unions believed, would probably cause reduction in demand and, therefore, a decrease in requirement for workers.

---

[7]Many of those who worked under the 'lump' system were tempted to avoid paying into pensions and national insurance. Unless these workers had other investments, many have found themselves destitute in old age or when they were unable to work due to injury.

[8]UCATT (Union of Construction, Allied Trades and Technicians).

[9]TGWU (Transport and General Workers Union) and NUGMW (National Union of General and Municipal Workers).

Whilst workers could earn large sums of money under the lump system, many commentators (and most particularly unions) recognised that this was bound to be severely impeded should construction activity decrease; this was exactly what happened as a result of the economic conditions caused by the regular oil crises that characterised the decade. The role of unions at this time can be viewed as being resonant with that played by the guilds. They wanted systems of regulation of wage rates and agreements on aspects considered important for the future well-being of members, such as training and apprenticeships. Agreement on rates of pay was considered to be the issue that needed attention. Unions argued that workers should receive a guarantee of £30 for a standard 35-hour week. Given that the stance of employers was that workers were able to earn far in excess of £30 by working significantly more hours than was being argued by the unions, it was always unlikely that agreement was ever to be reached. Accordingly, negotiations between unions and major employers broke down and a national strike was called by unions acting in coordination under the aegis of the NJCBI (National Joint Council for the Building Industry).

The strike commenced in April 1972 and was deliberately intended to coincide with the beginning of the spring/summer period when activity is usually greatest. The fact that it lasted for thirteen weeks was testament to the perseverance of each side in facing down the other. During the period relations between unions and employers became increasingly bitter and whilst mass picketing was normal in that period, the strike is now remembered for alleged incidents that occurred in Shrewsbury which resulted in the jailing of two men (one of whom was Ricky Tomlinson, the now famous actor). Whilst the strike ended in an agreement of sorts, employers were determined that their continued use of labour-only subcontractors and its attendant casualisation would continue unabated. The significance was that, as had been predicted by opponents of such methods of employing workers, as soon as workloads declined, lay-offs and reduction in rates of pay ensued. Unions which had been limited in their influence in buoyant times found that they were ignored during the downturn. Construction workers, in effect, were left to fend for themselves and find work wherever they could.

It is to be stressed that many construction organisations eschewed exploitation and attempted to maintain rates of pay for workers and were prepared to invest in training and education for new entrants. However, as work became increasingly scarce, there was a tendency for firms to offer lower tenders as the means by which to ensure that their resources were being utilised. Contractors offered lower rates of pay to subcontractors and suppliers. The former offered lower rates to those operatives who were lucky enough to get any work at all. The industry as a whole engaged in cost reduction wherever possible. Accordingly, expenditure on training and education was increasingly likely to be considered 'avoidable'.

Those who were willing to enter as trainees recognised that their prospects were likely to be uncertain. Workers with security in terms of employment were found only in those organisations which had reasonable expectations of future workload or in DLOs (Direct Labour Organisations) found in the public sector. The prevailing context became one in which, even though clients could procure work 'on the cheap', the standards produced and the process of achieving 'satisfaction' were likely to involve protracted contractual negotiation and dispute (contractors and subcontractors engaging in using clauses to extract additional payment). This was to become the norm for the remainder of the 1970s, the whole of the 1980s, and much of the 1990s. It took the exhortation of two influential reports to force the industry to confront its patterns of behaviour and attitudes to both people (throughout the supply chain) and customers.

3.10 Time for change?

By the 1990s the construction industry had become one typified by intrinsic characteristics such as risk aversion, lack of investment in its key resources, especially people, and a reduced willingness (and ability) to achieve what clients believed was possible. Even though many commentators argued that change was necessary, the difficulty seems to have been one based on a belief that change would be pointless. After all, it could be reasoned, if no one else does any different, and can submit lower tenders as a result, why should we. Moreover, the industry had been operating on short-term cycles for a long period in which its resource base was developed (invested in) only when this was considered to be essential. Why, many asked, should we look to the long term if clients are only interested in their own immediate interests (based on procuring their products and services as cheaply as possible). But as many commentators asserted, since industries such as manufacturing had been forced to change their approaches (especially to the way people were employed and utilised – see below) in order to become more customer-focused and competitive, why could construction not follow suit.

The long-term consequence of the approaches typically used by construction organisations was an industry in which there was a belief that the human aspect became less important than formerly. The fact that many used whatever methods they could find to cut costs in order to survive meant that construction did not offer long-term prospects or security. Those with requisite skills that enabled them to move elsewhere did so – frequently to manufacturing. In the 1980s and early 1990s many organisations involved in manufacturing products such as automotives and electronics were beginning to experience the effects of competition from the Far East, particularly Japan. When compared to British products, those produced by Japanese manufacturers were frequently not only cheaper but, significantly, also performed in a way that was superior in terms of reliability and quality (absence of defects and standard of finish).

Those who analysed how Japanese producers could solve the dilemma of how to reduce costs whilst simultaneously improving the quality of their products may have been surprised by what they discovered; most especially by the personnel practices that were routinely used. Instead of engaging in reduction of costs by abandonment of training and education, using short-term casualised labour, and creating a culture in which the importance of people was undermined (as construction had done for many years, indeed, decades), Japanese manufacturers focused on long-term improvement by using lifelong employment. Moreover, people were actively encouraged to contribute to continuous efforts to improve every aspect of every process used. Importantly, such efforts included all of those in the supply chain, particularly suppliers. Such approaches were collectively termed TQM (Total Quality Management). Though many construction organisations did eventually recognise and attempt to emulate the strategic principles that the Japanese had implemented with such apparent success, their stance was originally viewed as exceptional. Altering the established contractual attitudes and culture in terms of the treatment of people was difficult to achieve. As had been demonstrated previously, it would take construction's most influential client (the government, directly and through its agencies) to create the circumstances by which change became not just possible, but a strategic imperative.

3.11   Government intervention – the impact of the 'Latham' and *Rethinking Construction* reports

As is amply described in Murray and Langford (2003), construction is an industry that cannot claim to have been ignored. On the contrary, it might reasonably claim to have been over-analysed; certainly when one considers the number of reports that have been prepared in order to examine and report on various aspects of its performance and behaviour. Whilst some of the reports were explicitly to present empirical finds, many contain recommendations dedicated to suggesting how positive change may be effected. Sometimes, however, the climate in which a report is published is such as to be highly conducive to change. The reports' recommendations and/or conclusions are perceived as being so pertinent and essential that all but the most immutable agree that change must occur – not only for the benefit of consumers of construction but also for the long-term well-being of its participants. The Latham Report[10] (published in 1994) and *Rethinking Construction* prepared by the Construction Task Force (published in 1998) can be seen to have been 'right for their time'.

In the period leading to the commissioning and writing of the Latham Report, construction had suffered serious decline in its workload: almost 40%. The consequential loss of jobs (some half a million) and the closure of over 30,000 small enterprises meant that the industry was in a very poor state. The incoming Conservative government of 1992 believed that a review of the industry – and most especially of its contractual arrangements – was needed. Latham produced an interim report in 1993, *Trusting the Team,* and a final report *Constructing the Team* in 1994. Both of these reports were explicit in describing an industry that had become riven by disputes about contracts, was prone to make relatively poor returns on investment by engaging in underhand practices (commonly referred to as 'screwing' those with less power) and produced work that was less impressive than it might otherwise have been if more effort had been dedicated to working in a harmonious and cooperative way rather than engaging in conflict.

Latham argued that if the construction industry were to improve it was essential that there should be a willingness by all participants to behave differently. Crucially, whilst it was incumbent on clients to lead such a change, there was acceptance that clients – certainly those in the private sector – could not be forced to alter their behaviour. However, given that it was the government that had commissioned the report, Latham recognised that this represented an excellent opportunity. He recommended that any work procured through government departments or agencies should in future set an example to all other clients by engaging in best (model) practice. In order to do this and to propagate Latham's recommendations, a new organisation was formed: the Construction Industry Board (CIB).

The CIB body was established in 1995 on the basis that it should provide strategic direction for the industry by including representation from all parts of the industry: major clients (through the Construction Clients Forum which had been set up in 1994); professions (through the Construction Industry Council); employers (through the Construction Industry Employers Council); suppliers (through the Association of Construction Product Suppliers); and a group representing other interested groups, the Construction Liaison Group. Through twelve working groups the CIB published a number of good-practice guides to inform every organisation in the industry. Whilst this work continued up to and after the 1997 election, it was somewhat overshadowed by the consequences of another report commissioned by the incoming Labour government – *Rethinking Construction* (published in 1998).

The brief given to the group tasked with writing this report was to actively explore all aspects of construction and, using comparison of best practice found in other sectors, to recommend change that would ensure radical improvement to raise levels of performance, efficiency and satisfaction of clients. The fact that Labour came to power with the avowed intention to invest in large infrastructure projects and to develop the provision of facilities for public services, such as education (schools) and health, created an imperative to investigate how a more vibrant and healthy environment could allow the Labour party's manifesto to be facilitated.

Rethinking Construction (1998) may be viewed as having a major impact in terms of recognition and influence through the various initiatives it spawned. The report was prepared and written by a group known as the Construction Industry Task Force (CITF). This group was made up of influential clients, representatives from other industrial sectors (because of their expertise in change, innovation and excellence) and, significantly, in so far as developing a new consensus between employers and workers was concerned, a representative from a union (but one that had neither high membership in construction nor had been involved in the 1972 strike). The chair of the CITF was Sir John Egan, someone who was believed to possess particular knowledge and

---

[10]Eponymously named in reference to the author, Sir Michael Latham, a Conservative who had lost his seat in the 1992 election.

experience in leading change during his time as Chief Executive of Jaguar Cars and British Airports Authority (BAA); this is the reason why *Rethinking Construction* is very often referred to as the 'Egan Report'.

Whilst many believe that *Rethinking Construction* was a logical extension of Latham's report of four years earlier, it was intentional that it should be far wider and be both radical and adventurous in its recommendations. Among these, was the proposal that the industry should be willing to develop alternative forms of contractual relationships (partnerships and alliances). CITF believed that improvement in terms of ability to give its customers satisfaction should come through a greater willingness to learn from outside the industry – especially from manufacturing in the use of techniques such as quality management, standardisation, 'lean' and supply-chain management. Having learnt how sustainable change had been produced in other industrial sectors, the CITF acknowledged the importance of people. Even though they were careful with their words, it was hard not to deduce that there was veiled criticism of the way that many construction organisations had implemented strategies that were explicit in their misuse and undervaluing of those who carry out day-to-day tasks (both operational and managerial): people.

Like *Latham*, *Rethinking Construction* spawned many groups and initiatives which, it was believed, would encourage and support change. Among these, was the Movement for Innovation (M4I) which was established to discover examples of how organisations were using novel solutions to produce improvement and, by so doing, producing greater success in terms of client satisfaction, delivery in timely periods and increased efficiency. Because objective measurement of success was believed to be a crucial component of change, Key Performance Indicators (KPIs) were devised by M4I. Complementary to the work of M4I, the Construction Best Practice Programme was established using resources and expertise at the Building Research Establishment. In order to ensure that the efforts commenced in the aftermath of *Rethinking Construction* continued, a body known as the Strategic Forum for Construction was formed in 2001 which published another report, *Accelerating Change,* in 2001.

One issue that had been identified as being crucial to the development and implementation of TQM was people. As the so-called 'quality guru', Dr Deming, had frequently pointed out, the effort to improve process must be based on the willing support and active engagement of those carrying out day-to-day processes. This was an essential part of the doctrine of SPC (Statistical Process Control) that Deming taught to Japanese managers in their efforts to develop industry based on the desire to produce high-quality goods. However, when the authors of *Rethinking Construction* considered how people in construction were treated, they were not impressed by what they found. Unsurprisingly they found that in large part the importance of the human contribution to construction was considered to be not of significance. Significantly for a group that included the interests of clients and major employers in the industry, there was an explicit acknowledgement of the need to address 'people issues'. Unless this was done, they argued, commercial interests would suffer:

> *In the Task Force's view, much of construction does not yet recognise that its people are its greatest asset and treat them as such. Too much talent is simply wasted [...] We understand the difficulties posed by site conditions and the fragmented structure of the industry but construction cannot afford not to get the best from the people who create value for clients and profits for companies* (Construction Task Force, 1998, p.14)

Accordingly, the importance of people was accepted as being fundamental to the key objective of *Rethinking Construction*: that improvement was both long overdue and would require a change in attitude by every organisation. The key question was, of course, how. As with all of the other recommendations that had been proposed, people required the formation of a group dedicated to achieving precisely this objective. As a consequence the 'Respect for People' initiative came into being, one whose task would be to explore how the input of every individual could be harnessed so as to produce long-term and sustainable transformation in construction. Two reports were produced which presented a number of recommendations intended to assist organisations in considering how to become more people-focused. To do so, it was suggested, attention should be concentrated on three Rs:

- Respect;
- Recruitment;
- Retention.

Despite the emphasis on the importance of people, the group was keen to stress that their advice was not meant to be merely benevolent. A cyclical diagram was presented which made the link between enhanced respect for people, customer satisfaction and the commercial benefits (profits/economic use of resources[11]) that could be enjoyed by those organisations that adopted the principle of better treatment of employees. The logic, it was explained, was that those who are trained and educated can more actively contribute to improving the processes by which the products and services of construction are delivered to the client. Such a sentiment precisely resonated with what has become known as the 'Deming cycle' in which improvement in quality of goods leads to more of them being sold which, as well as producing higher profits, increases the reputation of the organisation, secures jobs and, in turn, leads to new opportunities being created. Ultimately this becomes a virtuous cycle – entirely the antithesis to that which had become all too prevalent in construction.

3.12   So where is construction currently and what is next?

The so-called 'Egan Agenda' has been analysed, explored and, by some commentators (Green, 1999a,b, 2002 for example), criticised. As advocates of the recommendations of what was contained in *Rethinking Construction* believe, the change in attitude represents an opportunity for the industry to permanently shift its attitudes (culture) so that the decline that was so clearly identified in the report of 1998 would become a thing of the past. Construction would enjoy the sort of benefits that successful Japanese companies such as Toyota and Sony had experienced. Detractors of 'Egan' suggested that what was being suggested was simply capitalistic interests realigning themselves so as to regain control over productive resources. For example, because of lack of investment in people, a severe skill shortage had been created which gave too much power to some workers who were capable of effectively holding employers to ransom when negotiating wage rates as the guilds had once been able to do. As those who criticise Egan assert, long-term agreements are simply the best possible way to redress the shift in power. After all, should the industry suffer a decline, it can always go back to its old ways (see below)!

The reality has been that organisations have indeed been more willing to invest in people in a way that would have been unthinkable a generation before. Investment in construction in the period from the mid-1990s has been steadily rising. A very benign economic environment has ensured that some construction sectors, especially residential housing and commercial office development, have been especially buoyant. Public-sector investment created an apparent dilemma for the 'New' Labour government. The manifesto had been clear about the commitment to invest in what were viewed as essential areas of social provision such as education and health. Additionally it was widely accepted that much of the country's infrastructure – utilities, especially water and sewage, and transport systems such as the railways – needed urgent attention in terms of essential maintenance and a vast amount of renewal and updating.

---

[11] Organisations may be non-profit oriented, as are those in the public sector.

New Labour, however, was an administration that wanted to maintain the low rates of taxation that had been a feature of the previous Conservative government led by John Major. The solution was to adopt an arrangement which enabled schemes intended for public use to be funded by private sector organisations. This arrangement, what is known as PFI (Private Finance Initiatives), had been developed under the previous government but effectively meant that any organisation capable of raising sufficient finance could build a project (like a hospital) and rent it to the government over a fixed period at agreed rates.

In a period when the amount of finance available for such schemes increased rapidly, many consortia (frequently including contractors) engaged in PFI work. The result was that construction demand increased dramatically in both the private and public sector. Construction organisations in every area of operation have experienced a period of sustained growth since 1997. This has required investment to ensure that they can respond to the expectations of clients who, as *Rethinking Construction* so amply stated, had experienced enhanced service and delivery from other sectors. Employees benefited from both an increased demand for their expertise, which raised wage rates, and more investment in their training and development. The importance of the need to 'respect people' in construction could hardly have been more apposite.

The difficulty for construction has been in coping with such demand. Ensuring an adequate supply of suitably trained people at professional and manual level has created difficulties. As those against the widespread use of contracting and casualisation had argued in the 1970s, lack of investment would cause long-term structural problems that could not be easily remedied. The response of the industry has been to encourage new entrants through enhanced salaries. Some operatives with skills which are highly prized have seen their rates of pay increase dramatically. Apprenticeships to train the next generation of craft operatives are once again in vogue. And not for the first time, immigrants have been encouraged to fill the 'skills gap' – especially those from Eastern European countries such as Poland.

British construction, it seemed, was finally emerging from decades of under-investment and could look to a bright new future. It offered opportunities that rivalled other industrial sectors and, it appeared, a dynamic environment that enabled people to experience variety and challenge. Those who advocated change and the need to learn from others – especially manufacturing which had, in turn, learnt from the strategic intentions of the Japanese – felt vindicated. Until the late summer of 2007 the prospects for construction appeared to be extremely good. But as Chapter 1 explained, the consequences of the financial crisis, caused by problems of liquidity in the banking system (stemming from the availability of finance to fund the so-called 'sub-prime' market), have led investors in some parts of construction to re-examine their intentions. Property prices are for the first time in a generation decreasing. Accordingly, developments that would have been carried out on the assumption of rising values are no longer taking place. Larger schemes requiring freely available finance are now being questioned because banks no longer have sufficient liquidity; indeed, the once unthinkable, de facto nationalisation of the banking system has taken place. As the effects of the impending recession take hold, consumption will reduce; unemployed people spend less which means that all commercial development and construction is either 'put on hold' or abandoned. There is a corresponding impact for public spending because tax receipts go down as people have less to spend, especially if unemployment goes up. In the case of the latter, the government discovers that, as well as having reduced income through tax receipts, expenditure must increase to cover benefits payable.

Construction workers are suddenly discovering that their services are no longer needed, especially those working on city-centre residential developments such as apartments. Because overall demand is reducing, so are wage rates. Companies that have made commitments with respect to training and education are finding that the key strategic objective may simply be garnering sufficient work to survive. However, as some of the case studies contained in Chapter 12 describe, those organisations that have developed strategic markets not dependent on rapidly rising property markets are better able to survive. Schemes dedicated to public-sector provision appear to provide stability. Indeed, not for the first time, investment in construction is being used as the means by which to 'accelerate' the economy. However, the question is being asked as to what will happen if the present government (or a future one) decides that it can no longer afford such expenditure?

The overall effect may be that an industry left to the ravages of 'market forces' may, like a recidivist criminal, revert back to its traditional methods of coping. This would mean a return to hiring only when essential (and at rates which are relatively unattractive compared to what has been enjoyed up until recently). Casualisation could reappear. Indeed, some commentators argue that it has never really disappeared and that exploitation is rife among immigrants who were attracted to an apparently vibrant industry that appeared to offer easy money. Instead, people who may have been recruited by 'gangmasters' find that construction provides them with work that is poorly paid and that their conditions are poor in comparison to what would be regarded as acceptable. All of this is not what advocates of 'Respect for People' envisaged.

3.13   Conclusion

As this book makes clear, understanding the influences and factors that create the environment is essential to developing strategic intentions that are timely and appropriate. Moreover, ensuring that valuable resources are used in a way that respects their importance (particularly the human element) and attempts to be effective and efficient will be no less important than at any time in the past. As this chapter has shown, too many managers responsible for implementing strategy in construction organisations have, certainly in the past, been too keen to simply reduce costs by whatever means available and, significantly, regardless of the human cost. Interestingly, the Labour government of the late 1970s considered the British construction industry to be so vital to national importance that it believed nationalisation was the only way to maintain its resource base at a sufficiently high level and in a way that could respond to national needs. In the aftermath of the defeat by the Thatcher administration that was so vehemently against such a strategy, nationalisation was all but forgotten. Indeed, until recently, the mere mention of the word in any context seemed utterly ridiculous. However, recent events have demonstrated that some industries – like banking – are so crucial that taking control over them is essential. Whilst no one has mentioned nationalisation of construction in Britain, it may be worth revisiting some of the arguments for and against. In particular, it is vital that governments recognise the role that construction has played throughout history in providing the environment in which a fully functioning and efficient society can take place. The problem is, of course, as Santayana states: failure to learn from past mistakes will indeed mean that they are likely to be repeated in the future.

Chapter 4
Understanding the environment – markets and competition

'Opportunity is a haughty goddess who wastes no time with those who are unprepared.' George S. Clason (1874–1957), soldier, businessman and writer on business affairs

4.1 Objectives of this chapter

As individuals we are all influenced by the world in which we live, work and play. Everything we experience, particularly major events in our life, plays a part in making us into the sort of people we are. Organisations are also influenced by events that occur in what is usually referred to as the 'environment', a term that Sims *et al.* define as being the 'social, political and cultural context' (1993, p.247). As previous chapters have described, those managers that have the task of making strategic decisions must deal with all of the potential influences that will affect the conditions in which they operate. This requires analysis of things which are intrinsically uncertain and, as a consequence, carries the inherent risk that assumptions will turn out to be inaccurate. Additionally, alterations in one influence may engender changes in others, all of which may alter that environment in ways that were unanticipated.

Every day we think about possibilities and, following a period of (reasonably) careful reflection, we make the best choice. Sometimes we have to make decisions that have longer-term implications, such as whether to take a job that may mean making big changes but with the potential for career development. In doing this, as well as the risk involved, each situation will be likely to involve positives and negatives. The final decision will depend on a multitude of influences and, ideally, should be the one we have strongest belief in. So, even though we make a choice that appears to others to be illogical, it instinctively feels right for us. This is important. If we have passion about what we have decided, then it is more likely that we will be successful (certainly when compared to doing something we are only half-hearted about). Organisations frequently make decisions that, even though there is risk, with sufficient belief and commitment, will turn out to be correct.

4.2 Appreciating the context of construction

As Chapter 1 explained, work carried out by the construction industry covers a very wide spectrum of clients and involves a tremendous variety of skills and expertise to meet the requirements of those who procure work. For those organisations that wish to analyse the environment in which they will work, it is important to appreciate that there are certain **Key Factors for Success** (KFS) for, in the first instance, the industry and, more specifically, its constituent parts. Lynch defines KFS as:

> ... those resources, skills and attributes of the organisation in the industry that are essential to deliver success (2006, p.92)

For those familiar with benchmarking (McCabe, 2001), KFS looks similar to KSF (key success factor). Given that the words are exactly the same, it may be taken that the objective is the same: to identify the things which, if achieved, will contribute to success. Some of these may be obvious (see list below). For any organisation, though, consideration of the specific factors that they (and their competitors) will have to contend with is crucial. If you can deal with these more successfully (efficiently and effectively) than others, you will be more competitive.

When thinking of those KFS that apply to construction, some care is required. Remember that the industry produces a wide range of products and services. Its workforce is very diverse which can be both an advantage and disadvantage. The location can almost literally be anywhere (barring inhospitable environments such as mountains and deep seas). Nonetheless, there are certain KFS that will be common, such as:

- They involve the assembly of components in almost any location;
- There are a usually a number of professionals involved;
- They rely on a wide range of trades;
- The workforce tends to be mobile;
- Wage levels are governed by the amount of skilled labour available;
- Speed of construction is important (though not to all).

The difficulty is in knowing what markets actually exist which involves understanding the potential demand that particular clients may have. Having consulted previous research by Newcombe (1994) and Rowlinson (1999), Boyd and Chinyio suggest that there are three categories of construction client (2006, p.7):

1. Uninformed;
2. Partially informed;
3. Well informed.

The first of these will, they explain, tend to be 'naive' because they procure on a one-off or very irregular basis. Because the second category will procure with greater regularity, they will have **some** idea of what is involved in the construction process (albeit their knowledge may be somewhat dated – depending on the length of time since their last involvement). The last category of client will be 'sophisticated'. This is because they will be involved in procuring construction on a very regular basis. In order to better understand potential markets, they contend that construction organisations tend to consider them in terms of their primary business or 'sector categorisation' (2006). Doing this, they believe, is sensible in that it allows construction organisations to better understand the needs of potential clients and how they can 'provide a better service'.

The following diagram presents a three-way categorisation of public, private and mixed clients:

Table 4.1   Some examples of types of construction customers and consumers

Public	Private	Mixed (some not-for-profit, some regulated)
Schools	Manufacturing	Rail
Social services	Agriculture	Universities
Police	Research	Housing associations
Roads	Retailing	Airports
Social housing	Energy	Religious
Defence	Commerce	BBC
Prisons	Hospitals ⎫	
Transport	Prisons ⎬ PFI	
Administrative facilities	Schools ⎭	

Importantly, they warn that too many involved in construction (both clients and those in the 'supply chain') see the process simply in technical terms and that results in a completed project which has 'physical reality [… and] substance' (2006, p.79). They argue that construction organisations should consider what they do as being not only about the product that the client receives but also about the way in which relationships develop. Understanding what clients want and what gives them satisfaction is crucial. Therefore, it seems the ability to deliver **value and quality** (my emphasis) is subject to clarification:

> … buildings are a lot more than their physical existence .Our industry is expert at delivering physical buildings: they have been trained to do it, they enjoy doing it, and the organisational systems have been evolved to make this happen. Whether this is satisfying for the client is another matter. (2006, p.80)

Crucially, no one organisation would expect to operate in every construction market. Rather, it will tend to concentrate its efforts in sub-markets (or sectors) of construction in which it has highest expertise and, of course, best potential for success. The important thing that must be remembered is that the organisation will hope to implement a strategy that achieves success and which is based on certain assumptions about the way that particular markets will develop. Experience also demonstrates that markets can sometimes alter. For an industry like construction which relies on the economy to be buoyant, this can mean that change in markets can occur very rapidly – such as when interest rates are raised rapidly.

For advocates of the prescriptive approach to strategy, this is something of a dilemma. How should organisations make choices based on assumptions that may suddenly change due to unforeseen events (and over which they have no direct control)? Whilst this may provide a very strong argument for an emergent approach which explicitly recognises and embraces change, there is the problem of making the organisation only capable of reacting. Construction is an industry that has used what is known as 'the contracting system'[1] to cope with fluctuating markets. As such, it possesses sufficient resources to cope with anticipated short-term requirements (day-to-day), but calls upon suppliers and subcontractors to provide additional labour should the need arise. Whilst the virtue of this system is that it enables an organisation to cope with increases or decreases in demand without incurring additional employee costs, it may mean that the available labour is not ideal. Following criticism that the industry received concerning lack of quality (consistency) in both the end product and the people employed, construction is attempting to remedy some of these issues by adopting a more proactive approach to the identification of potential markets and how they will alter. This chapter explains the way that process is carried out. As such, the following elements of environment will consider how the overall environment can be analysed and the appropriate tools which exist to carry out analysis.

4.3   How are markets considered?

Wickham makes the point that the term 'market' can mean different things to different people:

> To the economist [it] is an institutional system through which producers and consumers exchange value. [...] To a manager a market is the sum

---

[1]A system that is not without its detractors who typically argue that there is a temptation to reduce investment in crucial aspects of human resources such as training, education and safety.

total of demand for the [organisation's products or services ...] plus those of all competitors[2] (2000, p.48)

Identifying (and understanding) markets is of critical importance to those managers involved in strategic decision-making for the following reasons:

- Developing an intimate understanding of the potential behaviour and tastes of consumers (dealt with in detail in Chapter 5);
- Anticipation of potential demand which will consist of total (aggregate) and likely individual share;
- Appreciation of the profile(s) and strategies of any competitors;
- Consideration of the growth rates that may be possible if factors alter (either in terms of price, demand or supply);
- The ability of those within the supply chain to provide materials/components or labour (which is especially appropriate to construction).

According to Wickham, there are five factors that 'contribute to the overall attractiveness' of a market (2000, pp.49–50):

- Size;
- Growth;
- Supplier Concentration;
- Potential to differentiate;
- Margins available.

The first of these, **size**, refers to the potential consumers that exist. Usually, the larger the market, the more attractive it is likely to be to potential suppliers. The actual number of suppliers will depend on the complexities of the processes involved and the barriers that may exist which deter new entrants (see Porter's Five Force model in a subsequent section). In some markets there may be domination by a few key producers (such as in car manufacture). The size of the potential construction market is highly dependent on many factors and can be subdivided according to the needs of particular customers. So, even though there may be some very large organisations involved in the construction market, they will usually only carry out large complex projects which when considered in relation to the total market are relatively small.

The next factor, **growth**, depends on the prospects for increasing market share. This, in turn, depends on the maturity of the market. So, in a newly developing market it is highly likely that all organisations can increase market share. However, in a mature market in which all consumers are identified and their needs understood, there will be little point in supplying more unless it is to use the market 'mechanism' to reduce price and potentially undercut competitors (what is frequently referred to as a 'price war'). Even in a declining market there may be potential for increasing share by attracting custom away from competitors: usually by lowering prices although possibly by giving something better (value, quality, cachet).

Supplier concentration is an effective measure of the dominance by 'key players' (usually the largest). These dominant organisations will have advantages that make it difficult for new entrants to challenge – often because of cost advantage (as in the case of those who mass produce). They will often enjoy high brand association because of their ability to advertise and market themselves more effectively (expenditure on high-profile campaigns is much more cost effective). Smaller suppliers must therefore concentrate their efforts on those parts of the market that are

---

[2]The word organisation is used rather than Wickham's 'firm' in recognition of the fact that markets may consist of provision of products or services that are 'not-for-profit'.

ignored by the dominant few, usually because the costs involved are prohibitive to make return on investment worthwhile.

Potential to differentiate is based on those aspects of the product or service that mean potential customers believe that they gain something unique or special from one supplier that others cannot emulate because of their lack of expertise or their inability to supply using existent production methods or cost base. Accordingly, differentiation can be based on cost (being cheap), brand (being identified as popular or the best), service (having excellence through ability to provide 'back-up' and systems – but trying to ensure defects are reduced to a minimum [aiming for zero]). Certain suppliers are able to avail themselves of technology and features that others do not currently possess (or cannot afford).

The final factor, **margins available**, is especially important for those considering potential markets where profit is the motivation[3]. Margin may be expressed by the following relationship:

$$\text{Margin (M)} = \text{Sales Revenue (SR)} - \text{Production Cost (PC)}$$

In any competitive environment, if high margins are being made, this will act as an incentive to potential suppliers to enter (assuming they have the productive capability and sufficient resources). **Sales revenue** is determined by the price charged and by market conditions (supply and demand). Consequently, under what may be regarded as 'normal market conditions', if demand goes up but supply remains fixed (see explanation of production cost – below), the price will usually rise. If the price remains high, margins also remain high and there is an incentive for new suppliers to produce. As a consequence, supply will match demand and price will normally fall (which will reduce margin).

Production cost is determined by the costs of resources required. In any market where resources are scarce, their value rises and, of course, their costs rise. If sales revenue remains fixed, the margin will be reduced. The willingness of suppliers to pass any increases on to customers will be influenced by the desire to maintain margins but usually in the knowledge that other competitors may be willing to tolerate lower profits. Indeed, others may be able to charge less and still maintain margins because of their ability to use novel techniques or procure resources more cheaply.

Construction is an industry in which there is a great degree of flexibility in terms of resources. However, over the short-to-medium term capability is relatively fixed. Factories will usually have maximum levels of output. Only so many workers with requisite skills will exist. Any demand above maximum capacity will, unless alternative supplies can be found (such as from overseas), cause the price of the resource to increase. It is precisely this situation that has pertained in the UK in the period since the late 1990s. Not for the first time, workers were attracted to the UK by higher wages than they could earn at home, such as in eastern Europe. However, as demand declines then so do wages.

4.4   Appreciating the dynamics of markets and competition

We generally think about a 'market' in such a way as to suggest that it is a fixed entity and has a sense of permanence. However, markets frequently display characteristics that suggest they are far from static. Nonetheless, those who take key decisions concerning strategy are concerned with identifying the extent of current or future markets and, of course, the potential opportunities

---

[3]Whilst managers in non-profit-making organisations may not have profit as their main focus, they will still be assumed to be dedicated to calculating the real costs of providing goods or services and whether it is economic to do so; i.e. whether the funds received are adequate to pay for sufficient resources.

that may allow their organisation to gain an advantage over competitors: 'competitive advantage' (see below). What this requires is that managers responsible for strategic thinking to be able to identify who are the existing or potential customers, and how they might be persuaded to buy their product or service. In existing markets, especially those that have achieved what is known as 'maturity' (have been operating for a long period of time[4]), such analysis of the established demand patterns may be relatively straightforward. As such, calculating the number of existing customers may be found by consultation of published data. An equally important consideration will be the matter of who supplies these customers. Again, identifying them may be very easy. Accordingly, if an organisation wanted to enter an established market it would need to consider how it could persuade customers to switch allegiance. Alternatively, it might look at ways to develop new customers (and it could attempt to do both). Therefore, the key question is how **dynamic** the market is.

4.5   Overall analysis of the environment

When considering the environment it is important to remember that it can be dynamic in nature. Therefore, any analysis must acknowledge that events may cause change to occur which, of course, will have an impact on the strategy that is being implemented. Strategy, therefore, should be developed and implemented in such a way as to be capable of being adapted should the circumstance require. In order to assist in analysing characteristics of the environment, Ansoff and MacDonnell (1990) proposed a model in which there are two fundamental measures:

- **Changeability;**
- **Predictability.**

Changeability relates to the ways in which the environment is considered likely to alter in response to events and/or shifts in the behaviour of 'players' (customers, suppliers, etc.). Predictability relates to the amount of certainty (or uncertainty) that exists in the environment that pertain to the stability of a particular market. Construction, as an overall market, may be thought of as somewhere in between. Radical change does not happen very often very rapidly and, indeed, change is usually slow. However, when construction is considered as sub-markets, there may be greater variations in changeability and predictability. This means that even though analysis may be generalised at 'industry level', organisations must consider the factors that are peculiar to the areas of interest for them. So, for example, the producers of major construction products (such as bricks and cement) would consider the overall market as being an aggregation of many smaller markets.

Ansoff and MacDonnell (1990) recommend that the two measures contained in the model are subdivided so that changeability is considered from the perspectives of **complexity** and **novelty**, and predictability is considered in terms of **rapidity of change** and **visibility of the future**.

Complexity is a consideration of the factors that affect the organisation's environment and would include issues such as internationalisation (and how events elsewhere can create influence), the technology involved, and social and political influences.

Novelty concerns whether and how much the organisation should anticipate situations that are different from the norm.

Rapidity of change is self-explanatory.

---

[4]As to exact length, this depends on the product or service and other influences. Housebuilding, for example, may be considered to be a mature market but demand alters depending on economic conditions and customers' confidence in future prospects.

Visibility of the future is considered in terms of how much information (and its veracity) can be gleaned on what the future holds for the organisation.

Importantly, there is what is known as **Degree of turbulence** which, depending on its magnitude, suggests how much the effective analysis of the environment can be carried out, and what confidence can be associated with it. This is a five-point scale where point 1 is very low and point 5 is very high. Accordingly each point on the scale suggests the likely environmental conditions (and most appropriate responses):

Environmental condition	Response
Repetitive – 1	Continue doing what worked before.
Expanding – 2	Increase output on the basis of forecasting.
Changing – 3	Need to consider alterations to strategy (which will include organisational, technological, innovative) to respond in order to cope.
Discontinuous – 4	Organisational thinking should be orientated to enable rapid alterations in markets and conditions to be dealt with.
Surprising – 5	Nothing can be easily predicted and innovation becomes a typical characteristic of organisational life. The organisation (people) should develop an ability to learn and adapt to unexpected and unanticipated events.

In addition to what Ansoff and MacDonnell propose, there are a number of techniques that can be used to specifically analyse the environment. These are:

1. The market options matrix;
2. SWOT analysis;
3. PESTEL;
4. Key drivers of change;
5. Scenario-based planning to analyse the environment;
6. Porter's five force framework.

4.5.1   The market options matrix

This model provides a method by which an organisation can consider what it provides to a market in terms of either development of its products or services or increasing its market share by attracting new customers (see also Chapter 5). Whilst four main choices are available, a fifth, withdrawal, may be a viable option if the prevailing circumstances dictate that continued involvement is uneconomical. Such a choice will normally have dramatic consequences, in terms of personnel, if the market was significant.

Market penetration   This will involve the organisation in attracting customers from competitors by adopting strategies that are explained in greater detail in subsequent sections. Much will depend on the 'maturity' of the market and how much it is possible to use marketing and advertising to attract custom from competitors. If this is not possible (or economically viable), the organisation may consider either market development, or development of the actual product or service (usually by incorporating innovation or additional service/quality features).

Market development   This involves the attraction of customers who might not have previously been seen as natural or obvious. For instance, some customers may be unaware of the benefits that can be gained by using the product or service. As a consequence of advertising, these benefits may stimulate demand that did not previously exist. Another example might be a small organisation that expands its potential market by operating beyond its traditional territory.

Product or service development   The intention of an organisation which engages in development of its products or services is to ensure that existing (and potential) customers perceive that they are getting something better than before. As will be described in a later section, the objective is to **differentiate** from competitors which possibly may allow a higher price to be obtained. The manufacture of products such as cars and electronics, especially by Japanese companies, clearly demonstrates that offering superior quality (in terms of reliability and features) may be a very effective way to improve market share – especially if the price is no higher than that of competitors offering alternatives that are not as good.

Diversification   An organisation may decide that its current market base is either limited or in decline. Therefore, it might be useful to develop a strategy that means it seeks alternatives that are outside its normal area by moving into either related or completely new markets.

Withdrawal   In a market where there is over-supply it may uneconomic for all suppliers to remain. The hope is that others will withdraw before you do; much depends on how long you can sustain losses. However, if there are a number of products or services that are being offered (some of which may be substitutes for each other), withdrawal of such choice may reduce the supply sufficient to increase price. Sometimes, of course, there is no other alternative than to make the decision to abandon a market and, provided that the costs of restarting are not prohibitive, waiting until conditions improve (or developing a more effective way to deliver processes).

4.5.2 SWOT analysis[5]

SWOT is the acronym of 'Strengths, Weaknesses, Opportunities, Threats' and is probably one of the most widely recognised models for understanding strategy. Its beauty is its simplicity in that there are two dimensions to be considered: the internal and external environment. In the first instance such analysis will be what the immediate situation is. As with all strategic analysis, the objective is to consider the various situations that are likely to occur in the future environment. In considering these situations, it is possible to explore whether they will present the means by which the organisation can develop (an opportunity) or potential circumstances in which the organisation's well-being and/or survival may be threatened. Allied to this, will be consideration of the internal aspects that the organisation possesses which provide either advantage or disadvantage when compared to other competitors. This technique should be considered as being dynamic and, therefore, used in conjunction with another approach such as scenario-building analysis (see below).

Strengths and weaknesses   An organisation will consider aspects such as technology, human resources, expertise and specialised knowledge, and partnerships with others that are under their

---

[5]Given the level of recognition that SWOT enjoys, it is surprising that there is no agreement on the exact provenance of this model. Whilst Albert Humphrey who was involved in a research project at Stanford University in the 1960s is often credited with developing SWOT, there is no evidence to support this (although he did create what is known as the TAM, Team Action Model, as a way to cope with change). King (2004) and Haberberg (2000) believe that the term SWOT was commonly used by academics at Harvard University during the 1960s. Turner (2002), however, suggests that it was Igor Ansoff who developed SWOT.

control and can be altered (internal).These may be things that give it an advantage or superiority over others. If this is the conclusion, the organisation enjoys **strengths**. Alternatively, the organisation may identify aspects in other organisations (competitors) which give them an advantage. Accordingly, it would indicate that the internal resources are potential **weaknesses** and that they should be altered to allow it to attain the ability to compete. Resources used by organisations are considered in more detail in Chapter 6.

Opportunities and threats   Strengths and weaknesses are considered to be internal to the organisation and, therefore, under the control of managers. **Opportunities** and **threats**, however, are considered to be factors external to the organisation and, consequently, much less likely to be capable of being influenced by managers. For example, an opportunity could be a new market which is particularly attractive, most especially if the organisation possesses resources that make it well suited to providing products and/or services to that market (and ideally that competitors do not possess). On the other hand, a threat would be the potential decline of a traditional market or a shift by consumers to another product or service that the organisation has neither the expertise nor capability to provide (but which others may have).

Clearly, in carrying out a SWOT analysis, the objective of managers is to scan the environment in order to ascertain the potential opportunities and threats and the likelihood of their occurrence (some probability analysis would be useful). By re-orientating the internal resources, the organisation should aim to respond appropriately by increasing strengths and reducing weaknesses so as to be able to take advantage of any opportunities and, of course, avoid the threats.

Lynch provides guidance that would be likely to 'enhance the quality' of the SWOT (2006, p.451). An amended list includes the following:

- That the 'big picture' is important rather than too much detail;
- Understanding of the peculiarities of the context is crucial;
- It should aim to focus on 'specific' statements that will assist in providing clear direction;
- Whilst safety (prudence) is important, adventurous thinking might identify factors that others have failed to identify and, therefore, provide competitive advantage;
- Timeframes are important;
- That resources should be used in a way that emphasises flexibility (to cope with rapid change).

The difficulty of using SWOT, according to Lynch, is that too many managers try to envisage every 'conceivable issue' which provides a sense of overload and makes effective decisions on strategy more difficult. The ability of those managers carrying out the process to sense how the environment and markets will alter is critical to success. The difficulty for managers charged with the task of carrying out strategic decision-making is that they are often too remote and disconnected to enable them to be truly effective. By the time such managers gain a sense of any shifts in consumer tastes or behaviour, it may be too late – particularly if others have been more dynamic in providing an appropriate response.

4.5.3 PESTEL

The word PESTEL is the acronym of Political, Economic, Social, Technological, Environmental and Legal which are the six main influences that will occur in the environment (certainly at a national and international level). These are the main influences that any organisation, regardless of context, would have to consider and, of course, anticipate. Each of these influences usually includes a large number of factors, each of which should be considered in developing a strategy. Crucially, the timeframe over which the strategy is being considered should be borne in mind and it is probable that as the period increases so will the likelihood that change may occur (with the consequential need to alter the strategy). Examples of the factors (and how they might impact on construction) are as follows:

Political   This considers what the government(s) of countries in which an organisation operates is likely to do and which will impact on the strategy. Because a significant amount of construction work is procured by government at international, national and local level for a multitude of activities (e.g. transport, health, education, justice, military), the impact of political decisions should be carefully considered. Despite the fact that such work may now be carried out by non-government organisations, such as those operating under what is known as 'PFI' (Private Finance Initiatives), the influence of the incumbent (or incoming) political administration should be understood and incorporated into the strategy.

Legislation that is proposed by the 'standing' administration (as well as signals contained in the manifesto of opposition parties). This also includes proposals on taxation. Both of these will have a major influence on the way that construction is carried out (e.g. health and safety, employment practices, incentives to engage in work).

The influence of major representative bodies and would include economic analysts and influential consultancies, professional institutions (RISC, CIOB, RIBA), those who articulate the opinions of employer organisations (large contractors, subcontractors etc.), health and safety executive.

Economic   The general state of the economy will have a major impact on the desire of clients, especially in the private sector, to procure work. Therefore, any organisation carrying out analysis for strategy development will need to consider the following:

- Economic trends, such as the price of oil, which can have a major influence on international markets and the economy of every industrialised country;
- Interest rates which determine buying behaviour at every level which, in turn, will have an impact on consumption;
- The sense of well-being of the general population;
- Rates of employment which will determine rates of pay across sectors;
- Spending plans as determined by government (national and local).

Much of this information will be generally available through the analysis of economists and will be published through the business press and in reports by consultants with a specialised interest in construction.

Social   The way that we live and cooperate with each other as a society can have a major impact on the patterns of behaviour and, of course, in how we spend our money which will have a major influence on the demand for certain types of construction product. An increased demand for property by single people (as has happened in the UK since the 1960s) creates a market for housebuilders to meet such a demand[6].

Demography (the study of trends in population, which will include birth and death rates) will provide an indication of the size, growth, density and distribution of particular groups in the future. It is to be noted that as the population of a country becomes wealthier (a broad measure), the birth rate drops which, unless there is immigration, will raise the average age and, in the long term, create a demand for products and property appropriate to its needs.

---

[6]The fact that city and town centres are well provided for in terms of apartments may be seen to be part of this trend.

Technological   Construction is an industry that is constantly affected by changes in the methods used to carry out tasks and the materials that are used. Some commentators criticise the rate of change at which construction alters (most notably the authors of *Rethinking Construction*). The belief is, if other sectors have been able to incorporate changes in the way that technology is used, then so should construction in order that it too can enjoy similar benefits in terms of greater speed, reduction in cost and improvement in quality. Technology can also include the way that processes are carried out. The influence of Japanese production techniques has been very strong for all organisations involved in manufacturing to accept that they should implement techniques that will enable them to compete. As advocates of such techniques would typically argue, unless construction can provide customers with a product (and service) that matches clients' experiences elsewhere, it will lose business. Moreover, there is a view that, as clients potentially become more utilitarian in their use of buildings, they may be more prepared to use structures that have less permanence and can be produced far more cheaply by prefabrication and modularisation. Some contractors have demonstrated that it is entirely possible to use technology in this way to serve particular markets; examples are the use of modular bathrooms in repetitive-design apartments, budget hotels and prison building.

Environmental   The need to more carefully consider the way that we interact with the environment has become an important consideration for every person and, more especially for organisations engaged in industrial processes. The so-called 'Green Agenda' is one that impacts on every part of society and every sector of industry including construction. ISO 14000 is the international standard that is used to assess an organisation's commitment to the environment. Accordingly, any organisation should carefully consider whether the materials and processes used to carry out every task are as good as they might be in terms of environmental concerns. Whilst legislation is an important aspect of driving change, many major clients have developed strategies that are based upon an explicit commitment to addressing environmental concerns. This would potentially include the use of only those contractors who are able to demonstrate their commitment to the environment.

Legal   There are a number of legal concerns that every construction organisation should be aware of and be willing to adapt to if change occurs. Health and safety legislation and contractual law are two major factors that influence the way that construction organisations operate. However, any change in legislation – particularly if it concerns employment or government policy – will have a potential effect on construction. Failure to respond in an appropriate way will be likely to undermine an organisation's ability to compete effectively.

4.5.4   Key drivers of change

Whilst this approach to developing strategy may not be considered to be a particular model, its consideration is helpful to an organisation's ability to contemplate the influences that impact on it. Indeed, the consideration of factors in SWOT and PESTEL should be carried out in such a way as to recognise that they are highly dynamic. As Johnson *et al.* believe, using PESTEL analysis simply as a checklist is mistaken. The key to success is in thinking about the interaction of forces:

> It is important to identify a number of **key drivers of change** [their bold] which are forces likely to affect the structure of an industry, sector or market. […] Although there will be many changes occurring in the macro-environment of most organisations it will be the *combined effect* of just

some [their italics] of these separate factors that will be so important, rather than all of the factors separately (2005, p.69)

Change, a word that will be considered in greater detail in subsequent chapters, has become a constant for organisations in every market. The important thing for those in construction to be cognisant of is the overall effect for the sectors in which they operate. As Johnson *et al.* make clear, the 'specific drivers will vary by industry or sector' (2005, p.71). Those who are aware of the 1998 report, *Rethinking Construction*, will be familiar with the expression 'drivers of change'. These were proposed as 'fundamentals' which construction should be encouraged to apply in order to enjoy radical change similar to that of other industries (such as manufacturing and service, especially supermarket retailing). Many initiatives have been implemented as a consequence of the publication of this report and, as explained here, organisations throughout the industry have altered their approach (strategy) to attempt to comply with the recommendations.

Consistent with the desires of the authors of *Rethinking Construction,* many influences of change that have affected all sectors of industry have also impacted on construction. For example, globalisation is now taken for granted and, in order to remain competitive, organisations are willing to source their input of materials and labour from literally anywhere in the world. Consequently, it is quite common for both the labour force on site to be multi-national, and for a significant proportion of components to be produced outside the UK. Additionally, customers in other sectors have become used to buying foreign goods and are more likely to demand higher standards for the same money (the impact that China has made in the last decade has undoubtedly been important in reducing prices in many markets). Construction has come under pressure to match the simultaneous reduction in price and increase in quality. Paradoxically, whilst it is taken for granted that goods and people can be transported across the globe, the issue of sustainability and the environment has become a major concern. Customer behaviour is increasingly predicated by the desire to make ethical purchases. Construction, a major user of industries that contribute to the emission of 'greenhouse gases', will be expected by potential customers to demonstrate its commitment to implementing changes that result in greater sustainability.

4.5.5   Scenario-based planning to analyse the environment

Lynch defines a scenario as 'a model of a possible future environment for the organisation whose strategic implications can then be investigated' (2006, p.85). As he advises, the objective is not about prediction which assumes that it is possible to use what has happened already and to extrapolate; rather, a scenario is about considering a range of current situations and to then consider what might be likely to happen and, crucially, what the organisation would do, at the very least, to cope. Significantly, the use of different scenarios allows the decision-makers involved to be more creative in their strategic thinking than would be the case if they simply tried to make simplistic assumptions based on the belief that rationality is the guiding principle (see the list of points for guidance below). As a consequence, something radical may emerge (so-called 'thinking out of the box') as well as being able to, as Lynch states, 'understanding the dynamics of the strategic environment'.

For any construction organisation, this would be an extremely useful way of getting people to 'brainstorm' how they believe the market(s) that they serve may alter or be developed. An example would be a housebuilder that wants to consider changes in critical things like interest rates (and by implication the general state of the economy), demographics, cultural expectations (such as how space available is used) and energy requirements. A very radical, and admittedly highly unlikely, scenario would be that which occurred in car manufacturing in the 1980s, i.e. the importation of mass-produced and cheaper pre-constructed units. Assuming that purchasers were to choose these units in a more utilitarian way than the traditional belief in 'bricks and mortar', there would be a devastating effect on markets. Having considered the unthinkable, no matter how unlikely, would enable the managers in the organisation to develop contingencies should such a scenario come to pass. Clearly, the key would be in 'reading the market' and spotting any significant signals.

In using scenario-based analysis, Lynch suggests that there are certain things that should be borne in mind (2006):

● Start from an 'unusual viewpoint' that is very different from the likely scenarios normally proposed.
● 'Develop a *qualitative description* of a group of possible events or a *narrative* that shows how events will unfold.' As Lynch believes, the need to provide a quantitative dimension is not necessary.
● Having developed particular scenarios, it is necessary to 'explore' the potential 'outcomes', as he recommends two or three scenarios (more than three will become overly complicated).
● Recognising the inherent uncertainly of each scenario, and the effects on the organisation, is important. In each case there will be a range from most optimistic to most pessimistic.
● 'Test the usefulness of the scenario by the extent to which it leads to *new strategic thinking* rather than the continuance of existing strategy'.
● Remember that the objective of this process is to '**cope with uncertainty**, [my bold] not to attempt to predict the future'.

4.5.6   Porter's Five Forces Framework

Michael Porter has become one of the seminal influences on strategic thinking. In particular, he believed that organisations strive to achieve **competitive advantage**, which is the ability to create circumstances and utilise assets and resources which enable them to offer greater value (through price or quality) than others. The result is that such an organisation will enjoy a position that gives it an effective 'head-start' over others. In order to provide guidance as to how this might be done, Porter proposed a model that incorporates five forces which exist and which managers in an organisation should actively consider. These forces are:

● The bargaining power of suppliers;
● The bargaining power of buyers;
● The threat of new entrants;
● The threat of substitutes;
● The extent of competitive rivalry.

It is recommended that this model is used at the level at which activity takes place. As explained earlier in this chapter, appreciating the differences in influences (forces) in more specific markets is crucial. Therefore, in an industry like construction, whilst there may be forces that have an impact across the industry (such as government introducing new legislation), any organisation using this model must apply it in a way that recognises the particular issues and influences that exist in the environment in which it operates. It is also important to be aware that the five forces may be interdependent which will mean that a change in one may potentially cause alterations in others.

The bargaining power of suppliers

In every productive process there will normally be those who supply goods or services: materials or components to construct the building and the labour to do so. As we know from experience, when something becomes scarce its price tends to rise. Therefore, the bargaining power of a supplier is based on their ability (or the collective ability of a number) to control what they provide[7]. In times when construction activity is high, there will inevitably be a greater call upon those who supply materials and labour and, as a consequence, prices will rise. In a market where there is so much dependency on suppliers, this can be a major problem for those who are working on a 'fixed price'; any rapid price rises may wipe out potential profit and, worse, mean that they work for a loss. It is for this reason that the popularity of long-term agreements (commonly referred to as 'partnering') has increased, both with clients and suppliers.

Construction is different from other industries in that there is less dependence on particular suppliers. Indeed, if there is scarcity, the industry is adept at being able to source its supplies from elsewhere, such as labour from overseas (a tradition that has existed since the Industrial Revolution). Additionally, it is now very common for some materials, especially components, to be supplied from abroad (especially where labour costs are low – such as China). One trend that has become noticeable is for those suppliers of materials and goods which are used in construction to compete with builders (certainly those who carry out smaller-scale work), to sell directly to the general public. This is called '**forward integration**'. It is also notable that some subcontractors have become so dominant that they are more powerful than the main contractors on whom they are believed to depend.

The bargaining power of buyers

The bargaining position of buyers is determined by their relative power in comparison to those who buy from them. In an industry that is as large and as diverse as construction there may be many variations in the power of buyers whose decisions, of course, will be strongly influenced by the prevailing economic climate. For example, in the case of speculative developments, the desire to procure the services of particular trades will be dependent on the rate at which building takes place which, in turn, is influenced by sales. Much will also depend on how specialised is the product or service that is being sought. If the product or service can be easily substituted, and if there is no risk involved, the bargaining power of the buyer is increased. Sometimes, a buyer may believe that because the risk or cost of being dependent on key suppliers is too high they need to take action. Accordingly, they engage in what is called '**backward integration**' which involves their taking over subsidiary companies. For example, some contractors have specialist divisions for certain trades; this ensures an internal guaranteed supply and enables them to be able to negotiate more effectively with other suppliers.

The threat of new entrants

In any competitive market, new entrants will be attracted by the potential to make profit. In every economy, increased value of property, especially in the residential sector, will create the incentive for others to access the market which, for a period of time, can force up prices even further. This may create a disincentive to those who cannot afford the higher prices and, as such, act as what is known as a **barrier to entry**. Such barriers can exist (or be created) in order to act as protection against the entry of others which would inevitably dilute the influence and power of existing providers. Typical barriers that exist in certain markets are:

- The scale of production that is needed to be economic (especially in terms of investment in land and/or equipment).

---

[7]The use of collaboration to form what is known as a 'cartel' (an alliance formed to control production, competition and prices) may be subject to investigation and punishment by either national government or international commissions.

- Qualifications or licences that are required by new entrants.
- Availability of raw materials or skilled labour.
- The way that a product or service has a unique brand and, therefore, is differentiated.
- The ability to gain access to key suppliers.
- Loyalty of existing customers.
- Experience and intimate knowledge about the products and services that is known only by 'insiders'.
- The willingness of existing organisations to engage in retaliatory action by, for example, reducing prices.
- In the past, government action by the use of regulation or tariffs to ensure the survival of an industry (this is now almost unknown and is considered to be an unfair restraint of trade); for example, many markets that we now consider to be extremely competitive used to be, effectively, state monopolies (gas, electricity, telecommunications).

Markets will vary with respect to the ease of entry (in terms of barriers that exist). Construction, though, mainly because large investment is not usually required in capital equipment to finance production facilities, is relatively easy to enter. Indeed, there has been a long tradition of organisations and individuals entering (and leaving) the industry. However, whilst this might be an advantage during times of increased demand, there may be few controls or checks on the ability or competence of those who offer their services[8].

The threat of substitutes

In offering a product or service, there may be the possibility that others can provide something which is perceived to be a perfectly acceptable alternative (particularly if it is cheaper). Frequently, the buyer may consider a switch to the substitute if the original product becomes scarce. Alternatively, the decision may be guided by more general influences such as societal or cultural aspirations and/or general changes in economic well-being. For instance, the demographic trends of society will impact on the demand for particular aspects of construction.

The extent of competitive rivalry

Competition in a market will be influenced by the strength of the other four forces acting upon it. Where there are a number of providers in competition with one another, their ability will be affected by a number of factors:

- The relative size and number of organisations (the assumption being that the more there are, the smaller their market size will be which will, as a result, normally increase rivalry).
- The rate of growth (and decline) of the overall market (if a market is stable, any increase must be as a result of having 'acquired' some of the rivals' share).
- The proportion of fixed costs that must be borne by producers. This can result in two potential outcomes. On the one hand, an organisation rapidly increases sales (by price-cutting or offers) to gain a greater market share and, possibly, drive competitors out of business[9]. On the other hand, if costs are high, there is a motivation for competitors to engage in sharing them to both reduce risk and, potentially, to exclude new entrants (see above): mutual cooperation as opposed to mutually assured destruction (MAD).

---

[8]A criticism that is frequently levelled at housebuilders despite the existence of standards or guarantees such as British NHBC (National House-Building Council).
[9]A dangerous strategy as others may react by cutting costs which results in a 'price war' in which every organisation makes a loss which, of course, cannot be sustained limitlessly.

- The degree to which products or services can be differentiated. If this is the case, then it is likely that customers will be price conscious and rivalry will be less than if differentiation is difficult and market share can be increased only by price reduction.
- The emergence of new economies (as was the case in the 1980s with Japan and, more recently, with China).
- The costs of exiting the market are high because of, for instance, redundancy costs associated with widespread closure. This will mean that producers will be inclined to continue (regardless of the price received).

It is important to recognise that Porter's Five Forces model has not been without its critics. Therefore, using this model requires care and acknowledgement of the potential problems that exist. The main criticism that has been levelled at the model is its apparent lack of dynamics in terms of the magnitude of the five forces and their relative importance. It follows that every organisation will need to contextualise the forces in a way that is both appropriate and meaningful. If analysis is carried out, it must be done in such a way as to identify the veracity of data (and assumptions) that have been used. Accordingly, it is believed by advocates of the model, the results of analysis will provide the basis of extremely informative indications of potential scenarios that may emerge.

4.6   What are the dynamics of competition?

Competition can be usefully considered to be like concentric rings. This assists in appreciating the level of competitiveness offered by others. The simplistic logic is that the closer another organisation is to the centre (where the operative organisation is), the more competitive it will be. Hence, those in direct competition, and who offer products and/or services which consumers perceive to be a perfectly acceptable substitute, are potentially more threatening than 'near' competitors. Distant competitors are even less of a threat. The simplicity of this diagram does not show the intentions of competitors and a change in the strategy of one or more competitors, for example, may alter the level of threat that is offered. If a potential competitor (who would not normally be seen to be a major threat) were to develop a product or service that could provide consumers with an alternative they found attractive, the impact could seriously undermine the operative organisation.

Construction may be considered as being different from other industries in that whilst the 'end product' is made up of standardised components – bricks, plaster, concrete, paint and so on – the configuration will usually differ depending on the wishes and requirements of the client (apart from where there is an intention to create homogeneity). Accordingly, it might be assumed that competitive rivalry is less prevalent than in sectors where many of the factors above are found. However, as explained earlier in the book, firms tend to be clustered according to a factor such as working in a particular geographical area and they compete very vigorously for work which they believe they can muster sufficient resources to carry out (often guided by value). So, in any particular 'segment' of the construction market, there will frequently be a number of contractors attempting to compete for the same work (especially in periods where economic activity is reduced and there is a paucity of contracts).

In considering competition, it is useful to ascertain whether there are particular strategies that can be used to classify the approaches adopted by particular organisations. One particular model that may be used to do this is Michael Porter's 'generic strategies' which he first presented in his book *Competitive Strategy* (1980) and updated in *Competitive Advantage* (1985). This model suggests that competitive strategy is based on consideration of two dimensions:

- Strategic target;
- Strategic advantage.

As Porter believes, the first dimension considers whether the organisation attempts to provide its products or services to the whole market or simply a part thereof: what is known as 'focus' and assumed to be based on having particular expertise or resources. As in the diagram below, the second dimension shows that there are two broad approaches which are either by attracting customers by being cheaper than rivals, known as **cost leadership** or by offering different features or service from competitors (**differentiation**). The use of cost leadership and differentiation may also be used by those organisations implementing a focus strategy. Even though such organisations will have chosen to target a narrower part of the potential market, unless they are the only supplier with expertise or resources, they will be likely to need to make what they offer more attractive than their competitors can.

Porter advocated that it was important to concentrate on only one approach. As he argued, trying to cope with different strategies will dilute effort and expertise. Moreover, he thought that fighting others who have concentrated on one approach will undermine the likelihood of success. Any organisation that adopts more than one approach, he suggested, risks being caught in a situation where it cannot effectively compete using any approach: what he termed 'being stuck in the middle'. Some commentators, most notably John Kay in his book, *Foundations of Corporate Success* (1993), believe that adopting more than one strategy is perfectly feasible. Indeed, as the example provided in the next section indicates, it might be a very appropriate way to deal with the complexities of a market such as construction.

4.6.1 How would such strategies work in construction?

Organisations can choose whichever approach it is believed will produce the most likelihood of success. In a market in which there is intense competition, such as jobbing building, it may be necessary to keep costs as low as possible to win work; to compete by being 'cost leaders'. However, within this broad market it might still be possible to attract clients by being differentiated; such as being able to produce high-quality work or having a larger skill base which can be utilised (perhaps to give a speedier service). Finally, some construction clients have very particular requirements which will mean that only those with specialised resources or expertise can effectively be involved. Such construction organisations will focus their efforts on providing clients with what they need. In the case of those that adopt a differentiated or focused strategy, the key economic consideration is whether any additional finance needed to procure or develop the specialised resources or expertise will justify such investment (usually by higher margins).

There is nothing to stop an organisation adopting more than one of these strategies to deal with the potential markets that exist. For example, a large contractor might tender for general works on a competitive basis which involves being cheaper than competitors. In effect, such an organisation is using cost leadership. As previous chapters have explained, this 'traditional' approach has been one that has not served the industry well if the objective of being cheapest means that investment in training and development of productive processes is reduced. However, it is possible to produce work that is 'as good' as others but can be carried out more efficiently and effectively than competitors (perhaps by the use of quality management techniques to reduce waste and ensure 'right first time'). Such an organisation might at the same time have specialised resources that allow it to be able to provide a service to the general construction industry that is attractive to more discerning customers (who will be assumed to be willing to pay more). This organisation might also have developed resources to serve particular clients and therefore be focused[10]. Indeed, as analysis of fluctuating markets for construction (like all

---

[10]The author has personal experience of working for a large multi-national contractor involved in many different types of work and for a regional organisation that carried out general contracting, civil engineering, speculative housebuilding and some property development. The latter, in particular, gained the advantage of having a varied portfolio which allowed flexibility and the ability to cope with vagaries in market conditions.

other sectors) demonstrates, reliance on one approach carries risk in the event that demand declines.

4.7   Analysing competitive behaviour – a consideration of adaptive strategy

There are various styles that can be adopted that will give some clue as to the likely behaviour to be encountered. These, according to Miles and Snow, depend on considering appropriate ways of managing the environment, setting goals, and creating structure and implementing processes. The four types are:

Defenders who seek stability in the market. They usually occupy a well defined niche and will attempt to put others off by aggressively using their advantage (cost, marketing and promotion). Defenders have the following characteristics:

- Efficiency;
- Use of bureaucracy;
- Minimisation of costs through control.

Prospectors who seek innovation and want to create new customers by breaking into existing markets. They are often seen as high risk (may use others' money) and will be prepared to undercut to gain a share. This could be a potential threat to those who wish to defend their markets. Prospectors have the following characteristics:

- Willingness to change;
- Highly adaptable attitude by management and workers;
- Use of formality (through procedures) is minimal;
- Culture of entrepreneurship and risk-taking is normal.

Analysers who seek to survive by minimising risk and maximising profit by imitating what has been successful elsewhere. They may have secure markets for products or services but seek to adapt them to gain market share elsewhere. They could also be a potential threat to those who have established markets: defenders. Analysers have the following characteristics:

- Desire to be stable;
- Although willing to change, want to achieve it in a planned and orderly way;
- Will use controls to ensure that processes are carried out in a logical way;
- Innovative in that they explore others' practices/technology or ideas.

Reactors who are unwilling or unable to behave in a way that is consistent with the previous three. They change only when forced to do so or in crisis (when it is probably too late). These, potentially, are less threatening (they may even be 'picked off' by prospectors or analysers wishing to develop market share). The trouble is, in the short term they may cause market fluctuation (by, for example, dramatic price reduction to survive) which may cause a short-term drop in sales revenue. Reactors **may** have the following characteristics:

- Not predictable (and therefore it is difficult to prejudge their behaviour);
- They will attempt to get into new markets without having sufficient expertise: a 'have-a-go' mentality;
- Their structure and processes may be disorganised and reflect a sense of lack of clear focus.

4.8   How do markets grow?

In the previous section the rate of growth (and decline) was highlighted as a factor that will influence competitive rivalry. Markets for products or services have a tendency to rise and fall (and with them the fortunes of companies providing them). As a consequence it is possible to plot a curve that shows market share over a period of time and which passes through four stages:

- Introduction;
- Growth;
- Maturity;
- Decline.

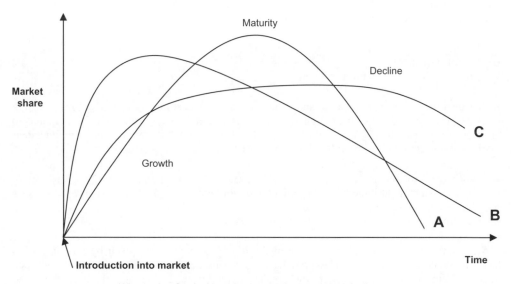

Where   A = Gradual introduction, peak and gradual decline
        B = Rapid introduction, peak and gradual decline
        C = Fairly gradual introduction, stable sales and very gradual decline

Figure 4.1   Three cycles of market life cycle.

In the early stages of the introduction of a new competitor there will be growth. The rate of such growth will be influenced by 'take-up' by customers and driven by factors such as cost relative to other substitutes or competitors. Cost may be determined by the investment that was required to bring it to market and, therefore, there is an attempt to recoup such expenditure. The rate at which the market increases will be influenced by demand which, in turn, may be a product of affordability. For example, personal computers, when first introduced in the 1980s, were seen as luxury items that only the very wealthy could afford. Competitors who saw that there was a viable market were able to enter. This will increase the overall market but, potentially, reduce the individual share that a particular producer enjoys.

The key question for those considering production and sales of their products or services is what stage the market is currently at (if at all). For the 'pioneers' that enter early there is great risk in introducing something that may be novel and has no brand identification associated with it. This means that, as well as all the costs associated with production and delivery, money will need to be spent on advertising and marketing. However, those that take such risks are in a position that allows them to demand a premium price which gives greater return on investment. An expanding market will potentially be to their advantage because of their experience and knowledge of what is required in production and sales. As has been experienced for as long as markets have been used to trade, others who see that there is profit to be made will be tempted to enter in the hope that they can equally benefit. In so doing, they will increase supply in a market that may already have reached the stage of maturity (and even decline). This will reduce everyone's market share and force all producers to either cease or engage in rivalry by, for example, cutting cost.

Once a product or service reaches a state of 'maturity', it will be necessary to decide the most appropriate strategy. Much will depend on the costs and economics of the situation. If revenues are good, it might be worth attempting to maintain by regular advertising. If there is increased competition, a price reduction could potentially ensure that sales remain at their present level. However, once decline occurs, some difficult decisions may be necessary. As sales fall, then so do revenues. As well as meaning that there is less money to spend, it could mean that there are serious economic considerations associated with over-production. This leads to surplus on the market which may cause prices to fall and revenues to reduce. Over a short period losses might be absorbed. However, longer decline may need to be addressed by considering whether investment is worthwhile to improve the product or service. This could increase its features or innovative capability. Such an evolutionary approach may cause the sales curve to shift up again. Such investment may merely prolong the period of decline and an entirely new product or service should be developed (a revolutionary approach). Having a number of 'lines' which are at various stages of sales will allow some degree of balance in terms of sales and, of course, revenue.

The key questions that should be considered are the time periods over which each of the four stages occur. They may be relatively long giving a period of months and, possibly, years which allows decisions to be made. However, increasingly, markets are prone to change very rapidly which means that any consideration of change (whether proactive or reactive) must be acted upon far more quickly. The ability of existing producers to engage in what is known as 'lean' production[11] means that even a very complicated product involving many components (such as a car) can be designed, manufactured and on sale within months (rather than years). This enables these producers to react quickly to new opportunities that emerge, in effect, to be more **competitive** than others (see next section). It potentially excludes outsiders who do not possess the industrial expertise and intimate knowledge from entering (although they may still be tempted by the possibility of being able to imitate at lower selling prices – as the Chinese are able to do because of lower comparative wages).

As your experience will have taught you, competition is something that we have come to accept, and indeed expect, in many aspects of our lives. In modern society we are bombarded with messages (some subliminal, some very explicit) which seek to convince us that we should give our custom to a particular organisation or supplier of goods or services. If the Internet has assisted us to anything it is to make such comparisons more easily than would have been the case only a few years ago. As a consequence, we are more aware of the 'competitive forces' that exist between those who vie for our custom. Construction has always been an industry that is inherently competitive; the evolution of the 'contracting system' was based on the premise that one contractor could carry out work faster and cheaper than another (McCabe, 2006)[12]. Therefore, a deeper appreciation of how competition occurs and the implications for those involved is crucial

---

[11] A concept originally developed by Toyota in Japan.

[12] Many have suggested that such competition, whilst being good for the client, was destructive to the industry and, most particularly, for those employed to carry out work on site whose conditions and pay are uncertain and, when recession occurs, relatively poor compared to others.

to understand how organisations can attempt to develop (and sustain) their competitive advantage.

4.9   Competitive advantage

Managers face the constant challenge of being able to respond to everything that goes on both within their organisation (strengths and weaknesses) and without – particularly those with which they are 'in competition'. The objective is normally to be at least as competitive as all others. In effect, what the organisation offers to potential consumers should be perceived as being no less 'good' than any comparable alternatives. However, an even better position is to be judged to be superior to others which makes choice very straightforward for consumers. Therefore, those managers concerned with strategy are constantly searching for ways to achieve superiority: what is known as **competitive advantage**. My own definition is:

> Competitive advantage is the existence in an organisation of an aspect(s) of performance or characteristics in their product(s) or service that are perceived by potential consumers to offer superiority in comparison to any other organisation with which it is in competition

Therefore, achieving competitive advantage is highly advantageous in that it gives an effective 'head start'. For example, if customers believe a product is better, they will need less convincing to buy it. This could potentially reduce the need to advertise, which reduces cost and, therefore, increases either profit or the ability to invest in development and improvement. It was precisely this 'virtuous cycle' which has enabled certain Japanese manufacturers to achieve the association between what they produce and excellence. Alternatively, an organisation might reduce its selling price in order to attract more custom. Whilst doing this might reduce profit, if the desired effect is to get customers to switch allegiance, it may be worth doing, especially if the organisation is attempting to gain access to a mature market. However, such a strategy is inherently risky in that competitors may match any price reduction which negates any advantage and over a longer period could lead to what is known as a 'price war'[13].

Competitive advantage cannot be assumed to be something that an organisation will enjoy permanently. Indeed, the corporate world is one which is replete with the imaginary 'carcasses' of businesses that once had dominant positions. Their failure may have been due to a variety of causes, but the common factor is that none would have contemplated such a future and, in all likelihood, did not sufficiently react to changes in market conditions and, most especially, to strategies by competitors. If they can, organisations wish both to develop and to **sustain** competitive advantage which will allow them to maintain their 'head start'. To be able to do this, though, requires the organisation to be in possession of advantages that others cannot easily copy or implement.

Lynch (2006, p.118), on the basis of work proposed by Aaker (1992), believes that there are certain potential sustainable competitive advantages (SCA) which are dependent on the particular area of 'business' in which an organisation operates. Accordingly, in 'high technology', organisations with SCA might possess technical excellence, reputation for quality customer service, financial resources and low-cost manufacturing. Organisations with SCA in 'service' might have a reputation for quality of service, high quality and training of staff, customer service, well-known

---

[13]Whilst customers will rarely complain about cheaper prices, the long-term consequence of a 'price war' can frequently be less availability of choice if some organisations leave the market because of reduced profits or inability to endure loss.

name and being customer-oriented. Those organisations with SCA in 'small business' might be able to demonstrate quality, prompt service, personalised service, keen prices and local availability. Finally, those organisations with SCA operating in manufacturing who are believed to be 'market leaders' would be likely to have low costs, strong branding, good distribution, quality of product and customers enjoying value for money.

There are a number of 'sources' of SCA that an organisation should consider:

- Differentiation;
- Low cost;
- Niche marketing;
- High performance of technology;
- Quality;
- Service;
- Vertical integration;
- Synergy;
- Culture, leadership and style of organisation.

Whilst all of these sources may be considered as being equal, each has a different basis and difficulty in terms of development and sustainability. The first three, differentiation, low cost and niche marketing, tend to attract a great deal of attention because they might appear to offer the simplest way to gain advantage. However, any advantage which was easy to produce would also be easy for others. If an organisation really wants to achieve SCA over others it would be better to do things that cannot be so easily imitated and that are embedded within the collective thinking of all those involved. The fact that Japanese organisations were able to demonstrate superior levels of excellence was not accidental; it required a level of commitment that others who engage in culture change find incredibly daunting. Moreover, as Kay (1993) suggests, SCA is best achieved when there is stability and continuity in all parts of the organisation. Crisis may be the catalyst for making change but it does not usually allow for calm thinking and reflection on the longer-term consequences. Kay asserts that there are four aspects of an organisation that contribute to its ability to enjoy sustainable competitive success:

- Strategic assets;
- Innovation;
- Reputation;
- Architecture.

Strategic assets, the first aspect, are any items that an organisation possesses and can use to create dominance. Such items could include things such as distinctiveness of the product or service, its unique knowledge of the market and its customers, production methods, ability to deliver on time and budget, research and innovation capability. Recent experience of contemporary business and management strongly suggests that advantages based on assets can be temporary. Unless an organisation has been able to protect its assets, knowledge or techniques (by the use of patents or copyright, for example), they will be replicated. Therefore, the only way to stay ahead is to constantly **innovate** and create (as has been a trademark of Japanese producers), which is the second aspect. Whilst this involves continual risk in terms of investment in research, development and marketing, it has been demonstrated that there is a clear advantage to using innovation as the basis for SCA. In particular, organisations which continually innovate are more likely to be able to develop a culture which fosters the desire to search for new ideas which have potential for market development.

The third aspect, **reputation**, is one that is extremely valuable. The so-called 'quality revolution' of the 1980s demonstrated that there is immense advantage to organisations that are able to consistently provide customers with quality in terms of cost, delivery or excellence – and often all three. As a result, they enjoy a reputation for trust and expertise which enhances relationships with both consumers and suppliers. This is important in that reputation reduces the need to invest in advertising and marketing (which is expensive and notoriously 'hit and miss'). Potentially, more can be charged for goods or services that are strongly associated with quality. Because of the mutual advantage that is enjoyed, suppliers who have a close relationship with organisations of repute are more likely to invest time and effort in ensuring that this aspect continues. Such relationships (frequently described as partnerships) are a particularly notable characteristic of 'world-class' organisations (see Chapter 5).

Finally, the last aspect, **architecture**, represents the way that the organisation structures itself within and without. An organisation's ability to achieve SCA is often dependent on how it creates its internal structure and develops its network of external relationships. The world in which contemporary organisations operate is one in which the ability to call upon any alliances that have been created may give a significant advantage in terms of being able to deliver, what may be for others, impossible outputs. These networks, together with the internal structure, will facilitate a degree of intimacy and understanding of the environment and market – especially opportunities for development – that competitors cannot easily achieve. As explanations of other aspects would suggest, strong and intimate understanding between all those involved in all processes is key to success. Therefore, cooperation can be an extremely important element of SCA.

4.10   Conclusion

In considering what creates the environment, it is important that an organisation recognises that demand patterns are created by the aggregate actions (and presumed behaviour) of both customers and consumers. As the next chapter describes, the former will make the decision to purchase the products or services of an organisation. The latter are those who actually consume. In many cases, customers and consumers are one and the same. However, this is not necessarily so. For example, governments spend money on construction such as roads, schools or hospitals; whilst we normally don't pay for these directly, we may use them on a regular basis: we are consumers. As any successful organisation will happily acknowledge, knowing to whom you provide your products or service is crucial.

Chapter 5

Strategy and its connection with consumers and customers – the arbiters of success

'The single most important thing to remember about any enterprise is that there are no results inside its walls. The result of a business is a satisfied customer.' Peter Drucker (1909–2005), key management academic, writer and commentator

5.1 Objectives of this chapter

In Chapter 1, the issue of understanding the needs of those who consume construction was considered. Chapter 4 considered the extent of construction markets and, in particular, how it might be possible to carry out analysis. As was explained, there is frequently a difficulty in making the necessary assumptions about construction clients' needs. The importance of appreciating these needs is to be able to respond accordingly by adjusting the supply of products or services. Therefore, developing an understanding of how particular consumers of construction are likely to behave, particularly if it is the result of influences such as the economy or shifts in markets, is invaluable to those that are given the task of managing and implementing strategy. This chapter considers the differences in definition between a consumer and a customer, terms that are widely believed to be interchangeable. Additionally, it will consider theoretical models that have been developed in order that data pertaining to consumers can be analysed and strategic decisions taken.

5.2 Defining consumers and customers

In considering strategy, there is an assumption that whatever the output of an organisation, whether it be product(s) or service(s), there are individuals or groups whose needs are fulfilled as a result. Such individuals or groups are known as consumers. They, in effect, become the final part of the production and selling process that all of us regularly engage in[1]. That some commentators use the expression 'end customer' is entirely consistent with considering the overall process as being reliant on interconnected (and inter-dependent) events and people. However, it introduces the possibility of confusion between consumers and customers. Whilst the former are concerned with satisfying needs (whether they have paid for the privilege

---

[1] Dr Deming, who is credited with assisting post-Second-World-War Japanese industry to develop its approaches to quality management, is said to have stressed that consumers are **the** most important part of the production process (see Deming's Chain Reaction, in McCabe, 1998, p.32).

or whether others have done so on their behalf[2]), the latter are, according to Wickham, defined as follows:

> An individual or organisation who buys a product [or service] with the intention of gaining economic value from it by adding value to or passing it on to other buyers unchanged (2000, pp.86–87)

As Wickham advises, customers can be either producers who are 'down the value chain' or those involved in distribution. Accordingly, whilst it is normal to consider customers as being those who buy goods and services in order to derive what is known as **utility** (value and benefit), this is not strictly correct. Customers are part of the value chain that leads to the satisfying of the utilitarian needs of consumers (the requirements of whom will have been presumed to have been accurately identified).

Consumers and customers, therefore, are essential components of strategy[3]. This is especially so in an industry like construction in which there are so many key 'actors' in the process who ensure that the end product we all use (consume) is available for activities as diverse as living, leisure, education, health, entertainment and transport. Accordingly, it is crucial that all organisations have at the heart of their strategy the desire to ensure consumer satisfaction and, as some stress is needed, delight at what they receive.

Those organisations that operate in construction are no different. This chapter will describe what this involves and how application of techniques can be used to attempt to ensure that enhanced satisfaction results in consumer loyalty[4]. Research (see Doyle, 1997) indicates that consumer loyalty can be an extremely worthwhile investment. For example, savings can be made because advertising is less necessary (remember that it costs five times more to attract a new consumer than to retain an existing one). It has been demonstrated that retention of existing customers is beneficial (a 5% increase resulting in 85% more profit). However, in order to do this, it is essential to be able to know what potential (and existing) consumers want.

5.3   Customers – who are they and what do they want?

Theodore Levitt, who was the Professor of Marketing at the Harvard Business School, argued that attracting and maintaining consumers is the key objective for any 'enterprise' (1960). As such, he maintained, developing an understanding of precisely what customers expect is central to strategic success. However, he stressed that even though it was usual to carry out analysis in broad terms (how they behave in a general way), it is helpful to try and consider consumers as individuals who have choice which will be influenced by what competitors offer. For many organisations, this is a matter of considering what is offered (or potentially may be offered), how good or bad it is when compared to others, and any other innovations or technical developments that might offer advantage.

As previous chapters have suggested, the competitive environment may change very dramatically (and rapidly), which alters the basis of all decisions taken previously. For example, those

---

[2]The public sector is dedicated to the provision of services (such as health and education) to those who do not pay (even though they may do so indirectly through taxation). Additionally, in organisations we use a range of products or services which are purchased for us to use in pursuance of strategic objectives.

[3]It is interesting to note how many leading textbooks fail to explicitly address the importance of either consumers or customers.

[4]It is widely accepted that increases mean that less effort needs to be spent remedying problems and, therefore, considering improvement.

who created what is known as Eurotunnel, the rail link under the English Channel, made their decisions on the basis of the existing ferry operators. They did not anticipate the growth of budget airlines which has reduced the price that travellers are prepared to pay (that construction costs increased so much that, on the basis of previously estimated charges, Eurotunnel could never make sufficient revenue to pay its accumulated debt). Clearly, in attempting to predict consumer behaviour, there will be inherent risk which deters new entrants (and protects those already serving their needs).

Most organisations will discover that the identification of customers is relatively straightforward and, according to Lynch (2006, p.159), involves consideration of three 'strands':

1. Understanding their needs;
2. Responding to changes in needs;
3. Providing value for money.

As he advises, consideration of each of these three strands, in a way that includes and integrates all of the stakeholders involved, will lead to the development of a strategy that is far more likely to be 'customer-driven'. So, for example, as far as understanding of needs is concerned, the organisation should have mechanisms to be able to investigate them on a regular basis.

An example of customer 'tracking'   In the case of regular purchases, such as in supermarkets, they have developed sophisticated methods of tracking our purchases which allow them to monitor their regularity. One of these, a loyalty card, will gain such information on the basis that the consumer will get something in return. The reality is that the amount is insignificant in comparison to the value of the information which is used to allocate stock levels and, more crucially, to carry out profiling of consumers (see below). Such profiles are particularly useful when considering new lines of products and services which, it is believed, will align with needs. This information will be highly likely to identify any changes in product lines and allow remedial action. Some supermarket consumers are especially price sensitive and, therefore, supermarket chains regularly monitor what they charge in comparison to others.

The challenge for other sectors, especially where there is not the regularity of purchasing by consumers (as in supermarkets), is to be able to achieve the same degree of market intelligence. Even though many elements of the construction market are certainly not as repetitive as supermarket purchasing, there is an increasing tendency to engage in longer-term arrangements – frequently referred to as 'partnerships' – intended to produce repeat work. Moreover, large clients, such as supermarket chains, have been keen to propagate such arrangements which, they believe, will result in better understanding of needs and enhanced ability of suppliers throughout the supply chain to achieve.

5.4   The use of 'customer-profiling'

Clearly, long-term inter-organisational arrangements are like those we experience on a personal level. You get to know others better, both their strengths and their weaknesses. Moreover, it is possible to know their expectations and preferences and, significantly, how to please them. Even though such knowledge may be known both informally and tacitly, it will be important to record and analyse it regularly. This can be done using what is known as **profiling** and will be especially important to allow newcomers (or those who have transferred) to know the basis for having made certain decisions concerning particular clients or suppliers. Equally, even where there is no previous history of relationships, using such a database to provide a record of analysis will still be extremely useful. Profiling can be defined:

> A profile is a tool that provides a record of the major characteristics or influences that determine the purchasing and procurement decisions of customers[5]

This record should be completed in way that systematically provides information that enables consistent analysis to be carried out and that clearly demonstrates what consumer needs are. Importantly, it should be possible to explain the basis of their current purchasing and procurement decisions and why they used one (or more) organisation rather than others. For the organisation carrying out the analysis, there is the incentive to know why they were successful (and thus maintain their advantage) or, if they are attempting to become so in the future, what they should consider doing. Additionally, providing consistent data will allow the organisation to carry out comparative analysis of the range of different organisations it seeks to either sell to (consumers) or procure from (suppliers). The profile should include:

- Name;
- Nature of business;
- Size of organisation (and relative proportion of market they account for);
- Identifiable needs;
- Aspects of consideration in purchasing and procurement decisions;
- Uniqueness of requirements;
- Individuals (and their position) taking decision;
- Matching of needs with organisation's ability to supply;
- Trends that can be easily identified (and consequences).

This list should not be seen as being exhaustive and if there are other aspects that can be usefully incorporated they should be. As with all data bases, the 'art' of analysis is in presenting information that has the virtue of simplicity and ease of presentation. Therefore, short expressions and key words are especially good, particularly where it is desirable to show a spread of organisations. This will allow other information to be presented:

- Relative costs involved in securing custom (including any costs that would be involved in 'switching');
- Buying power of key consumers;
- Interconnections that currently exist (such as partnerships or alliances) and which would preclude entry.

In order to deal with the variety of consumers and competitors involved in profiling, it may be necessary to use another level of analysis to test out the implications of altering strategy on consumers. This involves the use of what is known as the **customer–competitor matrix**.

5.5 Customer–competitor matrix

This is a technique that provides for two aspects of analysis:

- Customer needs;
- Competitor advantage.

---

[5]As such, it could be used for those in the supply chain, such as subcontractors and suppliers.

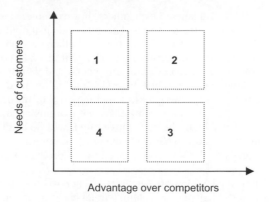

Figure 5.1 Matrix comparing customers and competitors.

The first aspect of analysis, what customers want (i.e. their needs), is considered from two perspectives. They may be consistent across the whole potential consumer group and be **homogenous**. Alternatively, consumer needs may be considered to be specific to particular customers (or small groups thereof). Accordingly, the latter type of need is considered to be **heterogeneous**. The second aspect, competitor advantage, is based on consideration of the totality of features or characteristics that are provided and how difficult it is for others to imitate them. In carrying out this analysis, it will be normal to consider economy of scale and the difficulty of imitation. There are four strategic situations that are possible:

1. Fragmented;
2. Specialised;
3. Volume;
4. Stalemate.

5.5.1 Implications of the matrix for construction

Depending on where a product or service is placed on the matrix will determine the most appropriate strategy that should be pursued. Therefore, a **fragmented strategy** is based on being different according to the particular consumer/customer (or group thereof) that is served, but accepting that others may be able to imitate such differences and, therefore, eliminate any competitive advantage. A small design practice, for example, may serve different clients in ways that allow them to be either competitive or preferred. However, unless their expertise is unique, others will be likely to attempt to 'poach' their clients.

A **specialised strategy** is one that will be appropriate for an organisation offering a product(s) and/or service(s) which contains unique or distinguishing characteristics that are capable of satisfying the needs of consumers. Crucially, the advantage that exists is significant and not easily imitated. For example, a specialist subcontractor, might possess expertise based on long experience (especially if it has a stable workforce), innovative technology and intimate knowledge of the processes required that competitors find extremely difficult to imitate. Such a strategy is highly appropriate for those organisations that are able to use patents or copyright to protect their ideas or technology.

Volume strategies are appropriate for organisations whose product(s) and/or service(s) is consistently needed by a large number of customers. Importantly, the organisation can use economies of scale to give itself the advantage over others. For example, standard component manufacturers normally produce very large quantities. This makes their individual cost very low and allows them to be sold to a wide variety of consumers (whether they are firms within the trade or individuals carrying out repairs or modernisation). This will increase consumers' perception that what they are being offered is differentiated form others (and therefore better value).

The price and/or features included in the product or service may allow its development into a 'market leader'. Price is certainly a useful weapon in attracting custom in most markets. Even though those pursuing volume strategies will acknowledge that margins are small (price minus cost), as long as sales are sufficiently high, the cumulative effect will be healthy profit/surplus. But if you can offer something at low cost, what is to stop others doing the same? The hope for organisations using such a strategy is that the costs of setting up production facilities are prohibitive in view of the low margins.

As the next strategy to be described, **stalemate**, suggests, using cost can prove both temporary and illusory. Indeed, organisations that pursue volume strategies in other industries, like the automotive sector (which some, such as the authors of *Rethinking Construction*, cite as being an exemplar), have come to recognise that outright hostility is mutually destructive. Rather, it is acknowledged, if it is possible to engage in some degree of cooperation, as in the costs of research and development, survival is more likely; new technological development can be shared (and passed on to the consumer) without incurring prohibitive costs. Japanese organisations, most notably Toyota, have shown that there is a demonstrable competitive advantage to constant innovation and improvement in the overall performance of their cars, particularly in their reliability. As they have shown, their success is entirely based on the dedicated effort of all who contribute to the supply chain (this involves strategic quality issues which are explained in greater detail in a subsequent section of this chapter). The challenge, as the authors of *Rethinking Construction* exhorted, is for all organisations to learn how such practices can be emulated and similar benefits enjoyed.

The final strategic situation, **stalemate**, as the word implies, is not a good one for an organisation to be pursuing. In this market, customer needs are very similar. Any development that is made by one competitor to differentiate its product(s) or service(s) is unlikely to be revolutionary (and will be inhibited by low margins – see below). Therefore, it will be relatively easy for others to incorporate changes or to imitate. Consequently, there is a strong likelihood that organisations will attempt to use selling price as their competitive advantage. The problem will be that any strategy based upon price reduction will simply be matched by others. Customers, believing that what they will receive is standardised, have a tendency to be price sensitive and can be easily convinced to switch their custom. Therefore, as well as wiping out any margins, it will create the dilemma of how to encourage them to return their custom. Undercutting competitors will potentially create losses (which will be presumed to be sustainable for a limited period).

Attempts to vigorously cut costs (such as reducing labour and material costs) will be disruptive and painful. If every organisation in the market does the same and creates a 'price war', the end result is likely to be good only for those that can survive longest (survival of the fittest or those with the ability to sustain reduced margins or losses)[6] and, of course, the consumers who get the same for lower cost. However, such an assumption has a longer-term consequence of reducing productive capability and, should overall demand increase, of creating shortages. This has been precisely the problem for construction in the 1990s. Following a long period during which construction activity was reduced (because of recession and reduced government spending on construction), many firms and construction bodies had severely reduced investment in training and development. The paucity of skilled labour and some vital materials has caused

---

[6]This was a characteristic of construction in the 1970s and 1980s when it was common for firms to offer to carry out work at zero profit and even potential loss. The objective was, as well as enabling resources to remain active, to carry out work faster than the target programme (and thus make saving through reduced on-costs) or to negotiate lower prices from labour (normally provided by subcontractors) or materials (from suppliers).

construction costs to increase which does not endear itself to those who have to pay for the end product: clients.

For organisations that are 'in stalemate', it is likely that survival will be the best short to medium-term aspiration. Ultimately, in the longer term, they should consider developing approaches that will create a strategy that results in more effective outputs. As a result, their products and services would be such as to move them into one of the other three quadrants (better still being in either specialised or volume markets). Consequently, the organisation would enjoy the benefit of having more significant advantage than competitors. Being able to develop advantage over competitors provides the rationale for the importance–performance model.

5.6 Importance–performance analysis

The purpose of this analytical model is to allow an organisation to consider the 'competitiveness' of what it offers by comparing it to potential substitutes provided by competitors. What is most powerful about this approach is that the perspective used is that of the most important person(s) in the value chain – the customer. The basis of analysis is carried out by considering two dimensions:

- Importance of features that are offered;
- Performance of these features.

In order to carry out the requisite analysis, it is necessary to be able to gain access to those customers (existing or potential) who should be requested to list the features that they consider to be fundamental in providing the ability to satisfy their needs. Those being surveyed should then be asked to define how **important** each feature is. Using a score-based rating is a useful way to provide measurement. The second dimension, the **performance** of each feature, is then provided by those being surveyed but, crucially, they do so in comparison with products or services offered by competitors. Again a score-based rating would be useful.

Whilst this information will be extremely useful as provided, it will be likely to have far greater impact when presented in diagrammatic form. The best way is one that shows the two dimensions; each of the dimensions is measured on a scale from low (say, 1) to high (say, 10). Accordingly, using 5 as the median for each dimension allows the creation of four quadrants:

- **High importance-poor performance** in which what the customers consider to be important to them is not being delivered as well as they think it should. The message is that effort is needed to achieve higher performance to get into the next quadrant.
- **High importance-high performance** in which the customers believe that the product or service has the right features and they perform well. Whilst this is good for the organisation, it should not be complacent. Competition may be (or become) intense and maintaining loyal customers through repeat custom is a key objective. Markets such as electronic products have demonstrated that constant innovation and development, together with extremely high levels of quality as the norm, lead customers to be averse to those who cannot consistently deliver comparable standards.
- **Low importance-high performance** means that the customers are possibly indifferent to the features but feel that they perform well. It is worth asking how much all of the superfluous features cost and whether the product or service would sell just as well without them. Reducing the features might allow the overall cost to be lowered which may increase sales.
- **Low importance-low performance** is certainly not where an organisation should aspire to be. Customers don't consider the features to be important and, anyway, they don't perform that well. However, improvement in the performance of some of the features might be worth considering if it increases sales and generates more revenue than the investment (see below for a detailed treatment of the financial considerations of such decision-making).

5.7   Market segmentation

These forms of analysis have an assumption that customers are consistent. Most organisations acknowledge the fact that customers tend to be more varied than some of the simplistic analytical models suggest is the case. In construction, large contractors, for instance, will be fully aware that every large client has particular differences which they will seek to satisfy. Even in more general markets, it makes sense to classify customers into sub-groups with needs, expectations and desires that differentiate them from other sub-groups. Carrying out such analysis will enable strategic effort to be dedicated to ensure that resources are sufficient to satisfy customers within a sub-group. Market segmentation is usually considered to be of three characteristics:

1. Product/service;
2. Producer;
3. Buyer.

The first, **product/service**, considers aspects such as price (which will tend to be relative to its features and comparable to other alternatives offered by competitors). Dependent on the market, the organisation will dedicate resources (and finance) to advertising and marketing as well as the need to innovate (research and development). Increasingly, some products can be altered to suit the sub-markets in which they are consumed. **Producer characteristics** will be determined by the nature of what is being offered, the type of organisation (history, culture, leadership), and the type of markets it has traditionally served. The producer characteristics will have a large influence on the choice of market segments that it believes it can best serve. The difficulty may come for an organisation which attempts to shift from one segment to another. For example, where a contractor has shifted from dealing with refurbishment of traditional or historical building to completion of new building for 'demanding clients' (perhaps because of a decline in the former), they might find that their expertise (and people) does not make the transition as easily as anticipated. Understanding different customers is fundamental to success. Being aware of the different characteristics of sub-groups of **buyers** is vastly important. So, for example, it will be important to consider the following questions:

● How buyer selection is made;
● Who is responsible (individual, committee, user, financier);
● The basis of decision-making (careful analysis as opposed to impulse);
● Relationship with organisations (including the one carrying out the analysis);
● Other considerations that will assist understanding.

Market segmentation which incorporates all of the three characteristics will allow the organisation to constantly monitor its position relative to others and, of course, customers. Being able to identify shifts in, for example, taste or buying behaviour earlier than others may allow a reallocation of resources or development that provides competitive advantage which enables an organisation to invest sufficiently (especially in innovation) to continue to stay ahead. This may require a significant shift in terms of strategic approach and, by implication, competitive behaviour.

5.8   Revisiting Miles and Snow's analytical model of competitive behaviour

As was explained in Chapter 4, Miles and Snow's characterisation of four types of competitive behaviour (1978) is a useful way to consider what an organisation does and its appropriateness to a particular segment. More importantly, understanding the basis of the competitive behaviour of others is essential. Accordingly, **defenders** will wish to use whatever 'weapons' they can to deter newcomers (**prospectors** or **analysers**) from attempting to gain access to their customer base. **Prospectors**, which tend to be dynamic and innovative in inclination, will look for segments that, it is anticipated, have the capability for exploitation. **Analysers**, on the other hand, even though they may be established in one segment, will be exploring the marketplace and, especially, other segments for opportunities. They may do so by offering additional features or price discounts which existing organisations will be presumed to resist and react to. **Reactors** are the easiest organisation to deal with in that, whatever segment they are in, they will simply follow what others do.

5.9   Using the BCG matrix to analyse customers

Many organisations provide more than one product or service; they have what is known as a 'portfolio'. This makes sense in that it means they are not totally reliant on a single product or service. In any competitive situation markets may shift because of external influences, such as a change in the economy which reduces spending (such as on new housing). Alternatively, as has been described above, customers can be persuaded to switch their purchasing to competitors who offer different (more attractive) features or a lower selling price[7]. The relative performance of the products or services is an important consideration for the organisation. Development of strategy will be assumed to be dependent on analysis of such performance and, it is assumed, continuing with those products and/or services that are successful and making decisions about those that are not. In making any decisions, there will be inevitable resource implications.

Analysis of the portfolio can be carried out using what is known as the BCG Portfolio Matrix. This was originally developed by Bruce Henderson of the Boston Consulting Group in the early 1970s. The intention is to categorise the portfolio of products or services that the organisation supplies in accordance with two dimensions:

- The relative market share of each of the products or services in comparison to the whole market;
- The potential growth rate that each product or service has.

As this diagram shows, there are names that are given to the four categorisations that can be applied to the portfolio following analysis. Each of these has significance in terms of how the products or services should be incorporated into future strategy and, more especially, the implications in terms of investment in resources.

1. **Star** is one that an organisation will want to develop because it has high relative market share and high potential growth rate. Even though finance may potentially be required to invest in either production or marketing, because of its high market share, it can be priced at a level which allows good return. Therefore, the higher margins will recoup investment. As long as this product or service can maintain its dominance long enough to pay off such investment, it will become a 'cash cow' (see below).
2. **Problem child** (also referred to as a 'question mark') is a product or service that has relative low market share but there is high potential growth. There may be reasons for this, for

---

[7]Construction contracting traditionally operated on the basis of lowest cost (all competitors tendering on the basis of the same drawings and specifications). Experience demonstrated to clients that the final 'product' and the service that accompanied it were far from consistent; this is a reason why many new clients prefer to use alternative arrangements which are based on collaboration and continuity of work.

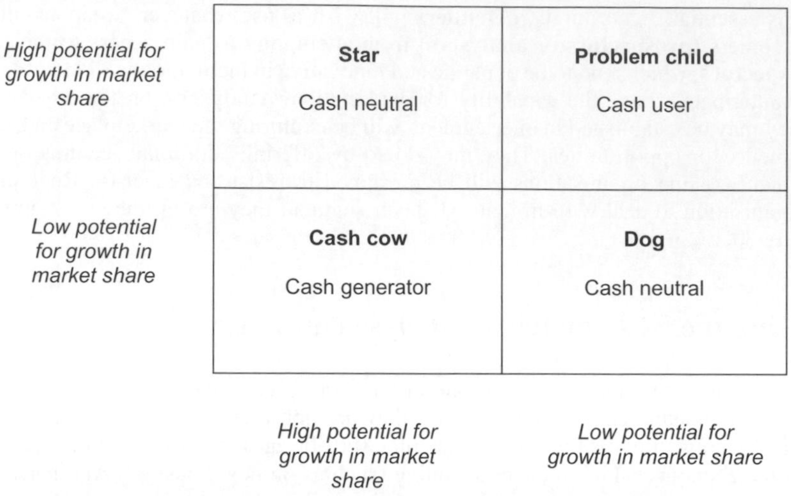

High potential for growth in market share	**Star** Cash neutral	**Problem child** Cash user
Low potential for growth in market share	**Cash cow** Cash generator	**Dog** Cash neutral
	High potential for growth in market share	Low potential for growth in market share

Figure 5.2   The BCG (Boston Consulting Group) Matrix.

example, because the product or service is new to the market or due to others becoming much more aggressive in terms of their desire to make what they offer dominant (it may once have been a 'star'). Therefore, the decision that must be taken is how wise it is to invest in its development. If the likelihood of gaining (or regaining) higher market share is good, then it is wise to commit finance. On the other hand, it may be that the market is one in which potential growth is limited. In this case, the effects of investment may be questionable if they come too late to make any difference to market share. In such a case, the product or service is likely to become a 'dog'.

3. **Dog** is a product or service that has low relative market share and low potential for growth. As such, it is highly unlikely that any additional investment in development will have a positive outcome. Therefore, as long as it is not making a loss, it may simply be a matter of doing nothing. However, in the contemporary marketplace, not investing may mean that customer loyalty will decline and new purchasers are not attracted. A more obvious decision would be to phase it out or to sell it on to another organisation that may have the desire and capabilities to reinvigorate the product or service. This would be especially so if the new organisation's operating costs were significantly lower.

4. **Cash cow** is a product or service that has high relative market share but low potential market growth. This will mean that even though it may be popular (or, possibly because of its special or unique features, the best current choice) and sell at a higher price, it is not essential to invest finance in development and marketing. Depending on the dynamics and behaviour of the existing customers, it will be possible to continue to use a cash cow to generate finance that can be used to develop 'stars' or 'problem children'. The danger that lies in this strategy is that others may provide alternatives that have similar or better features at lower prices. If this happens, cash cows will lose their position and potentially become 'dogs'.

5.10   The McKinsey Directional Policy Matrix

One of the main criticisms of the BCG matrix is that the buying behaviour of customers or consumers is rarely based on two factors only. Moreover, there are concerns about how terms such as 'relative market share' and 'potential growth rate' can be accurately defined. As a way of trying to deal with such potential problems, an alternative model was proposed by McKinsey management consultants in collaboration with both Royal Dutch/Shell and General Electric. In the case of the latter, their difficulty was in how to understand the differences (and resource priorities) of their 43 areas of operation.

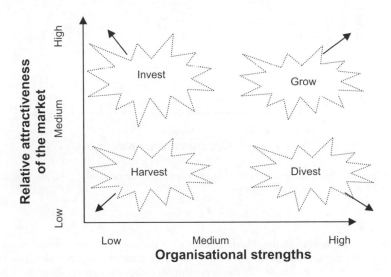

Figure 5.3   The McKinsey Directional Policy Matrix.

This matrix measures the relative health of the sector in which an organisation operates and its ability to pursue particular strategies. Analysis of results will suggest the way that future investment should be dedicated and, most especially, whether it is best to do one of the following:

1. Invest;
2. Grow;
3. Harvest;
4. Divest.

This matrix has only two axes which measure the **attractiveness** of the sector in which the organisation operates and its **business competitive position**. Attractiveness, as well as measuring potential growth, includes crucial aspects such as prospects for profits (or surplus), competition (who and where) and seasonal adjustments in the sector. If there are other factors that are particular (or peculiar), they can be incorporated into the matrix. Having decided what the aspects are, it is necessary to rate them and combine this into an overall score which, as the diagram below shows, can be plotted according to the attractiveness scale which is considered as being either high or low. The vertical scale, business competitive position, incorporates aspects such as how the organisation's products or services compare in terms of things such as price, reputation and quality. Additionally, aspects such as geographical position (and dominance), customer knowledge and understanding of other's strategy may also be analysed. As with attractiveness, each aspect must be rated and the overall score is considered against the scale which is high or low.

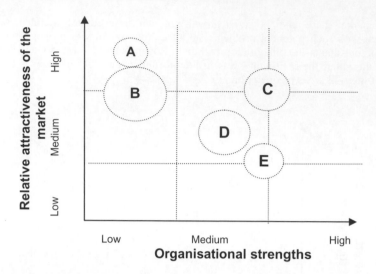

Figure 5.4   The McKinsey Directional Policy Matrix showing relative size of product market share for an organisation with multiple product lines.

What this diagram allows is to consider the relative position of a number of products and/or services which are plotted as circles where the size represents the size of the whole industry or sector and the wedge is the share that the organisation currently enjoys. Clearly, depending on which of the four quadrants a product and/or service is positioned will dictate the strategy that should be adopted:

- **Invest** which is the ideal position to be in and future investment should be dedicated to maintenance (such as by continued research and development) of the product or service.
- **Grow** where the market potential is high but the organisation does not have sufficient strength. The options that can be considered are either to gain as much from the current position as possible, or to invest in 'growing' the product or service so that it becomes more competitive.
- **Harvest** in which there are high strengths but low attractiveness. Therefore, whilst the short to medium term may continue to produce adequate returns, the longer-term prospects are unlikely to allow this to continue. Therefore, the organisation should look to harvesting what it can for as long as returns are sufficient.
- **Divest** in which the market is not attractive and the organisation's strengths are low. As a consequence, it is likely that getting rid of this product or service (divesting) is probably the best strategy. However, there may be a good reason for its continuance; for example, if it is a 'loss leader' in that it brings in customers who might otherwise not purchase at all, or where it is a complementary to other more profitable products or services.

5.11   Considering the value of these matrices

All of the models described have the objective of attempting to distil the behaviour of customers (and consumers) into simple aspects that can be readily analysed and measured. Unfortunately, the way that customers behave does not allow the sort of precise measurement that would be desirable for analysis. Moreover, the data that is elicited is usually historical which has the potential for leading to conclusions that may no longer be an accurate representation of what drives buying behaviour (and thereby markets).

There is a danger in trying to categorise particular products or services – especially if they are borderline and the choice of one or another impacts on decisions concerning investment. There is also a problem in the difficulty of trying to include perceptions that may have evolved as to how products or services really perform. As the next section describes, even though feedback from customers will assist in building an accurate picture of what is happening, interpreting information and data requires great skill – especially where there is a qualitative aspect.

5.12 Making connections with consumers – the importance of communication and feedback

Being able to accurately understand and predict what customers want is a key strategic tool. As students of even the most rudimentary study of management should be able to tell you, understanding requires all of the parties involved to be able to engage in forms of communication that will allow this to happen. Therefore, assuming that communication is axiomatic to the development of good strategic management, it is somewhat surprising to find that there is so little written about how it should be done. As Lynch is moved to remark:

> The whole subject is relatively poorly discussed in corporate strategy literature … Kay [1993] is the only recent strategist to deal in any depth with the issues [and even he in] *Foundations of Corporate Success* has only two chapters but they treat the subject from an economics rather than a marketing perspective and are rather simplistic as a result (2006, p.186)

Lynch contends that there are three reasons why any organisation should communicate with its customers. These are:

1. To provide information about the products or services it provides;
2. To 'persuade them to purchase or continue buying products or services';
3. In order to ensure that it can 'establish and secure' SCA.

As he goes on to explain, communication is about sending 'signals to the world at large', which will include, he stresses, stakeholders, individuals and any other groups – employees, suppliers, subcontractors, government and so on – who have an interest in the organisation. In so doing, therefore, the objective is to ensure that the strategy is developed in a way that recognises their importance and potential contribution towards achieving success. Clearly, this makes sense. To take a typical contractor as an example, how could they aspire to produce work of the highest quality without addressing the ability and dedication of those in their supply chain? This was considered to be a key objective which underpinned the recommendations presented by the construction task force (CTF) report, *Rethinking Construction* (1998).

5.13 The emergence of the concept of customer as 'king' – what the Japanese taught the west and how it has been developed

Whilst the importance of customer satisfaction is hardly a new concept, it has become something that is now perceived to be mandatory for all organisations. As the argument tends to be presented, constantly improving all processes is essential to ensure that an organisation constantly searches for new opportunities to enhance all customers' experience. Where organisations were aware of the importance of customers, there tended to be a belief that they would accept whatever was available. The most obvious manifestation of this was the often appalling quality of motor cars built in the UK in the 1970s. This apparent complacency was to change as a result of imports from Japan.

As car producers began to discover, followed by all other sectors of industry, if customers are offered products (and services) that give more value – certainly in terms of features but more especially as far as reliability is concerned – and they need pay no more, they will switch allegiance. The beginnings of the quality revolution may have been relatively muted, but the message that customers should become central to everything that goes on became pervasive. Many managers realised that they must learn how they too could implement quality management to ensure their processes were dedicated to achieving increased levels of customer satisfaction. As they soon discovered, what the Japanese had been able to achieve had required dedicated effort by all people involved in every aspect of the processes used.

The history of how the Japanese learnt to 'do quality' is well documented (see, for example, McCabe, 1998 and McCabe, 2001). Importantly, though, the influence of Dr W. Edwards Deming and Dr Joseph Duran was crucial in providing the fundamentals upon which those organisations that became pre-eminent in terms of their ability to provide superior products were able to develop. These fundamentals can be broadly summarised as follows:

- Making the customer central to the organisation's desire to achieve success (as Deming argued, 'The customer is the most important part of the production cycle');
- Developing techniques to intimately understand **precisely** what customers want;
- Using simple statistical methods to measure efforts to improve all processes (the way that inputs are transformed into outputs);
- Encouraging every person responsible for improvement in every aspect of what they do (the use of training and education to develop people's skills and confidence being essential);
- Developing an organisational culture that emphasises the desire to constantly engage in efforts to improve (the Japanese word, *kaizen*, which means continuous improvement, is now commonly used as a 'shorthand' for quality).

Even though many customers were initially reluctant to purchase Japanese cars and electronic goods, the experience of those who did helped to establish a reputation that you got more for your money and, most significantly, the goods were reliable. Those who had become accustomed to the experience of things breaking down were surprised to discover that Japanese goods did not do so with the same degree of regularity. Japanese goods – cars and electronic products being the main industries that had been chosen to be developed after the Second World War – were found to have levels of reliability that were vastly superior. As we all know, any customer experience – both good and bad – will be quickly disseminated and, if there is better value to be obtained elsewhere, others will follow. The consequence was that indigenous producers began to lose business and were forced to accept the need to learn what the Japanese had done and to emulate their examples.

American manufacturers, similar to their British counterparts, were losing business as customers switched allegiance[8]. Industry and academics explored the secrets of Japanese success. Not surprisingly, many offered their views as to what industry should do to emulate the success being

---

[8]This came as a shock as there were many who still harboured ill feelings about Japan's attack on America during the Second World War. Indeed, the author distinctly remembers an event in 1980 when an economics lecturer was asked about the prospect for Japanese cars in the UK. The response was that many had 'not forgotten' the ill treatment of British prisoners by the Japanese during their incarceration.

enjoyed by manufacturers able to consistently produce goods that performed outstandingly well and cost no more (often being cheaper to win market share). The belief that there was a particular formula meant that books on management sold in ever-increasing quantities and allowed Tom Peters (who co-authored *In Search of Excellence* with Robert Waterman in 1982) to become an internationally renowned consultant, albeit that some criticised his over-simplicity and apparent 'celebrity as consultant' image. However, the importance of the interest (some would suggest obsession) with the Japanese coalesced into what has become known as TQM (total quality management). TQM is defined by the International Standards Organisation in BS EN ISO 8402 as being:

> ... a management approach, centred on quality, based on the participation of all members and aiming at long-term success through customer satisfaction (1995, p.27)

Accordingly, an apparently new branch of management has been spawned which considers all the fundamentals of TQM. The panoply of techniques that come under the 'umbrella' of TQM (see below) provides a basis for applying concepts that can be used as the means by which to consider what customers want (or can be encouraged to consume). Importantly, TQM focuses effort on addressing the processes used to create products and services and, significantly, aims to constantly seek improvement. The result, it is argued by advocates, will not only be efficiency, but also a culture in which constant innovation and creativity are fostered.

Becoming a 'world-class' organisation (see section on benchmarking) was seen as essential in order to compete. However, as those who have tried to achieve this objective have found, it is far from easy to do – especially as Japanese producers have been doing it for so much longer than the new 'pretenders'. As Lynch points out, 'TQM is not without its costs and difficulties' but whilst 'benefits take time to emerge', the financial investment required to implement it is immediate (2006, p.435). Nonetheless, the belief that TQM offers the opportunity to dramatically improve the ability to increase customer satisfaction has become commonplace. As Lascelles and Dale contend, the benefits of TQI (total quality improvement) are crucial to success:

> TQI is concerned with the search for opportunities – opportunities for improving an organisation's ability to totally satisfy the customer. [...] the whole focus of TQI strategy will be on enhancing competitive advantage by enhancing the customer's perception of the company and the attractiveness of the product and service. This constant drive to enhance customer appeal through what the Japanese call *Miryokuteki Hinshitsu* (quality that fascinates) – the almost mystical idea that everything down to the tiniest detail has to be 'just so' – is integral to the concept of continuous improvement. Just like the concept of Total Quality *Miryokuteki Hinshitsu* is a vision, a paradigm, a value framework which will condition an entire organisational culture. This is the breakthrough, the stage at which an organisation finally breaks through to a new mind-set/paradigm: the autonomous and never-ending pursuit of complete customer satisfaction. (1993, p.294)

Two other Japanese words that are used in connection with the desire to become 'the best' are:

dantotsu which means the constant quest to be regarded as being the best.
zenbara which means the constant search for best practice in terms of management or use of technology, and then attempting to improve upon such practice to produce something that is even more superior (and often less expensive).

Whilst there are some who believe that the use of TQM and its associated techniques is appropriate only in environments where consistency of production processes is possible, i.e. manufacturing, there are many others who argue that they are equally applicable to construction. Indeed, many of the accounts of strategy provided in Chapter 12 are based on an explicit belief that making customer satisfaction an essential is a prerequisite for success. Moreover, these advocates of TQM stress the need for the active involvement and cooperation of people. Deming, the consultant sent to Japan to assist in the rebuilding of industry, and thus a major influence in their development of quality management, constantly made the point that improvement must be based on people: 'Quality is people, not products'. The remainder of this chapter will provide summaries of what are considered to be key concepts and/or techniques associated with the use of TQM. These are:

- Simple statistics, SPC (statistical process control) as the basis for continuous improvement of processes;
- Benchmarking, learning from the best and the quest to become 'world class';
- The EFQM excellence model;
- The SERVQUAL model for measuring customer satisfaction;
- Relationship marketing;
- Using lean to address customers needs;
- Partnerships and supply-chain management;
- Re-engineering of processes (usually referred to as BPR [business process re-engineering]).

5.13.1   Simple statistics, SPC (statistical process control) and Juran's quality control as the basis for continuous improvement of processes

The involvement of 'quality gurus', Dr W. Edwards Deming and Dr Joseph Juran, on the development of Japanese approaches to quality management is widely accepted to be seminal. As histories of Japan's post-war industrial development typically explain, Deming and Juran were sent to assist as advisors (see McCabe, 1998). In summary, Deming's message was that management effort should be concentrated on making systems for production as effective and as efficient as possible. Crucially, his belief was that all processes used in order to create outputs (either products or services) should be measured to determine the amount of variation: the degree to which there was deviation from what would be expected to be the norm, for example, a component being made – say, a light fitting – which is designed to be of particular dimensions to comply with an agreed international standard such as ISO. Whilst there is an assumption that every component is exactly right, measurement of actual dimensions may show that there is variation. There will be a certain degree of tolerance from the designed dimensions beyond which it will not fit (being too small or too large). The importance of the act of carrying out measurement is to provide accurate data on how much 'normal' variation is. This allows a control chart to be drawn which shows the degree to which 'normal' (and acceptable) variation occurs between the upper and lower control limits (see the following diagram):

The use of such charts is to provide an understanding of how much variation is occurring and, as the chart shows, to be able to identify what is in control and what is not. This allows the identification of so-called common causes of variation and special causes: the basis of statistical process control (SPC). By being able to identify the causes of variation, it is possible to assign the task of implementing changes, intended to produce improvement, to the right person(s). As the chart above shows, there are common and special causes. So, in the case of the light fitting, variation may be due to the machine not being correctly set. This would be easily identified by the operator and remedied by an adjustment. Such variation would be considered as a special

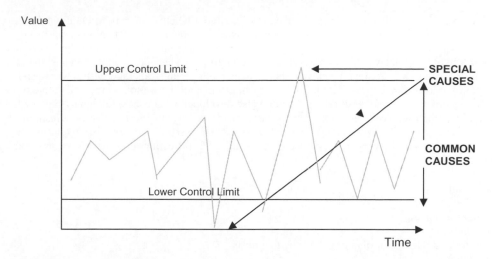

Figure 5.5   A typical control chart.

cause. However, variation in measurement may be due to the fact that inferior materials have been purchased by the organisation, possibly to save money. No matter what the operator does, they cannot remedy the variation. Such variation is within the system that has been created by management and is considered to be common.

Beyond stating the obvious, that variation is considered to be negative, the assumption is that the organisation will do everything possible to eliminate it. Therefore, the importance of using SPC and identifying the type of variation are crucial to continuous improvement by reducing variation. Indeed, as Deming argued, reduced variation in a process means that it is 'under control' and more likely to result in producing the intended results. He believed that variation due to special causes is usually very irregular (and certainly less than 10%) and its elimination should be made the task of those most intimately involved in the process. However, common causes provide most variation (90% plus). Such variation is inherent within the system that management has decided as the way of doing things (completing processes) and it is they who should create a system that encourages people to achieve success. What is important, Deming believed, is that managers cannot blame workers for problems that are the result of a bad system. The system should enable workers to achieve successful completion of tasks – and certainly not disable them.

The importance of SPC is that managers should be able to use their skills and expertise to cooperate with workers to demonstrably shift their focus and aspiration towards higher levels of quality by reduced variation and compliance with what the customer has ordered. Such a shift in the way that people (workers) behave is crucial to creating a radical change of culture. Even though Deming was a statistician, he emphasised the importance of the 'human input'. As Crainer states:

> Deming appreciated that no matter how powerful the tool of mathematical
> statistics might be, it would be ineffective unless used in the correct cultural
> context (1996, p.143)

Accordingly, the combination of 'culture and measurement' which the Japanese have used to such powerful effect in certain sectors led the Americans (and the west in general) to 'rediscover'

Deming and, according to Crainer, to invent 'what is now labelled (*sic*) Total Quality Management' (1996). Deming's beliefs are also found in what is known as the 'Deming chain reaction' which is based on a belief that, as quality improves, there is reduced cost in terms of work due to mistakes and delays. As a consequence, productivity is enhanced which means that the company can secure competitive advantage by being better than rivals (and with lower costs). It is possible to continue investment in improvement efforts and innovation which will create more jobs as market share further develops. Deming also developed what is now known as 'The PDCA cycle' (see below). As he asserted, once improvement becomes the norm in an organisation, its continuance becomes a virtuous cycle. The PDCA cycle is based on the objective to continuously improve each and every process used. As such, the following considerations are made iteratively:

- **Plan**
  Policy
  Development
- **Action**
  Possible change of plan based on the diagnosis
- **Check**
  Auditing
  Diagnosing
  Reporting
- **Do**
  Policy
  Deployment

Joseph Juran's approach to quality control

Juran is very similar to Deming in many ways. They worked together under the influence of Walter Shewhart (at Western Electric) and both travelled to Japan to advise managers involved in rebuilding industry. However, their approaches to improvement were different:

1. Deming concentrated on education whereas Juran believed that the key to success lay in actual implementation and he proposed what he called 'Company-Wide Quality Management' (CWQM) which will result in improvement at every level of the organisation.
2. Whereas Deming lectured to the *Kei-dan-ren* – the association of Japanese chief executives – Juran concentrated his message on those whom he considered to be in the best position to actually influence improvement: middle management and quality professionals.
3. Juran stressed that the key to managing improvement was to concentrate on the cost of quality; the best way to get the attention of management is to show how the 'bottom line' can be increased. Significantly, Deming thought that management in the west was too obsessed with cost (something he summarised as 'running a company on visible figures alone – counting the money' in his 'seven deadly diseases'). Juran, on the other hand, did not believe that this was a problem.

Juran's essential philosophy consists of four steps which lead to what is called the 'quality trilogy'. These are:

Step One – clearly identify specific things/projects that need to be done;
Step Two – provide definite plans for achieving what can be done;
Step Three – ensure that people are made responsible for doing certain things;
Step Four – make sure that the lessons that are learned during the previous three steps are captured and used in feedback.

The quality trilogy involves three essential aspects of quality:

- Planning;
- Control;
- Improvement.

Planning   Juran was forthright in his argument that quality does not happen by accident; it must be planned for. Therefore, Juran advised, a four-steps approach is necessary:

1. Clearly identify the needs of all of those involved in the process of producing the end product or service and get them actively involved.
2. Ensure that the needs that are identified in the above are put into language that is simple and unambiguous to those involved.
3. Once needs are identified and articulated, devise processes that are sufficiently robust as to achieve the desired outcome and satisfy everyone's needs.
4. Implement all processes and continually monitor them to ensure continued achievement and improvement which satisfies the customer's needs.

Control   Juran believed that, once processes are implemented, they must be continually monitored and measured. Like Deming, Juran's message was that it is only by controlling the processes that waste can be reduced and savings made. Also, like Deming, Juran, despite his focus on those who are directly in control of operations/processes, maintained the need for all managers to be aware of their responsibility to design a system in which every person can achieve the best result.

Improvement   According to Juran, improvement is the logical (and desired) consequence of planning and control. Once improvement starts to occur, it is incumbent upon senior management to consider ways in which all parts of the organisation can be improved by continuously developing its systems. Workers, provided that they are trained and supported, will benefit from improved engagement and morale. Juran's philosophy, like that preached by Deming, leads to the emphasis on the **total** of TQM.

Benchmarking, learning from the best and the quest to become 'world class'   Benchmarking is a technique that effectively developed from the recognition that some organisations were better at achieving customer satisfaction than others. The Japanese, in particular, had demonstrated their ability to mass produce in a way that was both cost effective (and economic) but, significantly, that the 'build quality' and performance (in terms of reliability) were far superior to anything produced by manufacturers in the west. If offered better value for the same money, customers will be tempted to switch. This was precisely the experience of Xerox who were pioneers of photocopiers but were discovering that they were losing business. When they carried out analysis of why this was happening, they found that their manufacturing costs were higher than what the Japanese were selling for. Worse, the reliability of their copiers was far more inferior. As a result, Xerox recognised that they needed to learn what it was that their competitors were doing that was different. Doing this is what benchmarking involves:

> [It is] A process of continuous improvement based on the comparison of an organisation's processes or products with those identified as best practice. The best practice comparison is used as a means of establishing achievable goals aimed at obtaining organisational superiority. (McGeorge and Palmer, 1997, p.83)

The word, processes, is important. As Deming had stressed, if you want to improve what you do, you must concentrate on all of the processes used on a day-to-day basis. Therefore, the key to benchmarking is to select particular processes that are used and to compare them to those used by another organisation (or part of the organisation) where their application produces demonstrably better results. There are various approaches to carrying out benchmarking:

- Internal;
- Competitive;
- Functional or generic.

Internal benchmarking   As suggested above, if an organisation has different parts (departments, offices or sections) and there is one that excels in the way it does certain things, it would be useful to analyse why. By looking at the way that this part implements processes, it is entirely possible that there may be valuable insights that can be applied elsewhere in order to give similar ability to excel. Because this approach to benchmarking does not require access outside of the organisation, it should be the most straightforward and should not create potential difficulties, especially if the comparison is with a direct competitor (see next section).

Competitive benchmarking   This approach to benchmarking is based on the desire to learn from those against whom you compete and, of course, to emulate any success (competitive advantage) they enjoy. The most obvious difficulty with this approach is in gaining the sort of intimate knowledge of processes that will be required to conduct benchmarking. Even though cursory analysis as an outsider will be a useful starting point, more detailed analysis is needed. However, those who are determined to learn the 'secrets' of others will normally find ways to gain this information[9].

Functional or generic benchmarking   This approach to benchmarking will potentially result in the most useful information. The objective is to learn how an organisation operating in an entirely unrelated industry achieves superior levels of customer satisfaction (either through the quality of the products or the service which it delivers). In particular, processes are the key to understanding. Therefore, context is not crucial; learning how they do what they do is. Moreover, as McGeorge and Palmer explain, this method allows an organisation to:

> … [break] down barriers to thinking and offers a great opportunity for innovation. It also broadens the knowledge base and offers creative and stimulating ideas. (1997, p.88)

The concept of benchmarking is deliberately intended to assist an organisation to learn how to improve. The analogy to sport is apposite in that any individual who wants to develop their skill and competitive ability will measure their progress against the best. They will learn from those who have demonstrated so by setting world records, winning competitions (such as World Cups) or gaining medals at events such as the Olympics. If you want to be the best, you have to do so by competing (and beating) all others. Organisations can demonstrate their status as being 'the best' by winning awards such as the EFQM Excellence Model (see below). As Dale *et al.* explain (on the basis of work carried out by Williams and Bersch, 1989):

> … strong, world-class product and service quality-related competitiveness can only be achieved when an organization has reached the stage of being able to compete for the top awards[10]… (1994, p.124)

---

[9]It has been accepted practice in the motor industry for manufacturers to buy any new models produced by a competitor and to carry out a 'tear-down' in which the way that the model has been assembled and any new technology will be looked at to discover if useful information can be gleaned – and applied to its own models.

As they assert, there are a number of key features that those organisations that win awards are able to demonstrate (1994, p.125):

- A 'leadership culture' that stresses commitment by all and in which every person is encouraged to implement innovative ideas that will potentially result in improvement (this is frequently termed 'empowerment').
- Organisational changes which have resulted in improvement through 'cross-functional management' and continuous development of processes.
- 'Strategic benchmarking practiced at all levels, in conjunction with an integrated system of internal and external performance measurement.'

As they explain, in such organisations, TQM and improvement become commonplace and constant improvement is the norm. However, what they also suggest is that, whilst winning an award is an explicit demonstration of being part of an 'elite', this label can only be given to those that are regarded as having achieved 'world-class' status. To be regarded as 'world class', they believe, is 'characterized by the total integration of quality improvement and business strategy to delight the customer' (Dale *et al.*, 1994). They provide a fuller explanation of what attaining 'world-class' status involves:

> The never-ending pursuit of complete customer satisfaction is a personal goal of everyone in the organization and an integral part of their everyday working lives. TQM is no longer dependent on top-down drives to improve motivation and policy deployment, but it is driven laterally throughout the organization [...] Customer desires and business goals, growth and strategies are inseparable; Total Quality is the integrative and self-evident organizational truth; the vision of the entire organization is aligned to the voice of the customer ... (1994, p.126)

Thus, becoming 'world class' means that an organisation is absolutely dedicated to the fullest possible achievement of customers' delight (Deming believed that the word, 'satisfaction', is too passive). Doing this is good business. For example, examination of the surveys carried out into customer experience of cars demonstrates that the majority of the top ten are still those Japanese companies that have implemented TQM. The experience of Toyota (see lean production) is seminal. The next section describes what the EFQM Excellence Model is and how it can be used as the basis of developing an organisation's journey towards TQM. Whilst winning awards and becoming 'world class' may be ambitious long-term aspirations, an organisation has to start somewhere[11].

5.13.2   The EFQM Excellence Model

The basis of what is known as the EFQM Excellence Model is to provide an organisation with the means by which to implement and achieve excellence. The eight fundamental concepts of excellence (see list below), which are interrelated, are based on a desire to attain 'Results Orientation' which leads to greater customer focus. This will require leadership and 'constancy of purpose' which, it is stressed, should be supported by management based on the 'use of pro-

---

[10] They refer to the Deming Application Prize, Malcolm Baldrige National Quality Award and the European Quality Company Award. The first was developed in Japan as an eponymous honour to the work of Dr Deming. The second exists in America and is in memory of a proponent of quality management (drafting the Quality Improvement Act 1987 but who was killed in a rodeo accident later that year). The final award is a European version of The Deming Prize and Baldrige Award and is now the EFQM Excellence Model described in detail in this chapter.

[11] The Chinese maxim of 'the journey of a thousand miles beginning with one small step' is highly appropriate!

cesses and facts'. The consequences of this will be greater engagement and collaborative working which will engender a culture of 'continuous learning, improvement and innovation' which will establish the basis upon which the desired results will be achieved.

In any approach dedicated to achieving excellence the key to success is constant improvement of processes. The EFQM Excellence Model is based upon precisely the assumption that an organisation will, following analysis, seek to implement solutions to existing problems and continuously improve every process used. As Porter and Tanner stress, it is important that every person in the organisation considers the objective of achieving excellence to be a 'strategic imperative' (1996, p.120). Oakland stresses the importance of people:

> The EFQM model [recognises] that processes are the means by which a company or organisation harnesses and releases the talents of its people to produce results performance. Moreover, improvement in the performance can be achieved only by improving the processes by involving the people. (1999, p.99)

The model is usually shown in a diagrammatic representation (see below). As this shows, there are two aspects to it: 'enablers' and 'results'. Within each of these, there are criteria that are used to assess the organisation's achievement of particular goals that are critical to the achievement of the eight 'fundamentals' of excellence:

- Leadership and consistency of purpose;
- People development, involvement and satisfaction;
- Customer focus;
- Supplier partnerships;
- Processes and measurement;
- Continuous improvement and innovation;
- Public responsibility;
- Results orientation.

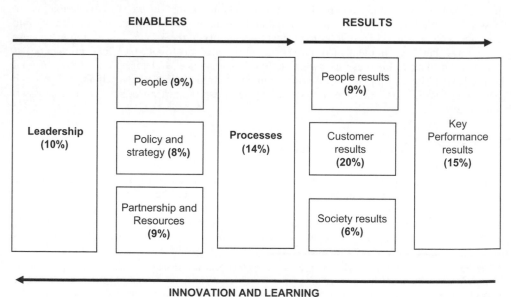

Figure 5.6   The EFQM Excellence Model ® (with permission from EFQM).

EFQM (which is the organisation responsible for the development and maintenance of the model) explain that the main benefit of using it is for an organisation to find ways to develop its strengths and weaknesses through 'self-assessment'. This involves analysing all of its approaches to carrying out tasks and considering ways of ensuring that its strategic objectives are successfully achieved through dedication to customer satisfaction. The main objective of self-assessment is to analyse best practice that exists in the organisation (referred to as **strengths**), and, perhaps more importantly, in order to develop the organisation's capabilities, to analyse any potential weaknesses that will undermine its abilities (referred to as **areas for improvement**). Importantly, the use of EFQM Excellence requires the following steps:

- Gain commitment of senior management;
- Plan the way that self-assessment is used;
- Select teams to carry out the exercise;
- Communicate what is going on to all employees;
- Implement the chosen approach;
- Subsequent to approach being used, analyse and disseminate results to those directly concerned;
- Create plans for teams to dedicate themselves to being involved in areas for improvement that have been identified, and for procedures to be written that capture any best practice which emerges.

Developing excellence also requires those involved in carrying out sufficient analysis to answer the following questions:

- What its strengths and areas for improvement are;
- How best practice can be applied (either that which exists within the organisation or from other organisations);
- How much needs to be done in order to create excellence;
- Where additional effort should be dedicated (particularly with respect to supporting the development of people).

EFQM provides the following five steps to ensure that the output of self-assessment is effectively used:

Step One Collation of the strengths and areas for improvement in a way that ensures there is a logic and rationale for action. This should be done using information that is coherent and as soon after collection as possible.

Step Two Decide upon the method and criteria to be used in order to prioritise the output of the self-assessment. In doing this, their impact on the following should be borne in mind:
- Strategic intentions;
- Critical success factors;
- SWOT analysis.

Step Three How the outcomes arising from self-assessment can be effectively used by consideration of:
- Changes that will be required;
- People that will be prepared to champion the changes;
- Employees who will support the changes.

Step Four Implementation of the priority actions by all concerned, remembering to:
- Allocate necessary resources;
- Agree plans;
- Coordinate the input of particular teams/departments or individuals.

Step Five Review and measurement of results to consider how effective the implementation of action has been:
- Monitor progress;
- Change resource input if required;
- Be responsive to external environment and differing circumstances.

As the diagram above shows, a major criterion for assessment is customer results (20%). Being able to accurately measure how effective the improvement has been on customers is therefore very important. The next section provides explanation of a specific model intended to measure customer satisfaction.

5.13.3   The SERVQUAL model for measuring customer satisfaction

One of the most widely recognised models for benchmarking customer satisfaction is known as SERVQUAL. This model was developed by Parasuraman *et al.* (1988). The SERVQUAL model attempts to measure the differences (**gaps**) that exist between what a customer expects and their perception of the level of actual quality they receive. As Parasuraman *et al.* believe, 'the criteria used by consumers is assessing service quality to fit ten potentially overlapping dimensions' (1988, p.17). These are:

1. **Tangibles**
   the physical attributes that exist in a product or service;
2. **Reliability**
   whether the product or service performed in accordance with what would have been expected;
3. **Responsiveness**
   the willingness of the service provider to respond to requests which may require particular alterations;
4. **Communication**
   the ability to talk to customers in a way that they can understand;
5. **Credibility**
   the honesty and esteem that the provider is held in by customers;
6. **Security**
   which can be physical, financial or concern confidentiality;
7. **Competence**
   how able the employees of the provider are to deliver the service;
8. **Courtesy**
   the politeness and respect which staff in the provider's organisation show to their customers;
9. **Understanding/knowing the customer**
   how able the provider is in appreciating what their customers really want;
10. **Access**
    the ease with which customers can communicate with staff in the provider's organisation.

Following research work by Parasuraman *et al.*, these ten dimensions were refined and reduced to five that appear in the SERVQUAL model. As these five (listed below) show, the first three (tangibles, reliability and responsiveness) are unchanged from the original list, whereas the last two are a combination of the other seven original dimensions:

1. **Tangibles**   Physical facilities, equipment, and appearance of personnel;
2. **Reliability**   Ability to perform the promised service dependably and accurately;
3. **Responsiveness** Willingness to help customers and provide prompt service;
4. **Assurance**   Knowledge and courtesy of employees and their ability to inspire trust and confidence;
5. **Empathy**   Caring, individualized (*sic*) attention the firm provides its customers.

As Parasuraman *et al.* point out, these five dimensions provide what they admit to be a 'basic skeleton' (1988, p.30). Therefore, they advise, it will need to be 'adapted or supplemented to fit the characteristics or specific research needs of a particular organization' (1988, p.31). As Parasuraman *et al.* also advise, the SERVQUAL model works best when it is used together with 'other forms of service quality measurement' (1988). The most important thing about SERQUAL, the authors stress, is that it is used systematically over a long period of time, and that any gaps that are identified are dealt with immediately. This, I believe, is the most important thing to emerge from their study: that there is a desire to find measurable gaps between customer expectation and their perceptions of the levels of service they receive. By constantly attempting to reduce such gaps, it is highly probable that the excellence this organisation provides will be improved.

5.13.4   *Relationship marketing*

If customers are important, it follows that effort should be dedicated to understanding their needs and developing methods that enable the organisation to make closer connections with them. This is the intention of what is known as 'relationship marketing' which, according to Christopher *et al.*, provides the link between quality, customer service and marketing:

> Relationship marketing has as its concern the dual focus of getting and keeping customers. Traditionally, much of the emphasis of marketing has been directed towards the 'getting' of customers rather than the 'keeping' of them. Relationship marketing aims to close the loop. (1993, p.4)

Relationship marketing is therefore much more focused on ensuring that the needs of individual customers are fully understood. This can be through a variety of techniques. However, the task is one that involves every person in the organisation and stresses their importance in developing 'bonds' with customers which gives a sense of commitment to giving the best possible service. There will be some people who have closer contact with customers than others and it is useful to identify the needs and support that each person has to ensure that they are able to provide a superior level of service. On the basis of Judd's work (1987), Christopher *et al.* present a matrix based on the level of contact that organisational members will potentially have with customers. This matrix includes four types:

- Contractor;
- Modifier;
- Influencer;
- Isolated.

A **contractor** is someone with most contact. They will typically work in a sales or customer-services environment. As a consequence, they must be fully conversant with the marketing strategy of the organisation. Additionally, they should be prepared and motivated to deal with customers in a way that ensures the perception given is of a professional and caring organisation. It is sensible that those appointed to this role should have skills that are suitable to giving the correct impression. A **modifier** is someone likely to be in contact with customers but on a much less frequent basis than contractors. For example, someone working in accounts, although primarily interested in financial data etc., will, from time to time, contact customers. Therefore, they should be able to understand the importance of maintaining an effective relationship. An **influencer** is someone who, even though not directly involved with customers, should be aware of the consequences of their decision-making. For instance, those who work in research and development should always have the interests of the customer as the key objective. Finally, the last type of person, those known as **isolated,** even though they have minimal or no contact at all with customers, should, nonetheless, be aware of the critical importance of ensuring that all processes are oriented so that customers are willing to repeat their custom.

Retaining existing customers is always a more effective way to conduct business than constantly trying to replenish them. According to Christopher *et al.*, relationship marketing provides a 'ladder' in which the emphasis is on ensuring that the relationship with every customer converts them from trying the product/service for the first time, what they call 'catching', to becoming advocates[12] (1993, p.22). The ladder provides the following guidance:

- That the experience that customers have is sufficient to ensure that, initially, they **support** the organisation by bringing repeat business, and in the longer term.
- That they actively **advocate** to others to consider purchasing the same product or service.

If an organisation can get to the top of the ladder, its need to advertise in order to attract speculative customers is much reduced. Instead, finance can be dedicated to maintenance of their custom (continuing the relationship) and, with logical resonance to what Deming suggested about the virtues of quality, it will potentially allow any money saved on advertising either to be passed on to the customer (who gets more for less) or to be dedicated to improving or enhancing the products or services provided.

How to develop a relationship marketing strategy

The following diagram summarises the main elements that are involved in creating the framework for developing a relationship marketing strategy:

---

[12]Some customers are so pleased with the service they receive that they happily advertise the organisation to others. For example, some of us carry bags that provide free advertising. Designer clothes can serve the same purpose.

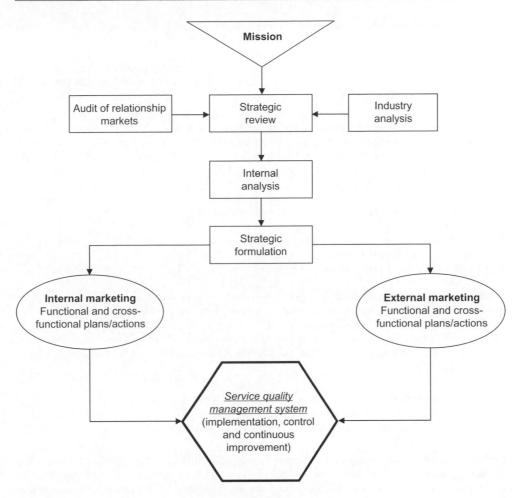

Figure 5.7 The framework for developing a relationship strategy.

Mission This requires those responsible for strategy to be clear about the organisation's aims and beliefs. The mission of the organisation provides a vision and focus from which the actions of all employees will occur. This should mean that everyone is in no doubt about the following: What the organisation seeks to achieve;

- How they can contribute to customer satisfaction;
- 'Reflect the distinctive competences' that the organisation possesses;
- Be realistic enough to reflect the strengths, weaknesses, opportunities and threats that the organisation currently faces;
- To be capable of adapting to align the organisation with rapid changes in customer needs.

As subsequent chapters maintain, the importance of gaining support and cooperation in developing the organisation's objectives is crucial. If they have helped to create it, they will logically support it.

Strategic review and internal analysis   These occur subsequent to having created a mission statement and having carried out a number of activities that aim to provide accurate indications of the competitive position that the organisation currently enjoys (these are described below). The objective of conducting these activities is to achieve, in every possible way, perceptions amongst customers that the product or service they have received was better than that which could have been purchased from a potential rival.

Audit of relationship markets and industry analysis   In order to give the customer 'more' than that which competitors provide, it is essential to make a comparison. Christopher *et al.* suggest that there are five things which have a major effect on the profitability of any industry, and, therefore, more crucially, the ability of individual organisations to achieve particular levels of commercial success. These are:

1. The difficulties that new entrants face in gaining access;
2. The power that buyers can exercise;
3. The potential for substitution of one product or service for another;
4. The power that suppliers have;
5. The actual level of competition.

What is important about these five elements is that, in a market where one or more are in existence, the resulting competition from the need to chase customers frequently means that profit levels are extremely low. Any organisation in such a position will find that all of its efforts are dedicated to continually having to attract customers (frequently by cutting costs). Often this cycle of decline can only be reversed by concentrating on adding value through customer satisfaction.

Strategy formulation   An organisation should decide what strategy is most likely to succeed. Making the customer the 'heart' of strategy is always going to be sensible. The consequence is that everything possible is done to add 'extra additional' value through every process used.

Internal marketing and external marketing   Essentially, this part of developing a relationship marketing strategy repeats one of the key principles of TQM: knowing what is most likely to achieve maximum customer satisfaction. Furthermore, it requires the organisation to ensure that high satisfaction levels can be consistently attained. Achieving high levels of customer satisfaction requires all those involved in every activity to understand the expectations from others within the 'production chain'. This principle is especially crucial where the next stage is reliant on the output of the preceding one. Accordingly, those receiving such output are considered to be 'internal customers' (and should be regarded in the same way as those who eventually buy or consume the end product or service). Like external customers, carrying out marketing to better understand the expectations of internal customers is a vital part of improvement.

Based on research work carried out by Levitt (1983), Christopher *et al.* suggest that 'offering' should be considered on four levels (1993, p.57):

1. **Core** which consists of the physical or tangible characteristics that constitute the product or service. Customers have certain minimum expectations which, in order that they are satisfied, must be achieved.
2. **Expected** which consists of aspects of the product or service which customers believe should accompany it. For instance, with most goods, it is expected that there will be a warranty period that covers failures or breakdowns.
3. **Augmented** which consists of providing additional features or service that allows the product or service to be perceived to be better than that which competitors can provide.

4. **Potential** which consists of the provision of features or elements of the product or service which the customer believes will allow them to derive greater levels of value than could be achieved by purchasing alternatives.

In attempting to achieve a relationship marketing strategy, the focus is on the last two of these levels, most especially the last. As Japanese producers of electronic and automotive products have demonstrated so successfully, customers who believe they have received products that provide them with high value tend to become enthusiastic advocates and, more importantly in the long term, very loyal in terms of repeat buying.

5.13.5   *The importance of lean – from Toyota to construction*

The expression, 'world-class organisation', tends to be associated with the Japanese. There are increasing examples of organisations that are judged to be world class that are not Japanese. They learnt quickly and emulated the lessons of those that pioneered the principles. In any cursory study of these pioneers, certain company names are notable. One, Toyota, is particularly worthy of consideration for the way that it implemented TQM through its use of what is known as **lean**, a technique based on improvement and, most especially, reduction of waste. Interestingly, the inspiration was a visit to what was regarded as the citadel of mass-production methods – the River Rouge Ford plant in Detroit.

Eiji Toyoda,[13] who in 1950 was a young Japanese engineer, visited the Ford production facility to learn how Ford produced vast numbers of cars. In the thirteen years until his visit, Eiji's company had produced only 2,685 cars. Ford, on the other hand, produced over three times that number in a single day! Given the desire of Japanese producers to learn, it was obvious that Ford's production methods provided the benchmark for success. Toyoda would have been expected to implement Ford's methods in his family's factory. Anyone visiting River Rouge would surely have been impressed by the rate at which cars could be produced. However, what Toyoda would have seen was a manufacturing system that treated its workers in a way that seemed to dehumanise them:

> [Ford] not only created a great, integrated machine for turning out cars,
> but pioneered a workforce as standardised and interchangeable as his
> automobiles and as dedicated to a single purpose as the thousands of
> machine tools they tended (Pursell, 1994, p.99)

Toyoda, together with a production genius he employed, called Taiichi Ohno, believed that what they saw was extremely wasteful, both in terms of physical aspects but also in terms of the human spirit. As they reasoned, there must be a better way to operate production systems. Consequently, the reduction of waste (the Japanese word is *muda*) has become a crucial component of the lean philosophy. Lean is an approach to organising that aims to develop methods of working that are both sensible and logical, but which are not imposed by senior managers who believe that their view of rationality is unquestionable.

Lean uses processes as the method of analysis to identify wasted effort with a view to creating improvement. By looking at the overall process, it is possible to consider what is known as the **value stream**. Instead of considering workload as occurring in discrete parts, it becomes essential

---

[13]Toyoda was a family company founded in 1937. Because this name meant 'abundant rice field' it was decided that a better alternative should be used. Following a competition in which 27,000 suggestions were made, the name Toyota emerged, a word that has no meaning.

to consider the interdependency of activities. The key concepts are that obstacles and blockages are removed in order to allow **flow** and that all activities are synchronised to **pull** the product or service from its inception, through design and production so that it meets end-customer expectations and creates what they perceive to be value. Innovation and timeliness are crucial to this technique and the overall objective is to continually strive for perfection.

Allied to this are other critical principles that were developed at Toyota:

- Flexible working practices *(shojinka)* by those directly involved in the production process which enabled workers to control what they were doing;
- Teamworking;
- Creative thinking *(soikufu)*;
- A rapid development cycle;
- Ability to change production lines extremely quickly to ensure that what is produced is exactly what customers want;
- Because of the ability of workers to change production equipment themselves, cars could be made in small batches which allowed regular changes in the models to rapidly meet changes in customer requirements;
- An emphasis on improving the quality of what was being produced *(kaizen)* rather than Ford's obsession with targets to be achieved;
- Extremely close relationships with suppliers (partnerships – see next section) so that they become, in effect, an extension of Toyota;[14]
- Rationalised stock systems which use the principle of 'pull' to draw from suppliers as and when components are needed – better known as 'just-in-time' – to reduce waste through deterioration, damage and potential theft. It also reduces space needed and decreases costs. Whilst the provenance of lean is most definitely placed in the manufacturing environment of cars, a 'derivative'[15] has been developed so as to be relevant to the context in which construction takes place. Accordingly, there is a set of principles called 'lean construction'[16] which have been proposed. Importantly, these principles fully accept that the way that construction operates – being project-based – makes it very different from manufacturing. However, the objective remains the same: the improvement of processes with a view to removing waste and maximising value so as to give the customer maximum benefit.

The early proponents of lean construction (Lauri Koskela and Glen Ballard) argued that the traditional method of planning activities in advance (and remotely) patently did not work and that it would be better to acknowledge the temporary nature of the 'production system' on site. Koskela (2000) and also Koskela and Howell (2002) contended that Toyota's approach to production could be appropriate to construction. Additionally, Bertelsen (2003) argued that construction, because it is project-based and relies on relationships (socialisation), should be considered by using chaos and complex systems theory. This combination of thinking led to construction being thought of as consisting of three things: transformation (T), flow (F) and value generation (G). (Bertelsen, 2003)

A very practical way of implementing the principles of lean construction is the technique known as 'the last planner'. This is operated on the basis that project planning should be carried out by those who know best, usually those most intimately involved and based on site – the 'last

---

[14]The word, 'symbiosis', which is defined as being a cooperative or mutually beneficial relationship between two people or groups, is absolutely apposite.

[15]In the aftermath of the 'credit crunch', this is now a word that is similar to bureaucracy in that it is seen as being pejorative.

[16]A term that came into being as a result of the inaugural meeting of the International Group for Lean Construction (IGLC) in 1993.

planners'. All activities should be coordinated and carried out using float so that risk is minimised. It is vital to understand the interdependence of activities and to clarify whether there are impediments to successful completion. Planning is carried out on the basis of being able to **promise** that an activity will actually occur. Once this has happened, the last planner responsible will ensure that the requisite standard has been achieved and subsequent work can take place. Thus, a so-called 'promise cycle' occurs.

Importantly, the key measure of success is based on the percentage of promises completed (PPC) and this will be carried out on site and continuously analysed so as to learn lessons and improve future practice, for example, by using other quality-management techniques, such as Taichi Ohno's 'five whys', to discover the root cause. As advocates fully accept, what is important in using this technique is that there is a desire by all involved to shift their thinking from a culture of blame and recrimination to proactive problem-solving and continuous improvement. It also emphasises cooperation between all participants so as to understand their contribution to reducing variation in activities and how this can increase customer satisfaction. Like Toyota, supporters of lean believe that value generation is the key to achieving customer delight.

5.13.6   Partnerships and supply-chain management

In addressing quality management, it is important to understand the needs of the customer (as was explained in relationship marketing). Inherent in the general description of TQM and associated techniques is an assumption that all processes are dedicated to improvement so as to deliver added value to the customer. However, it is extremely rare that one organisation will not rely on organisations that are technically outside its immediate control. At the very least an organisation will usually procure raw materials. More likely it will need to be supplied with components which it will assemble into a finished product, as is the case in car manufacture and construction.

Crucially, therefore, the customer's judgement of the quality of the end product/service will be dependent on how every component performs. In order that clients get exactly what they want, it is useful for them to have an extremely close relationship with the organisation that provides the product or service. In turn, it makes sense for the organisation which provides the product or service to its customer also to be close to those which supply it. The techniques of partnering and supply-chain management are valuable ways of ensuring that there is effective influence throughout the whole of the process.

Partnering and supply-chain management may be seen to be not only complementary but also indistinguishable. An example of not being able to separate one from the other is apparent in the way that Clive Cain contextualises the importance of partnering in his book, *Profitable Partnering for Lean Construction*:

> … if the construction industry is serious about delivering a radical improvement in performance and if it intends to import best practice in supply chain management (as every report around the world is demanding), the following definition of partnering ought to be adopted throughout the industry. (2004, p.7)

Unsurprisingly, the actual definition of partnering does nothing to distinguish it from supply-chain management:

> The formation of long-term, strategic supply-side relationships between the firms to make up a design and construction supply chain that is capable of delivering a comprehensive range of building types and construction activities for a variety of demand-side customers (small and occasional as well as major repeat customers).

> The primary purpose of such strategic supply-side partnering relationships is to enable the supply-side design and construction firms to work together at both project and strategic level to continuously drive out unnecessary costs (caused by the inefficient utilisation of labour and materials) and to continuously drive up whole-life quality.
>
> The output of such strategic supply-side partnering should be the continuous conversion of unnecessary costs into lower prices and higher profits, whilst improving the whole-life value of the building or the constructed product. (Cain, 2004, pp.7–8)

Using definitions provided by EFQM (the body responsible for ongoing development and maintenance of the Excellence Model), Cain does provide some differentiation between 'partnering' and the 'supply chain'. Respectively, these are:

> A working relationship between two or more parties creating added value for the customer. Partners can include suppliers, distributors, joint ventures, and alliances. (2004, p.219)
>
> The integrated structure of activities that procure, produce and deliver products and services to customers (2004, p.223)

Partnering, therefore, is about developing a collaborative relationship between the customer(s) for whom the product and/or service is created and the organisation that attempts to achieve this. The objective is for customers and the organisations that supply them to become intimately involved over a period of time so as to fully understand each other's needs and, equally importantly, potential difficulties. Like all of us, the more we get to know others, the easier the relationship is likley to be. Openness and honesty contribute to harmonious working relationships and a desire to get the best results which, depending on the agreement, may be shared: see Mari Sako's comparison of ACR (Arm's-Length Contractual Relations) and OCR (Obligational Contractual Relations, also known as 'arms around') patterns (1992). This relationship can be replicated between the supply organisation and those organisations that supply it with materials, components or people. Logically, the desire to partner can continue backwards up the supply chain until there is a logical end: being either those engaged in eliciting and processing raw materials or those who provide labour.

If partnering is about creating suitable relationships that enable organisations within the supply chain to cooperate harmoniously, supply-chain management is based on a desire to ensure that all processes employed by every organisation are as effective and as efficient as possible. The supply chain begins and ends with the customer. They provide the desire and it is all the organisations, engaged in the activities that eventually produce what they want, that will (or will not) achieve success in terms of satisfaction. Accordingly, the supply chain will include the entire cycle of activities: the extraction of raw materials, their processing into another form (such as clay into bricks), design, subsequent manufacture and/or assembly, distribution and, of course, final delivery. The key, as always, is to search for improvement and reduce wasted effort and cost. Using the *Building Down Barriers Handbook of Supply Chain Management*, Cain provides a full definition of what supply-chain management involves:

> Replacing short-term single project relationships with long-term, multiple project relationships based on trust and cooperation. These standing supply chains focus on delivering value as defined by their clients. Long-term strategic supply chain alliances can incorporate continuous improvement targets to reduce costs and enhance quality, and focus on the through-life cost and functional performance of buildings. The idea of continuous improvement, based on a systematic analysis of the weaknesses and strengths in existing design and construction processes, underpins every

aspect of the Building Down Barriers approach to supply chain management. Without this discipline, it would be impossible to reduce through-life costs significantly, or enhance quality, deliver superior functionality or any other design benefits, or improve levels and certainty of profits for the supply chain. (2004, p.224)

Cain asserts that there are a number of 'supporting principles' to ensure that supply-chain management is achieved (2004, p.225):

1. The need to precisely identify the end-users' 'values' using 'valued management tools';
2. Establishment of supplier relationships with all key organisations;
3. 'Integrate project activities' with the emphasis on developing a 'right-first-time' culture;
4. The desire to concentrate effort on managing costs 'collaboratively' so that all unnecessary expenditure is reduced and, eventually, eliminated;
5. 'Develop continuous improvement' at strategic level and throughout every activity and associated processes.

As should be abundantly clear already and which will be emphasised in subsequent chapters (especially Chapter 7 that deals with 'Organisation[al] Matters'), the importance of people cannot be overstated. As Cain argues, successful supply-chain management can only take place by developing every person through 'expert coaching, training and support' (2004).

5.13.7   Re-engineering of processes (usually referred to as BPR [Business Process Reengineering])

This is an approach based on the radical transformation of carrying out operations. As such, it is consistent with all of the techniques already described. Each process is analysed to ensure that it is most efficient (see also lean production) and that the configuration of all processes is such as to create customer value.

The overall aim is to redesign the way that the organisation operates and to alter processes in such a way as to enable rather than to disable.

In any organisation there will be accepted ways of doing things: custom and practice. It is a strange tradition that, even though people see that such ways may be outdated or inappropriate, they are either reluctant to suggest change or, worse, discouraged from doing so. Consequently, ways to improve are never implemented (indeed, never even considered). Therefore, it is asserted, people should be regularly consulted as to how they can 'reconfigure' the processes they use to carry out day-to-day operations. This will include the technical, managerial and organisational aspects. So, for example, the following would be considered:

- Product and service improvement;
- New innovations and developments;
- Logistics;
- Planning;
- Production times or cycles;
- Creation of stakeholder value.

Implementing BPR

1. Establish a clear view of the strategic purpose;
2. Set challenging goals;

3. Define core processes;
4. Redesign processes to create 'higher-level' processes which match the new expectations;
5. Use teamwork to create potential for change and to implement solutions.

Benefits of BPR

In organisations where BPR has been used, it has allowed significant increases in productivity, reduction in cost and slimming of times from order to delivery. Like lean production, it should be seen as a vital part of improving efficiency and quality – NOT as a method of reducing staff. To do this will be to undermine the dedication and motivation of the very people that are vital to improvement. If this happens, the people affected will hardly be likely to give the level of support that is required for success. Worse, they will be highly suspicious of any future initiatives. This, of course, will hardly be 'fertile ground' upon which further improvement can be developed.

5.14   Conclusion

Boyd and Chinyio make the point that understanding what construction clients want is difficult because of the fact that the 'area is multidimensional, dynamic and contradictory' (2006, p.297). In their book they recommend that there is a need for all parties to engage in a better model of thinking about communication of needs so that those in supply can adequately deliver satisfaction (they provide a 'toolkit for engagement'). What is important is that organisations involved in construction are willing to use appropriate tools or techniques which will allow them to 'surface' knowledge that enables them to understand what client expectations are. In so doing, it is more likely that those involved can deliver the levels of service and excellence that are manifest in other industries such as car manufacture and which are described in preceding sections of this chapter. However, whilst the use of systems is undoubtedly important, the processes of carrying out construction involve organisations and, as Chapter 7 emphasises, these consist of people who bring with them creativity and innovative thinking (as well as potential problems). Getting to know how people can collaborate to create value-driven solutions is how successful strategy can be achieved so as to give construction customers exactly what they want.

Chapter 6

Developing and maintaining organisational resources – the basis for delivering strategy

'The greatest achievement of the human spirit is to live up to one's opportunities and make the most of one's resources.' Marquis de Vauvenargues, French moralist and essayist, 1714–1747

6.1 Objectives of this chapter

In previous chapters there has been discussion as to what an organisation does in order to achieve its strategic objectives. Implicit within this statement is the assumption that an organisation possesses the ability to deliver its promises to those customers it wishes to serve. And in order to do this it must have resources which, as well as being the capability, may be defined as 'the means by which the organisation generates *value* [his italics]' (Lynch, 2006, p.189). Crucially, therefore, resources are the elements that distinguish one organisation from another, most especially its competitors.

In using the word, 'value', Lynch's definition makes clear that the objective of using resources is to create something that is greater than the costs of production. As a consequence, the organisation makes either profit or, in the case of non-profit-making entities, surplus (or maximisation of output for the input). Resources are the keystone of the organisation and securing the best available resources that can be afforded in the most appropriate mix will have a large impact on its subsequent success in achieving its strategy. It is important to remember that no organisation has access to limitless funds to finance its resources[1].

In any organisation there will normally be three main types of resource:

- Human;
- Financial;
- Productive capability.

The human 'element' of organisation is considered in Chapter 7, which explores aspects of organisation. In this chapter an overview of how the human input is effectively managed as a resource will be carried out. Most especially, the focus of analysis of human resources is on understanding how the input of people in organisations is a 'lynchpin' to making strategy work – through inspirational and effective leadership and a dedicated and motivated workforce.

---

[1]This is true of governments (although there may be occasions when budgetary restraint may be abandoned – such as during a crisis like a war) and even the largest and most profitable companies who seek to attain the highest possible return on investment.

Finance is examined in order to understand its importance in terms of making appropriate strategic decisions. Having sufficient money to be able to invest in any resources such as people, plant, machinery and technology is essential. Even if an organisation is working well, there is a need for constant development which requires innovation and investment. However, as well as consideration of investment, there is a need to judge the effectiveness of previous decisions. This requires managers to use financial information to provide the basis of quotients that enable them to ascertain the efficiency of either the whole or the parts of the organisation. Using financial data in this way is known as 'ratio analysis' and should provide indicators of the effectiveness of the organisational strategy being pursued.

The last of the three key resources, productive capability, is considered in terms of how the carrying out of tasks can be managed more effectively. Most importantly, by concentrating effort on the completion of tasks intended to supply goods or services, organisations can ensure that the goods and/or services can be produced more efficiently. The key objective is to look at operational processes to improve 'value-adding'. Managing resources in this way is also called 'operations management'. Importantly, technology and innovation, which are integral to carrying out operations, are considered in detail in Chapter 8.

Finally, construction's management of certain key resources has not always been good. Chapter 3 suggested that the input of people in construction has not always been valued as highly as it should be (see *Rethinking Construction*, Construction Task Force, 1998 and McCabe, 2007). As this chapter explains, managing any resource effectively will be more likely to contribute to the successful attainment of organisational objectives. Analysis of so-called 'excellent' (highly successful) organisations suggests that a key feature of their approach is dedicated human resource management[2]. Prior to considering the three key resources, it is important to understand the theoretical principles of resources.

6.2   Resource usage – an overview

In any industry, such as construction, there are key factors that are common to organisations (even though they will vary depending on the sector). It was Japanese strategist, Kenichi Ohmae (who was former head of influential management consultants, McKinsey, where Peters and Waterman worked before publishing *In Search of Excellence*, 1982), who believed that every organisation should be aware of the essential resources that will allow it to serve the marketplace in the best way it can depending on the prevailing circumstances – the 'competitive environment' (1983). As such, he argued that there are three principal areas which, given that they all begin with the letter C, have become known as 'Ohmae's three Cs'[3]:

- Customers – including price, service, quality and the technical expectations;
- Competition – including comparisons in terms of price, functionality and ability to provide a better 'than the rest' service;
- Corporation.

The first two of these have already been covered extensively in previous chapters. The last, corporation, considers the attributes that an organisation possesses which will give it the ability to, at the very least, match what competitors offer. Better still, of course, it should aim to use its resources to continue to create the opportunity to set the standard that others must emulate. In

[2] Frequently as an integral part of corporate social responsibility (CSR).
[3] Ohmae's three factors is not without criticism (see Ghemawat, 1991). In particular, the concerns are that identification is far from easy, that there is difficulty in making causal connections, that too much generalisation is assumed and that the dynamics of markets makes their use spurious.

industries outside construction, such as microelectronics, telecommunications and car manufacture, competition is highly intense and there is a need to constantly scan the environment to analyse how other organisations use their resources. Lynch provides a list of the factors that can usefully be considered with respect to corporation (2006, p.196) and which, in the context of an analysis of construction, are highly pertinent:

Cost differentials   In some industries organisations operate on the basis of lowest cost, the airline industry being a good example. As certain low-cost carriers (such as Ryanair and EasyJet) have shown, providing cheap air travel, despite criticisms concerning service, can enable you to prosper. Construction is an industry that has a long tradition of attempting to reduce its costs (see McCabe, 2007). However, the overall effect has been negative in terms of the ability of firms to match the expectations of clients and, most especially, has severely undermined the level of quality produced.

Economies of scale   These have been shown to be important in some industries in driving down costs and making goods and services affordable to a much wider 'customer base' (as Ford was able to do with the Model-T). However, if economies of scale rely on mass production there can be negative consequences. Henry Ford may have been a master of creating productive systems, but he failed to recognise the impact on people whom he simply regarded as being necessary to keep the machines going[4]. Ball (1988) describes how the post-Second-World-War period in construction was one in which employers saw industrialisation as being the way to reduce cost and to ensure that reliance on the skills of traditional craft guilds was undermined. However, whilst the criticism contained in cost differentials equally applies to this factor, it is important to recognise that there is the potential for clients to look elsewhere for a cheaper (and more utilitarian) solution to their construction needs which may become increasingly attractive – as McDonalds demonstrated in mass-producing the facilities to sell their food.

Labour costs   Certain industries have shown that moving their production or service facilities to countries with lower wage rates may be a way to reduce cost (as manufacturers and financial organisations such as banks have done). The overall effect is questionable in that whilst it may allow the organisation to provide cheaper products or services, there is the negative aspect of reducing indigenous employment and potential diminution of quality and standards. Construction has a long tradition of using migrant labour, going back to the large-scale building associated with the Industrial Revolution, to ensure that there is an adequate supply of workers. The consequence of doing this has been a legacy where many are sceptical of the real desire by employers to alter the belief that construction workers count for very little, despite the implementation of the *Respect for People* initiative which came about as a result of *Rethinking Construction* (see McCabe, 2007).

Production output levels   A constant criticism of construction has been its apparent inability to match the levels of output produced elsewhere, Germany and the US being cited as exemplars of superior performance (*c.f.* Ball, 1988). One consequence of *Rethinking Construction* and the quality debate (see next resource factor) is that the comparison is no longer solely between British construction and its counterparts elsewhere. As has been explained previously, British industries such as car manufacturing and shipbuilding have learned the harsh lesson that they cannot match production output found in other countries and, therefore, cannot ensure that long-term costs can

---

[4]As Pursell (1994) explains, Ford ensured that workers were discouraged from interfering with production by the use of what was called the 'Service Department' which consisted of 'thugs' who used violence to ensure compliance.

be reduced, so they become uneconomical. Charging more to ensure that costs are covered may work in the very short term, but customers will switch to those producers that charge less for the same product or service. Even harder to deal with was when Japanese could offer more in their product, in terms of additional features and warranty, for less money. As the authors of *Rethinking Construction* related, clients have become increasingly exasperated at the inability of contractors and their subcontractors and suppliers to emulate the customer experience they enjoy when dealing with other industries. One way to attempt to match the example of others is to 'benchmark' (see McCabe, 2001). Moreover, as the Japanese have so clearly shown, having a dedicated and passionate workforce is essential.

Quality operations   As in the previous factor, the level of quality achieved by British construction frequently leaves much to be desired. A notable example is that achieved by British housebuilders; it may be argued that this is the closest comparable to industries that produce large numbers of similar products ('batch' as opposed to 'mass', see Woodward, 1965). Despite the fact that the overall product is made of standard components (consisting of some 40,000 according to *Rethinking Construction*, Construction Task Force 1998, p.27), there is an apparent inability to achieve consistent standards of quality. Indeed, surveys of purchasers of new houses demonstrate levels of dissatisfaction with what they receive. When compared, for example, to bespoke sports cars (like the Morgan, for instance), the inadequacy of housebuilders to ensure quality in all aspects of their operations becomes all the more glaring. As Ball noted in 1988, lack of investment in both plant and people has been a major problem which, as *Rethinking Construction* asserted, needed immediate attention.

Innovative ability   Given what has been stated already about the traditional lack of investment and conservative attitudes that tend to exist in construction, it is perhaps no surprise that innovation is less prevalent than in other sectors. Organisations working in areas which are novel and highly dynamic are inherently more inclined to take risks. Accordingly, if the only way to survive or prosper is to do something radically different, it is worth having a go. Construction has a tendency to wait to see if it works elsewhere before embracing change (see Chapter 9 for a more detailed treatment). Once again, the 'spin-off' from *Rethinking Construction* has been to encourage thinking that leads to innovation. Moreover, as you should be aware, the desire of successive governments has been to encourage active participation and partnerships between industry and academia. Where this has happened in other sectors, there tends to be a highly symbiotic and virtuous relationship that generates benefits to all concerned.

Labour–management relations   This has been broadly covered already, but you are encouraged to consult McCabe (2007) which provides a socio-historical analysis of industrial relations in British construction. The 'Respect for People' initiative may be seen to be a long-overdue attempt to undo the damage that the national building strike of 1972 caused in terms of relationships between employers and major construction unions (which was an attempt to resist casualisation and the impact of 'lump' labour).

Technologies and copyright   See section on innovative ability (above).

Skills   At this stage, this should need no elaboration. Suffice to state, though, the levels of skill and expertise in the industry need to be constantly reviewed and improved by continuous training and education. This applies equally to manual and non-manual people in the industry.

All of the above are factors that may be considered as having validity across an industry (such as construction). However, it should be apparent that every organisation will seek to develop a blend of factors that will give it the uniqueness that enables it to enjoy competitive advantage. This approach is known as a resource-based view (RBV) and is considered below.

6.3   A contemporary consideration of how organisations use resources

Possession of resources is not without cost. As an analogy, think about car ownership. Anyone who owns their own vehicle will tell you that there are many costs that must be considered: tax, insurance, maintenance and, of course, fuel. As a consequence, in comparison to other modes of travel, such as bicycle, bus or train, it may not be the cheapest form of getting about. As well as providing more comfort, it allows the owner complete freedom as to when it is used. The same principle applies to organisational resources; they have consequential costs. Traditionally it was commonly accepted that control was essential and that all resources used should be owned. Additional costs were accounted for in the price that the customer should be expected to pay[5]. If you do own all of the resources that are used in production, there is an incentive to attempt to use them effectively.

A downside to owning all resources is that it tends to create large and unwieldy corporations which, in times of change and dynamics in the market, can cause decision-making to be far too slow to react appropriately[6]. In order to avoid this and to make organisations more agile, an alternative approach was used whereby some functions that, hitherto, had been carried out using resources that were internally owned, were carried out externally (either by specialist suppliers or contractors). Whilst many might believe that this technique of resource management to reduce direct costs incurred by the organisation, so-called 'outsourcing'[7], was novel, they might be surprised to learn that construction had been using it since the large-scale building schemes associated with the Industrial Revolution. Then the objective was, as Cooney explains, to create dynamic organisations that could, using the opportunities that capitalism provided, serve the needs of clients, who wanted their schemes operating (and returning dividend on investment) as speedily as possible, and also to make profit with minimum risk (1955)[8].

Outsourcing (or, as it is usually referred to in construction, 'subcontracting') has undoubted attractions. It allows an organisation to reduce its resources in terms of staff and machinery. Both capital investment and day-to-day operating costs are no longer the concern of the organisation that has outsourced. When demand increases it probably means that costs paid may be higher, unless long-term fixed prices have been agreed. Better still, though, should there be a reduction in demand, the costs of excess staff and having machinery operating at less than optimum are not its concern. As a result, there is a belief that having any resources (beyond absolute minimum) represents tied-up capital. Indeed, some organisations possess almost no resources beyond what is essential to trade. For example, Internet traders may not even own the computers that they use to receive orders. They will merely ensure that they can get the products from their suppliers via a courier to the customer. Their competitive advantage is to be able to do so at a cost that is lower than those of retailers on, for example, the high street.

Outsourcing assumes that those organisations that supply the goods or services do so at a cost that is 'acceptable'. Most especially, the price charged to the customer is greater than that which has been incurred in using the outsourcing organisations. However, as construction has found, the trouble with markets is that they can be unpredictable (and volatile). Some railway subcontractors, for example, frequently found that the rates they had to pay to attract sufficient labour

---

[5]Henry Ford, who was a 'control freak', bought rubber plantations (for tyres) and mines (for steel) to ensure regularity of supply of raw materials used in his cars.

[6]The state-owned car production corporation, BMC (which became British Leyland), is frequently cited as the best example of an organisation that became practically unmanageable.

[7]Lynch defines it as 'The decision to use the outside 'market' to buy in products or services rather than use the organisations' own to make them' (2006, p.198).

[8]Coleman (1966) provides ample evidence that whilst many contractors and subcontractors were unscrupulous in their use of contracting to make profit at the expense of workers (see, especially, his explanation of the system called 'truck'), there were notable examples of those who attempted to ensure that conditions were improved.

reduced their profit to a point that made continuance of the work uneconomic. Some carried on in the hope of 'clawing back' some profit from using whatever methods they could, frequently at the risk of serious injury and death to workers. Whilst some went bankrupt in the process, others merely waited until they had been paid and promptly disappeared, leaving those who had provided their labour in good faith empty-handed. The pain to individuals (and disruption to clients) that this causes (which still continues to this day) is perhaps the biggest problem that the industry has to deal with. As the next section explains, the key to strategic success is ensuring that the maximum utility is derived from resources used; the expression that is most commonly used is to 'add value'.

6.4   The objective of adding value to resources

Adding value to resources is based on the simple economic concept of being able to procure them at a price that allows them to be used to produce something that has a value greater than the sum total of its components. As Mr Micawber, a character in Charles Dickens' book, *David Copperfield*, reflects[9]:

> Annual income twenty pounds, annual expenditure nineteen six, result happiness. Annual income twenty pounds, annual expenditure twenty pound ought and six, result misery

Micawber's message is simple but critical; none of us should live beyond the means by which we can afford to purchase. This applies to all organisations, regardless of whether they are very small (such as a builder working on their own as a sole trader), or a large multi-national organisation carrying out work across the globe (as many contractors do). As previous sections will have described, ensuring that an organisation can remain competitive in an increasingly dynamic world is never easy. In the past it may have been able to achieve its processes effectively and efficiently (and still continues to do so). However, the lesson that many organisations have learned is that there is a constant threat from competitors (existing or new), who may develop ways of utilising their resources more effectively. British car manufacturers learned this lesson the hard way when they discovered that their Japanese counterparts were able to make their products with levels of efficiency that were undreamed of. Whilst this lowered the costs of production, they were also able to deliver a product that had higher levels of quality (in terms of reliability). Customers soon discovered that they could have greater value by switching their allegiance to such producers (even though they were aware that the consequence of buying 'foreign' undermines domestic producers).

Adding value in this way, therefore, can have the effect of radically shifting costs and revenues (and the 'economic rent' that resources attract – see below) of all those who operate in a particular market. It is incumbent upon organisations to constantly strive to add value to every process that they carry out. As the example of Japan's influence on car manufacture demonstrates, whilst cost reduction is a very effective way to add value, this does not have to lead to diminution of standards or quality. An abiding message that construction has received from commentators and clients is that the 'dog-eat-dog' approach creates dissatisfaction among all participants. Rather, it is argued, cooperation and collaboration are essential between all those who contribute to the

---

[9]Like many of his classic novels, this story actually originally appeared in nineteen monthly one-shilling instalments between May 1849 and November 1850 published by Bradbury and Evans, London. Whilst the novel is always referred to as *David Copperfield*, it was actually published under the title of *The Personal History, Adventures, Experience and Observation of David Copperfield the Younger of Blunderstone Rookery*.

end product or service. Two important concepts can be considered in pursuance of this objective:

- The value chain;
- The value system.

These have their provenance in the work carried out by seminal strategic thinker, Professor Michael Porter, in the 1980s at the Harvard Business School. The value chain considers all of the activities, both primary and secondary (defined below), that need to be organised to create the end product and/or service. The desire to understand value is regarded as being 'classical' to the analysis of resources. Indeed, as Porter recommends, in carrying out such analysis, in terms of what goes on, there are a number of crucial aspects concerned with production. These contribute directly to output and, normally, can be measured directly (as in site activities in construction). Accordingly, the primary activities shown are concerned with the following:

1. Inbound logistics – how all 'components' used in producing the end product/service (materials and labour) are received and used in a timely and appropriate way. The use of techniques such as 'just-in-time' would be very useful in ensuring that the overall quality of the construction is as high as possible[10].
2. Operations – this is the most crucial part of the construction activities carried out. Whilst detailed analysis of how contractors and subcontractors carry out operations on site is beyond this book, all students of construction should be familiar with the range of planning and scheduling techniques that exist to ensure production is as effective and timely as expected[11].
3. Outbound logistics – this is concerned with the distribution of goods or services to customers. Whilst this may not appear to be as relevant in construction as, for example, an Internet company delivering its goods, it is crucial that every opportunity to attract customers and give them exactly what they expect is pursued (in combination with the last two primary activities).
4. Marketing and sales – as well as including the advertising and promotion, this will necessitate consideration of all of the components of customer service that have been described in an earlier chapter.
5. Service – like the previous activity, this is an essential part of delivering a 'quality' experience to customers and, crucially, can contribute to ensuring enhanced levels of satisfaction and, as a direct consequence, repeat demand/custom.

Porter also identified 'support' activities which, even though their contribution to production is less easy to measure, are vital to the overall effectiveness. He contends that there are four elements which provide support:

1. Firm[12] infrastructure – this would encompass how all the elements of organisation are configured to create both its identity and purpose.

---

[10] Increasingly, prefabrication and modularisation are used to deliver whole parts of the construction such as bathroom 'pods' which, as well as being of an extremely high standard of finish (factory level), reduce the need to manage the interaction of trades as would be required for the traditional approach.

[11] Equally, though, every student should be aware that the nature of site activities is such that, whilst planners hope for the most optimistic outcome, they should expect the 'unexpected' (weather, unreliability of suppliers or subcontractors, design problems and so on), and include contingency both in terms of time and money (in CPM [Critical Path Method], the use of what is known as 'float' for non-critical activities allows some degree of flexibility).

[12] Porter's use of the word, firm, indicates that, at the point of publishing this work, he considered strategic analysis not to be applicable to organisations that did not seek to achieve profit.

2. Human resource management – this will be considered in detail later in the chapter but would include all aspects that are associated with procuring, inducing, developing and using every person within (and increasingly without[13]) the organisation. As such, it would include departments such as human resources (personnel), training and development, and safety, all of which should be assumed to be focused on getting the best from every person and, additionally, scanning the environment for new innovations and developments that will give it a 'competitive edge'.

3. Technology development – this will be considered in a later chapter but, in summary, it is important to recognise that the things we consider to be part of everyday life often had their origins in laboratories that were dedicated to finding new innovations and developing the 'next generation' of products. A very legitimate criticism of construction is its unwillingness to consider research into technological development of its products or services to be a vital part of expenditure.

4. Procurement – this is a word that will be familiar to all but the very new entrants to construction. It is seen to be crucial to the ability of organisations to achieve supply of goods, services and labour. Traditionally, every construction organisation had what was called a 'buying' department which carried out this function. Increasing complexity and inter-connectedness of the construction process (especially though 'partnering') have meant that procurement is seen to be much more complicated and critical than hitherto: requiring skills more than simply 'beating them down on price'!

Michael Porter proposed the value chain as a way to explore the **internal** aspects of what goes on within the organisation. Recognising that what any organisation does is usually part of a wider system of interconnected parts (especially if external subcontractors and suppliers are used), he presented what he called the 'value system'. Accordingly, the value system incorporates every organisation that contributes to the overall objective of providing the end customer with a product or service that, as a minimum, matches and, better still, exceeds their expectations. Porter's proposal of the importance of the 'value system' can be viewed as being entirely resonant and in sympathy with Total Quality Management (TQM) which was beginning to attract the attention of western managers, especially in northern America, who were aware of the increasing economic threat from Japanese manufacturers in the early 1980s. What was discovered by managers intrigued by the apparent sudden success of their Japanese competitors was the close cooperation and understanding of needs that existed between suppliers and producers. The aim was, as Sako put it, to develop 'arms-around relationships', as opposed to 'arms-length relationships' which, she contended, tend to typically exist in the west (1993).

If the supermarket sector is considered, Tesco's ability to demonstrate its competitiveness in terms of price and delivery is due to its obsession with improving its own internal processes in conjunction with its suppliers. Whilst its approach has not been without criticism (Lawrence, 2004; Blythman, 2005; Bevan, 2006; Sims, 2007), Tesco has created an extremely effective supply network that uses what they call 'continual replenishment' to continually restock their stores[14]. As well as ensuring that the produce on sale is usually freshly delivered and that stock levels are minimal (usually being held for a matter of hours) which frees floor space for sales, they use their database to build up a profile of demand patterns that exist in each of their stores. Like those engaged in manufacturing, such a distribution network relies on the willingness of suppliers to actively engage in what advocates of 'lean' describe as 'pull' (see Womack and Jones, 1996).

---

[13] If you can assist the employees of organisations that supply you, then this is likely to enhance the possibility of creating competitive advantage.
[14] The reason why there is often a lorry making a delivery when you visit a particular store.

Form a strategic point of view, understanding the linkages within any system is crucial to being able to determine what can currently be achieved and, potentially, how future improvement in delivery and service can be achieved. Bringing together the best combination of skills and abilities to deliver construction projects on time and within budget will strongly contribute to overall success. Having good relationships with subcontractors and suppliers, therefore, can create a system of 'partnerships' that can enable construction teams to rise to the challenges that were proposed in *Rethinking Construction*, Construction Task Force (1998).

6.5   How resources are valued – applying 'economic rent'

All resources have value. For example, we, as individuals, usually hope that the skills and qualifications we possess will allow us to earn as much as possible. However, how much are we really worth? This depends on exactly what our particular skills are and how plentiful (or scarce) are others who possess qualifications that we have. Gaining specialist knowledge that few, if any, others have will be likely to put us into a privileged position in terms of status and, therefore, earning power. This is true of all resources and, unsurprisingly, economics (in terms of supply and demand) provides the basis of determining the value of any resource.

In a stable market it is possible to judge the value that may be attached to a resource. In construction, this will mean that accurate costs can be produced for materials and labour. However, as the industry has experienced in recent years because of the increase in demand, the rates that have to be paid to particular trades have risen very dramatically. This is because there are shortages of trained operatives in those trades; plumbers are often cited as an example. It is reputed that some plumbers were earning in excess of £60,000 per year[15] (which we will assume to be £250 per day). As a result, operatives earning far less than this may have been tempted to retrain to become plumbers. If we consider the average construction wage to be £150 per day, then using the reputed rate of £250 per day, plumbers earn £100 more which reflects their relative importance and scarcity. For as long as plumbers remain scarce, the £100 is considered to be 'economic rent'[16], a concept that was originally proposed by nineteenth-century economist, David Ricardo (1772–1823), and can be defined as being:

> The excess that a factor of production earns over the minimum required to ensure that it continues in its current use

For example, if a contracting organisation suddenly finds it is losing operatives who are tempted by higher rates elsewhere (which often happens), they are faced with two choices. They can hire others who will work for the rate that the original operatives were being paid. However, this is risky in that there is no guarantee that these other operatives will either be available or possess sufficient expertise. The second choice is to offer higher wage rates to the workers tempted to leave which will take away their incentive to move. Such additional money would be 'economic rent'. Economic rent can be subdivided into two types:

- Ricardian;
- Monopoly.

---

[15] At the time of writing, it is reported that courses to train plumbers are over-subscribed which, of course, will increase the number available (supply). Assuming a relatively steady rate of demand, this will mean that plumbers will probably be forced to reduce their rates of pay.

[16] See Ricardo (1817).

The first of these is an eponymous reference to Ricardo and is used with respect to those resources that an organisation possesses which give it a competitive advantage. For example, if an organisation has developed special expertise or has a well-trained workforce that means it is capable of charging its customers more than its competitors, this additional money is considered to be **Ricardian**. Construction firms, especially those in specialist trades or professions, do this if there is a shortage – such as plumbers called out to emergencies during the winter. The second type of economic rent, **monopoly**, refers to resources that allow the organisation to offer products and/or services that are considered to be unique or distinctive and for which customers will pay more.

An organisation enjoying Ricardian or monopoly rent would be foolish to be complacent that it will maintain such advantage into the long term. As a later chapter that explores the importance of knowledge, technology and innovation will explain, the nature of competitive markets is that competitors should be assumed to be working on introducing developments that will allow them to match expertise and skill. Indeed, the concept of 'benchmarking' (see McCabe, 2001) is a management concept explicitly intended to identify gaps in expertise and to learn how to match the ability (and therefore resources) of proven market leaders.

Finally, as part of this section, it is worth making reference to what is known as 'Schumpeterian rent'[17]. Such rent is named after the seminal American ecomomist, Joseph Schumpeter (1883–1950), who identified that organisations that develop innovative products or services are usually able to charge a premium, if for no other reason than to justify the risk of the investment incurred. Accordingly, as long as the product/service remains innovative, the revenues generated may produce 'super-profit' which is the incentive to invest in development. However, as the history of innovation and technology has demonstrated time after time, unless an organisation can absolutely protect its creativity through the use of copyrights or patents[18], competitors will seek to imitate the developments that have produced super-profit. Whilst this means that such 'followers' will not enjoy the same level of profit, they will have avoided the risk of investment and will know what has been successful elsewhere. Construction, it must be noted, tends to be an industry in which research and innovation are limited. New developments frequently originate from other sectors and are adapted to suit the particular context[19].

6.6   How organisations develop capability based on their resources – an appreciation of the importance of the resource-based perspective (RBV)

Having better resources than competitors is axiomatic. History is replete with examples in both the military and commercial context of how having a particular strength or expertise enables one side (or organisation) to gain advantage. In contemporary terms, analysts seek to answer the question of how success was achieved, especially when against the 'odds' (such as where the victor has fewer resources or the general circumstances were unfavourable). The answer that analysis elicits is that whilst conditions in a particular context (market) may be helpful, if the

---

[17]See Schumpeter (1934).

[18]Which may only protect the exact configuration and application of technology.

[19]Given the relatively low rates of return that have traditionally existed in the industry (less than two per cent being common in the 1980s and 1990s), it is probably not surprising that organisations focused their efforts on reduction in the costs of labour and materials and, most certainly, were averse to the sort of capital investment seen in, for instance, manufacturing. This creates a 'vortex of decline' that means nothing ever improves until someone is brave enough to break the cycle of decline.

organisation is able to muster 'special' resources that it can utilise better, then it is more likely to enjoy success – even if, for example, the economic climate is challenging.

Accordingly, the emphasis has focused on consideration of how an organisation develops or acquires resource capability that enables it to achieve dominance or competitive advantage. This approach is known as the resource-based view (RBV) of strategy[20] and has been developed since 1984 by a number of authors (Connor,1991, provides an historical analysis and Hoopes *et al.*, 2003, edit a special edition of *Strategic Management Journal* which is entirely dedicated to the approach).

One problem that critics of RBV have identified is that to assume that markets can be analysed easily (in an attempt to identify greatest opportunities for profitability) is overly simplistic. The cliché, 'when the going gets tough, the tough get going', is apposite in this respect. Construction is a sector that is quintessentially prone to experiencing peaks and troughs which are largely governed by the economic conditions. Making profit is obviously easier in good times than in bad. However, it is rare that all firms operating in the industry make consistent profits or losses. Moreover, even in difficult economic conditions such as during the 1970s, 1980s and early 1990s some were able to increase levels of profitability. RBV, therefore, would attempt to explore how some construction organisations can use their resources better than others.

The question that may still be asked is how those successful organisations manage their resources in order to achieve sustainable competitive advantage. Whilst Lynch suggests that there is no agreement amongst strategic thinkers and commentators on RBV as to the 'precise source' of the advantages of successful organisations[21], he provides a number of key 'elements' that comprise RBV which are likely to produce SCA:

1. **Prior or acquired resources** – having been successful in the past, these will be useful in replicating accomplishment in areas where expertise and experience have been developed. Building on this success is bound to give a head-start on others who may wish to emulate such achievements. It is also highly likely that mistakes will have been learned earlier (which is always valuable, if frustrating!). For instance, very successful contractors will often have developed from small, family concerns that commenced in very limited markets (such as being 'jobbing builders').

2. **Imitability** – this considers the difficultly that competitors will have in copying the resources of those who are successful. Some will be much easier to imitate than others. Technology and design are notoriously problematic in stopping others imitating (especially if proprietary components have been used). Intellectual capital (what people know and can use to contribute to organisational success) is much harder to imitate. With respect to the latter, some organisations are able to harness their people in a way that enables them to cooperate and collaborate to a degree that others cannot achieve. It is for this reason that senior managers and, especially, executives are often paid very high salaries – both to recognise their organisational ability and know-how and to deter others from trying to poach them!

   Lynch suggests three ways in which an organisation can make the task of imitation of resources more difficult for competitors (2006, p.219):

   - 'Tangible uniqueness' which would include branding or being in a particular location (some contractors are based in regions, such as the south-west, including London, where others are reluctant to try and operate).
   - 'Casual ambiguity' where it is difficult for outsiders (competitors) to know what has actually provided the basis for success. This is especially so where complexity is part of

---

[20]Wernerfelt (1984) was the first to challenge the accepted wisdom that identifying those markets that were likely to be most profitable was the most effective way to develop strategy.

[21]The attempt to provide definitive answers is always going to be proved fallacious – as demonstrated by Peters and Waterman's identification of their 'eight attributes of excellent companies' in their book, *In Search of Excellence.*

the everyday processes and where specialist knowledge (such as an intimate understanding of customers) is required. Many successful contractual relationships have been formed on the basis of an intimate understanding between a number of key people. For those trying to 'break into' this market, it will be extremely difficult to know how these relationships operate or to know what creates success.

- 'Investment deterrence' in which the amount of money required and the risks involved provide too great a barrier to all but the determined (and perhaps foolhardy) competitors. In property speculation, for example, there is a need to be able to buy land and/or buildings and, until some or all of the development is sold, to pay for construction work to be carried out[22].

3. **Substitutability** – in some organisations there are resources that cannot be substituted without radically altering the essence of the organisation. This is particularly the case where there is a dominant leader who takes a 'hands-on' approach to all day-to-day matters concerning the organisation (especially the case in a family-owned business). Should they leave, the direction/culture is likely to change. This might equally be the case for a number of influential employees. This will be explored in greater detail in the next chapter.

4. **Durability** – the length of time that a resource can maintain its ability to provide advantage is important. Some will be transient, such as being involved in a market earlier than others. Some resources will maintain their advantage for longer, particularly those that are harder to imitate (see above). In many sectors, having a long-standing reputation – based on quality and service – can be very valuable.

In addition to these four, in order for a resource to generate SCA for the organisation, it is best if it provides benefit internally and not be shared with others. If the advantage is diluted by sharing with other organisations, its effect will be reduced by not allowing the full impact to be enjoyed by those who have most involvement. For example, if a new method or process is developed by a group of workers that indeed gives SCA, they should feel that they are recognised and, possibly, rewarded. Ensuring that surplus (profit) is reinvested in the people who created the advantage will give people the perception that they are valued and, as the exploration of 'people issues' in the next chapter explains, encourage them to redouble their efforts to ensure future success. Resources that deliver SCA are frequently those that are innovative and, even though others may eventually imitate them, they will allow the organisation to be at the forefront in providing 'breakthrough' technology and methods. Innovation is explored in greater detail in Chapter 8.

6.7   Resources and their place in the structure

Considering resources as if they are homogenous and have equal impact throughout the organisation is not accurate. The importance of resources will be dependent on their application within particular levels of the organisational hierarchy. As Chaharbaghi and Lynch (1999) identified, the higher the level within the hierarchy that resources are utilised, the greater the chance that they will contribute. The following list shows the four hierarchical levels at which resources can be used (shown in order of presumed contribution to competitive advantage):

- Breakthrough;
- Core;

---

[22]At the time of writing, the consequences of the so-called 'credit crunch' are beginning to be experienced which are making it extremely difficult for developers to raise finance from banks to carry out development.

- Base;
- Peripheral.

Clearly, the most important of these resources are those which allow 'breakthrough'. These are most likely to be highly innovative and, at least in the short term, enable the organisation to be ahead of its main competitors. The next level at which resources may be applied, 'core', will give the organisation its distinctiveness. Whilst these resources will be imitated, their advantage is likely to last longer than breakthrough resources which usually create a new standard or benchmark for an industry (or sector). The lower levels of resources, 'base' and 'peripheral', are less important but, as the explanation of productive capability demonstrates, they will be crucial in ensuring that the day-to-day processes are as effective as possible in delivering the outputs that will eventually contribute to overall strategy.

6.8   Using VRIO framework to decide on resource application

Making decisions about how resources should be applied is crucial to success. This, of course, is part of the role of a manager who will expect to be judged by the quality of their decision-making. As you should be aware, many theoretical models exist to assist managers in making better decisions (see, for example, Vroom and Yetton, 1973). Consistent with this desire, Professor Jay Barney from the Ohio State University has provided what is known as the VRIO framework, the letters of which are taken from 'Valuable', 'Rare', 'Cannot be imitated' and 'Organising capability'.

The first question, whether the resource is valuable, is one that should be considered with respect to how competitive it will enable the organisation to become in comparison to others. The second, rarity, is useful in that if the resource is plentiful, others will be able to obtain it with ease and any competitive advantage will be very short-lived. Thirdly, the greater the difficulty in imitation, the harder it will be for competitors to emulate any success that the organisation using this resource will have enjoyed. Finally, having resources that can potentially generate SCA is one thing, but using them effectively is quite another thing – particularly in a coordinated way that genuinely benefits the whole organisation. As the exploration of organisation that is carried out in the next chapter demonstrates, 'joined-up thinking' in organisations is more difficult than senior managers believe is possible (most especially where they are large and complex). Indeed, empirical evidence demonstrates that the larger an organisation becomes, the greater the tendency for entropy to occur. Accordingly, and consistent with such a tendency, there is a danger of different parts of an organisation acting in ways that are beneficial to the perceived needs of managers involved, but unless there is coordinated thinking for the whole system, with less regard for the impact on other parts than there ideally should be[23].

6.9   The importance of resources, distinctive capability and core competences

In considering resources, there is the difficulty of knowing exactly what are the elements or components that create the ability for the organisation to be what it is and, ideally, to add value

---

[23]Coordinated thinking and action using systems that were sympathetic to those who work within the organisation were an essential part of the success of the Japanese manufacturers who have demonstrated their pre-eminence in terms of customer care and superior levels of quality.

in ways that others cannot (and, by virtue, to be able to enjoy sustainable competitive advantage). Whilst this may seem relatively easy in that the organisation possesses superior technology and expertise, there are frequently aspects that are much harder to identify precisely and therefore put a value upon (see below). In attempting to make this identification, it is useful to consider resources in three broad categories (see Collis and Montgomery 1995). These are:

- Tangible;
- Intangible;
- Organisational capability.

As the word suggests, 'tangible' resources are those that have physical qualities such as plant and trained employees, based in particular locations. The second category, intangible, refers to things that organisation is able to definitely benefit from, but have no physical being. As such, brand names, expertise in service and delivery, and use of innovative ideas would be considered as being intangible. Finally, organisational capability would consider aspects of leadership and cooperation/teamwork amongst the various parts of the organisation.

Distinctive capabilities are three resource aspects of organisations that John Kay proposed in 1993 in his book, *Foundations of Corporate Success*, as the way that SCA can be delivered. These three areas are:

1. Architecture;
2. Reputation;
3. Innovative capability.

The first of these refers to the structure of inter-relationships within the organisation and with others who, even though they are external, are crucial to success (such as suppliers and subcontractors). Importantly, the ability to develop partnerships that operate on the basis of mutual cooperation and trust is a distinct advantage. Empirical evidence in sectors where this has occurred shows that, where those within the 'supply chain' provide the end customer with what they want (and frequently more), they enjoy brand identification that allows them to be perceived as superior to competitors. This is the second of the distinctive capabilities: reputation. The lesson of the quality movement of the 1980s and 1990s has shown that those who are able to gain a reputation for quality products and service can lead the market and frequently charge more for what is seen to be a premium service[24]. The third, innovative capability, has been described already. Importantly, the ability of an organisation to innovate will be strongly influenced by the culture that exists and which encourages people to suggest and develop new ideas. The issue of culture and strategy will be addressed in the next chapter.

Consistent with distinctive capabilities, Hamel and Prahalad (1994) developed the idea of what they termed **core competences**. These are the particular skills and abilities that an organisation has been able to develop (both in terms of people and technology/innovation) and which enable it to do things that competitors find difficult to match. The importance of core competences is that once an organisation becomes successful in certain key areas it is highly likely that their influence will be experienced throughout. For example, all construction contractors have similar skills. Some, though, are able to utilise their skills to carry out large and complex projects in a way that others cannot. Having developed such a reputation though achievement, such contractors will be likely to attract employees that want to work in an

---

[24] Sony, Toyota and Waitrose are exemplars of being seen to have an excellent reputation amongst customers.

environment that is challenging to them but is dominated by a 'can-do' culture (failure is not an option).

6.10   Developing an understanding of people, productive capability and finance

In the last three sections the intention is to provide an overview of the way in which the three key resources can contribute to effective strategy. The first, the consideration of people, which is normally referred to as 'human resource management' in contemporary practice, is fundamental to the success of any organisation. As subsequent chapters emphasise, those organisations which want to inculcate a culture of innovation and the fostering of new ideas intended to ensure competitive advantage must rely on people. Computers, regardless of how powerful they are, are no substitute for the creativity that the human mind brings to any situation. Accordingly, effective management of human resources is a key objective for those managers responsible for organisational strategy.

The second, finance, is essential in that all organisations rely on a steady flow of funds to invest in all of the key tasks that are required to carry out those operations (see final section) that will be completed in order to attain organisational objectives consistent with its strategy. Financial management consists of ensuring that all money is spent in a way that is effective and provides the optimum return for investors. In a profit-oriented organisation (business), this is about maximising returns. Equally, though, for a non-profit-orientated organisation, such as those that operate in the public sector, the objective is to use finance to maximise the effectiveness of resources used. As will be described, in order to manage the way it uses finance, an organisation may employ a technique known as 'ratio analysis' to evaluate the success (or otherwise) of strategic decisions taken.

Operations management is concerned with all aspects of the production of the goods or services that an organisation is involved in. Most importantly, the key objective is to configure the arrangement of labour, materials and plant or equipment in such a way as to maximise value-adding. In effect, all tasks should be carried out in the most efficient manner.

6.10.1   Effective HRM – the way to develop people

Whilst HRM is seen as a contemporary approach to the management of people, its predecessor was personnel management. The following definitions provide a direct comparison between them:

> Personnel management is most realistically seen as a series of activities enabling working man and his employing organisation to reach agreement about the nature and objectives or the employment relationship between them, and then to fulfil those agreements (Torrington and Chapman, 1979, p.4)

> Human resource management is a series of activities which: first enables working people and the organisation which uses their skills to agree about the objectives and nature of the working relationship and, secondly, ensures that the agreement is fulfilled (Torrington *et al.*, 2002, p.13)

The table below shows that there are significant differences between the two – most especially the desire to achieve long-term improvement through the people involved.

Table 6.1 The key differences between personnel management and human resources management (source: Guest, D. E. 1987)

Factors for consideration	Personnel management	Human resources management
Timing and planning	Short-term, reactive, marginal	Long-term, proactive, strategic, integrated
Psychological contract	Compliance	Commitment
Control systems	External controls	Self-control
Employee relations perspective	Pluralist, collective, low trust	Unitary, individual, high trust
Preferred structures/systems	Bureaucratic/mechanistic, centralised, formal defined roles	Organic, devolved, flexible roles
Typical roles involved in carrying out 'tasks'	Specialist/professional	Largely integrated into line management
Evaluation criteria (how is success judged?)	Cost reduction	Maximum utilisation (human asset accounting)

HRM is an approach to managing people within (and potentially beyond) the organisation in such a way as to achieve longer-term strategic goals. Employees are no longer merely employed for their ability to suffice in terms of qualification and experience. Rather, they should be involved in every aspect of the organisation's activities. People should be developed in a way that ensures they can contribute to improving every aspect of what the organisation does in pursuance of its objectives.

The origins of HRM can be traced to a course that was presented at the Harvard Business School in 1981 to explore the potential impact of an approach to collectively managing people towards known, desired (strategic) goals and which was referred to as organisational development (OD). Most especially, the course wished to consider how the input of employees (people) could contribute to the attainment of organisational objectives. The context in which this course was developed was that American companies were discovering that their ability to compete was coming under threat from Japan. Those interested in finding out why Japanese manufacturers were so much better (in terms of efficiency and quality) found by analysis of their approaches to management, most especially to people, that people were integrated in a way that emphasised their importance to contributing to improvement.

The Harvard course also emphasised the importance of people as the means by which an organisation can develop towards achievement of goals that create 'stretch' (those that are ambitious). *Managing Human Assets* (Beer *et al.*, 1984) propagated this message. Significantly it presented a model that put the person (employee) as being central to the desire to achieve high outcomes through human resource flow, rewards and work systems. The inter-relationship between these is commonly referred to as 'The Harvard Model'.

Thus, 'employee influence' considers how much authority people within the organisation have over the tasks that they are expected to carry out and the degree of power that is devolved by management to those less senior. The central assumption is that there should be agreement ('congruence' – see below) between managers and workers in terms of organisational objectives and how they are attained. 'Human resource flow' considers how people enter into the organisation through recruitment and selection policies and, subsequently, how they are managed/treated in order to best achieve the desired outcomes. Importantly, the organisation should ensure that it can recruit the people who are capable and dedicated to doing this. In other words, get the best people you can, and then do everything to develop their skills. 'Reward systems' refer to the way that the efforts of people are recognised. As such, they are assumed to be appropriate to ensuring

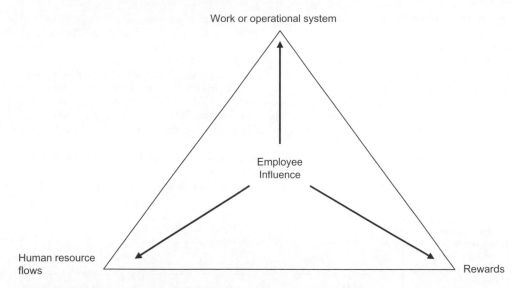

Work or operational system

Employee
Influence

Human resource
flows

Rewards

Figure 6.1 The Harvard Model for Human Resources (reproduced with permission from *Managing Human Assets* by Beer *et al.*, 1984).

that every member is given sufficient reward to create high levels of motivation towards achievement of their personal and organisational goals.

Motivation is, of course, at the heart of HRM and is seen as being the key to success in developing effective employees. Nonetheless, it is widely accepted that rewards can be both explicit (extrinsic) in terms of pay and conditions and implicit (intrinsic) in the way that people perceive themselves in terms of their relationship to their peers, their managers and their own ability to develop with respect to self-confidence, dealing with new challenges, and satisfaction with their careers. As Beer *et al.* stressed, the inclusion of people affected by the reward system in its design and application is an effective way to assist in creating a sense of ownership and personal value. The last part of the model, 'the work system', considers how all aspects of internal management and administration (such as procedures, communication and technology) are applied in the pursuit of attaining organisational goals as effectively and efficiently as possible.

For the model to work effectively, according to Beer *at al.*, it is crucial that managers in an organisation should orientate their decisions concerning HRM with direct relevance to the following 'Cs':

- Commitment;
- Competence;
- Congruence;
- Cost-effectiveness.

Accordingly, the first C, commitment, is concerned with what managers implement in terms of HRM (and beyond) to ensure that people are willing to dedicate themselves to the organisational goals. As such, there are connections with the 'culture-excellence' approach that was identified by Burnes (2004) and which acknowledged the influence of Japanese organisations in their ability to create organisations in which the workforce appeared to be obsessive in their belief that the organisation's goals were aligned to their own. The second C, competence, is concerned with attracting the sort of individuals who possess the requisite skills and ability to ensure organisational success. Alternatively, if they do not, those with the correct attitude and aptitude can be selected and developed from within by training and education. The third C, congruence, is concerned with attempting to create correspondence between what management want (strategic) and the personal goals of each employee. The final C, cost-effectiveness, is concerned with attempting to ensure that the expense of producing an effective HRM strategy within the organisation is not so great as to outweigh any of the benefits. However, as advocates of HRM point out, the costs of not having an effective and sensible approach to HRM will eventually cripple the organisation. Or, as the Construction Task Force argued in *Rethinking Construction*, absence of effective HR will seriously undermine the industry's ability to become world class.

In order to summarise the way that HRM is conceived and developed within an organisation, but is appropriate to the external environment, Beer *et al.* presented a diagrammatical representation ('a map') that combines the interests of stakeholders and situational factors:

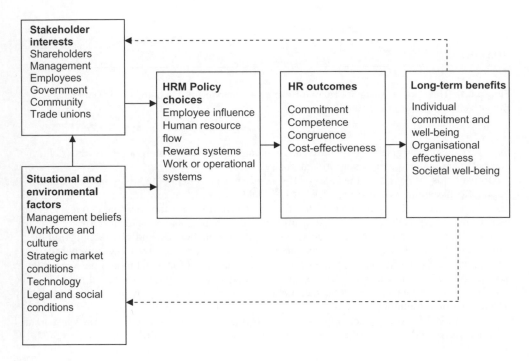

Figure 6.2 The Harvard 'Map of HRM' (reproduced with permission from *Managing Human Assets* by Beer *et al.*, 1984).

Guest (1987), whose work on human resources was carried out in the UK, developed an alternative HRM model to that proposed by Beer *et al.* His view is that the purpose of HRM in any organisation is assumed to be about the achievement of four key objectives:

1. Strategic integration (planning/implementation);
2. High employee commitment to the organisation;
3. The ability of the workforce to be able to adapt to changing circumstances;
4. A highly dedicated workforce that is committed to the achievement of consistently superior results.

In attempting to produce organisational success, Guest argues, it is crucial that management ensures that all of the HR policies are fully integrated into its business strategy. To explain this, Guest presents a model (table) for HR that differs in the fact that it has seven policy categories (the Harvard model has four):

Table 6.2 Guest's Model of HRM (source: Guest, D.E., 1987, p.516)

Policies	Human resource outcomes	Organisational outcomes
Organisational and job design		High work performance
Policy formulation and implementation/management of change	Strategic planning and implementation	High problem-solving
Recruitment, selection and socialisation into the organisation	Commitment	Successful change
Appraisal, training, education and continuous development	Flexibility/adaptability	Low turnover (known as 'churn')
Flow of expertise and creativity through all of the organisation	Commitment to objectives	Low absence
Reward systems	Culture of continuous improvement	Extremely low sense of grievance
Communication systems	Understanding by everyone of strategic objectives	Maximum effectiveness and efficiency by optimal usage of HR input; people feel valued

Thus, where Harvard proposed 'Human resource flow', Guest instead suggests 'Manpower flow' and 'Recruitment, selection and socialization'. What Harvard calls 'Work systems', Guest instead believes to be 'Organizational and job design'. Both models have 'Reward systems' in common. The other three categories in Guest's model, 'Policy formulation and management of change', 'Appraisal, training and development' and 'Communications' are, even though not inconsistent with the Harvard model, treated in a more detailed way.

The point to make about these models is that they stress that people are at the crux of an organisation's ability to create improvement and, as the 'culture-excellence' approach emphasises, its 'journey' towards becoming superior to other competitors. However, for an industry such as construction that has a long tradition of eschewing the belief that treating workers in an altruistic and benevolent way is beneficial[25], any suggestion that production should factor in such concerns was unlikely to be easily embraced. On the other hand, however, the industry faces greater pressure to improve its practices (most notably by the effects of the 'Respect for People Initiative'). As many ask: what needs to be done, and by whom?

Carrying out an HR audit

Any management audit is intended to consider the effectiveness of processes used. Accordingly, an audit of an organisation's HR is intended to examine the relationship between all aspects of people. Specifically it will consider how people are attracted to the organisation, how they are

---

[25] A sweeping generalisation in which there were undoubtedly impressive exceptions.

used and whether the policies being used are sufficient to attain the overall objectives. Lynch, on the basis of Tyson (1995) and Rosen (1995), suggests that there are a number of issues that an HR audit should consider. As he states, the main objectives of the HR audit are to provide 'basic information' on how people are used and to 'explore in detail the role and contribution of the human resource management function in the development of [an organisation's] corporate strategy' (2006, p.244). Accordingly, an organisational HR audit should include the following:

- Leadership and how its style affects others;
- The way that people are organised (structure);
- Skills and capabilities that are required by particular people (and teams);
- Morale and reward 'systems' used (and their apparent effectiveness);
- Methods used to select, train and develop people through, for example, induction, training and education and other courses intended to encourage innovation;
- Policies and procedures to guide actions and behaviour are appropriate to objectives (and up-to-date);
- Turnover rate (known as 'churn');
- Health and safety matters;
- Who (which department) is responsible for HR?

More specifically, he believes that the 'role and contribution' of the HR strategy should analyse the following (2006):

- Relationship with the corporate strategy;
- Key aspects of the HR strategy which are integral to the overall strategy;
- The ability to respond to changes in environment;
- Methods for review of the effectiveness of the HR strategy;
- The timeframes that exist for making decisions.

Overall, the input of people into the organisation should be viewed as being a never-ending task. The objective of improving this extremely vital resource should be one that strategists are constantly willing to dedicate time and effort to. Chapter 7, which deals with aspects of organisation, stresses the need for people to be able to work and cooperate in a way that ensures they are able to give their input. The way that this will most effectively occur is when there is specific responsibility given to a person(s) whose task is to manage the HR on a day-to-day basis; in effect, there is a HR manager (although frequently at a very senior level) to ensure that all of the issues (shown on the lists above) are being dealt with proactively.

6.10.2   *Finance as a key resource*

Whilst a detailed treatment of finance and financial management is beyond the scope of this book, it is nonetheless important to recognise the significance of finance as a key resource. Most especially, without adequate finance it becomes extremely difficult to achieve strategic objectives which, for example, may require the funding of new ideas and innovation, attraction of new employees with specialist skills or training of existing ones, or the investment in new technology and equipment. How to ensure that the organisation uses finance correctly is always a moot point. Should the organisation borrow money rather than use its own reserves (defined below)? Having used finance to 'drive' the organisation, are there techniques that can be used to monitor effectiveness? The answer to these questions are, respectively, dependent on cost and risk; and, yes, there is a technique that uses what is frequently referred to as **ratio analysis**.

Obtaining finance

Ensuring that the organisation has a regular supply of finance is axiomatic. It is effectively the lifeblood that keeps everything else going. Like the blood supply that any human being is utterly dependent on, without adequate finance, organisational activities will either have to be curtailed or potentially cease. Whilst spending can be reduced in various ways, for example, by cutting costs through wage cuts to employees or by seeking cheaper materials or subcontractors to be used, there may be potential consequences. If the same output can be produced without any diminution in customer (or consumer) satisfaction, there may be no overall negative effect. However, reducing costs frequently has an impact in ways that cannot easily be anticipated. For instance, people resent having their wages reduced which causes 'de-motivation'. Using cheaper sources for materials and subcontract labour can possibly result in lower standards which will impact on the overall quality of the product or service.

Finance may be obtained in various ways. However, it is important to make the distinction between profit-orientated and non-profit-orientated organisations. In the case of the former, they are largely dependent on the willingness of investors. This may be through invested capital (savings). It may be through loans from individuals (or groups), or from financial institutions such as banks (commercial or merchant). In the case of companies[26], investment may be through the raising of equity capital which effectively means that individuals or groups acting on behalf of others (such as pension funds) buy a percentage of the company which is based on the number of 'shares' (and their relative price). The important consideration in all of these is that those who provide finance will normally want a return on their investment (this is a key ratio used in analysing companies – see below). Significantly, those who buy shares may invest in a company in full knowledge of the likelihood that there will be no return in the short term or even medium term. Any return will be based on assumptions of growth in the future.

In the case of non-profit-making organisations, finance will still be crucial for carrying out day-to-day activities. However, the motivation for gaining finance will not be profit for investors (even though surplus may occur from activities). Instead, the desire of those who provide funds is to allow activities to take place that will ensure that products or services are provided to those who would not receive them otherwise. For example, governments raise money through taxation and excise that enables the national health service to operate. Alternatively, local taxation may be used to provide local services such as leisure or refuse disposal. Additionally, finance may come from sources such as donations received by charities, benevolent trusts or through international bodies such as the EU, UN or World Bank.

Managing financial resources

Requirements for finance alter as an organisation becomes larger and, usually, more complex. Whilst there are some well known, large organisations operating in the construction industry, the majority tend to be small (normally involving fewer than ten people). Finance for these small organisations is likely to be limited to paying for wages and materials on a weekly basis as work is carried out. Because there is no need to invest in capital equipment, start-up capital is likely to be small and limited to buying essential tools and, perhaps, a vehicle (and even this could be hired). As long as the organisation is able to secure regular work (and be paid in a timely way), it can pay for materials used (which will usually have been supplied on credit) and pay any employees or subcontractors who have been employed in the process of carrying out work.

As a contracting organisation becomes larger, its need for cash increases. It may have to take on additional staff to administer the greater variety of activities that are needed to compete. Office

---

[26]A company is a legally defined entity requiring registration in compliance with relevant company law.

space will be likely to be increased which, in turn, will increase the overheads that must be borne by the profits that each piece of work carried out (separate contract) generates. The dilemma that any organisation faces is how large it can become before overheads become too great a burden on the primary activities and make continued survival difficult. As explained previously, the construction industry is one that has a long tradition of operating on small margins which requires participants to be adept at keeping the costs of operation low. This encourages a culture of 'short-termism' and a tendency to avoid investing in anything other than the absolutely essential. Most particularly, there is avoidance of what are known as assets which are the resources that normally have long-term value. These could include items such as the premises and equipment that is used (including vehicles). Often, though, these are purchased with a view to paying for them over a longer period using proceeds from the profits (or in the case of a non-profit-orientated organisation, from the surplus). As such, there is a liability against the organisation which, in effect, is a debt that must be paid.

Using ratios to monitor progress

As Thompson *et al.* argue, 'the stronger a company's overall performance, the less likely the need for radical changes in strategy' (2008, p.97). The message they present is that, if an organisation is seen to be financially successful, it will be likely to have a 'well-conceived, well-executed strategy' (2008). The import of this statement is that managers should be able to make use of tools and techniques that will enable them to carry out measurement of financial performance. Whilst many techniques exist, the most widely used is that which is based on using information extracted from financial sources which, by being combined, can provide an easy and effective measure. This is the key objective of what is known as **ratio analysis**. So, similar to the way that measuring the efficiency of fuel consumption by a vehicle is achieved by dividing the unit of distance (miles or kilometres) by the measure of input (gallons or litres), so too can elements of financial data be used. These ratios can be used for internal purposes in order to judge performance from period to period to monitor whether there is any variation and, if so, why? They can also be used for external purposes in order to compare performance against others, especially competitors (assuming that the basis for making such comparisons – the financial data – is available).

Lynch provides a number of these (2006, p.304):

Liquidity which measures 'the ability to survive and avoid default' and usually consists of two key ratios:

- $\text{Current ratio} = \dfrac{\text{current assets}}{\text{current liabilities}}$

- $\text{Acid test} = \dfrac{\text{current assets} - \text{stocks}}{\text{current liabilities}}$

Gearing which measures 'the financial strength and the different forms of finance' and is usually based on two key ratios:

- $\text{Gearing} = \dfrac{\text{long-term borrowing}}{\text{capital and reserves}}$

- $\text{Interest cover} = \dfrac{\text{earnings before interest and tax (EBIT)}}{\text{interest}}$

Profitability which measures the amount of money made by the business and normally consists of two key ratios:

- Profit margin $= \dfrac{\text{EBIT}}{\text{sales}}$

- Return on capital employed $= \dfrac{\text{EBIT}}{\text{capital employed}}$

Investor ratios which measure 'the earnings available to those who own the company and can consist of the following ratios (note that profit is assumed to be 'net' which is after the deduction of tax and interest):

- Net profit margin $= \dfrac{\text{net profit}}{\text{sales}}$

- Earnings per share $= \dfrac{\text{net profit margin}}{\text{number of shares}}$

- P/E ratio $= \dfrac{\text{price per share}}{\text{number of shares}}$

- Earnings yield $= \dfrac{\text{earnings per share}}{\text{price per share}}$

- Dividends per share $= \dfrac{\text{dividends}}{\text{number of shares}}$

Trading activity which measures key aspects of the company's performance in terms of stock levels, the amount owed by and to the company and a measure of the value of the company being traded ('turned over'). Accordingly, the following ratios can be used:

- Stock cover $= \dfrac{\text{cost of sales}}{\text{stock}}$

- Debtor days $= \dfrac{\text{debtors} \times 365}{\text{sales}}$

- Creditor days $= \dfrac{\text{other creditors} \times 365}{\text{sales}}$

- Fixed asset turnover $= \dfrac{\text{sales}}{\text{fixed assets}}$

Whatever techniques are used, the importance of being able to judge effectiveness is important. This requires those using ratios such as those shown above (or indeed any other techniques that may be considered appropriate) to be able to spot variances from the norm (or what would be expected under normal conditions). To be able to do so will enable them to make decisions as to whether the particular tactics (or indeed the overall strategy) are still correct. If not, some alteration may be needed in either (or both). Alteration may be needed in the nature of the operations carried out to ensure greater efficiency and/or effectiveness. As the next section describes, analysis of tasks carried out to produce products or services may be needed (which can include the use of key ratios).

6.10.3   *Developing productive capability by improving operations*

Wickham explains that operational resources are those used to carry out 'operations' which will produce the 'output' in terms of the products or service that an organisation delivers as part of its strategy (2000, p.147). As with all other resources explained in this chapter, a great deal depends on the nature of the organisation (its purpose), the way that it achieves outputs (technology and processes), and the size of the organisation (which will depend on the magnitude of the 'market' it serves and its relative share). Included among resources used in operations, according to Wickham, are the following (2000, pp.147–8):

- **Machinery** which is the 'equipment' used by an organisation to 'manufacture' products or services. Clearly, some processes require high investment in machinery (like car manufacture) and some do not (like individuals who use their knowledge and intellect). Machinery used in construction varies greatly (from that of the large contractors involved in complex projects to that of individual trades personnel, such as a painter, who may have very simple tools).
- **Buildings** which are required to provide the 'envelope' in which operations occur. For some processes, buildings are essential. The relatively cheap goods we enjoy consuming in large quantities will almost certainly have been produced by mass-production techniques under cover in buildings that are commonly referred to as 'factories'. Construction, apart from offices needed for administration (at head or regional office level, site being temporary), does not carry out production **in** buildings. Quite obviously, the objective is to produce buildings, some of which it then becomes reliant on. If you consider what a building consists of, its various materials and components[27], you will realise that almost without exception all will have been produced and/or processed in facilities (usually factories) that will have been built by construction.
- **Storage and distribution facilities** many processes require places (usually buildings) in which to store materials that will be used in production (or have been produced) and for the purpose of transference to eventual destination (customers/consumers). Large supermarkets and Internet retailers (like Amazon) are absolutely reliant on this form of 'resource' which enables its outlets to be replenished on a regular basis (some stores having a number of daily deliveries). As with all buildings, construction is the industry that provides these facilities; some are truly gargantuan in size (consisting of thousands of square metres or being the size of many football pitches).
- **Research and development assets** some types of business are dependent on developing their products and service by innovation and creativity (see later chapter). All major biochemical, foodstuff and microelectronic manufacturers tend to have large research departments, often located in dedicated facilities, whose function is to develop their products for future markets and, potentially, new markets (if the research leads to radical innovation)[28]. Whilst research and development in construction is not as extensive as in other industries, it does exist. Organisations such as the BRE (Building Research Establishment) provide research for the construction industry. Whilst some contractors carry out research into materials or processes, this is usually left to specialists (many of whom are manufacturers with better expertise and wider markets).
- **Vehicles** few organisations are able to operate without vehicles to transport materials, people, equipment or the finished product and/or service. In the case of companies that are dependent on the ability to distribute regularly (daily deliveries or movement of other resources), there

---

[27]Bricks, timber, glass, tiles, concrete, steel, elements of services for electrics and gas, and secondary fixings (sanitary ware, ironmongery, control panels, light switches, paint and so on). This list is not complete and, as you should be aware, the construction of even a simple house consists of many thousands of individual elements (see *Rethinking Construction*).

[28]The sales of such products may be measured in millions which justifies the large amounts of finance required for research and development.

may be large fleets of vehicles (vans and lorries). Increasingly, the task of movement of resources has been 'outsourced' to companies who take responsibility for the provision of sufficient vehicles and the associated direct and indirect costs (drivers, fuel, repairs and maintenance, taxation, insurance). Construction organisations have traditionally operated sufficient vehicles to conduct their daily business. The nature of the processes being carried out will dictate the necessity for particular forms of transport. So, for example, a small jobbing builder may require only a small van (indeed, many use cars but carry their tools in the boot). However, a large civil-engineering organisation would need vehicles to carry out the work. These may be procured from subcontractors or from hire firms (see below) and, in turn, require lorries to transport them from the storage depot (see above) to wherever the work is to be carried out.

- **Office equipment** any modern organisation is likely to be dependent on all of the paraphernalia that is part of any administrative system and bureaucracy. This will include telephones, computers, printers, photocopiers, furniture for staff, and filing systems to deal with the inevitable paperwork that is generated (which, despite the predictions made about paperless offices, shows no sign of abating).

There are three key considerations with respect to resources:

1. The investment decision – to buy or otherwise?
2. Analysing costs of resources;
3. Using integration as a tactic to procure resources.

Investing in productive resources – a question of whether to buy outright

The first of these concerns whether it is sensible to purchase the resources required for operations outright or, alternatively, like outsourcing of transport described above, to pay for them on a 'needs' basis. Everything an organisation requires can be procured without having to purchase[29]. This normally involves hiring or leasing; the former tends to be for a relatively short period, days or probably a few weeks, whereas the latter is for much longer, possibly even years. The main advantage of ownership is that it ensures control and the resource can be used at whatever time suits. However, it involves what is known as 'capital expenditure' which requires access to finance for either the whole purchase price or, more likely, a deposit followed by instalments (which may include interest).

Once a resource is purchased, all responsibility for costs associated with upkeep shift to the owner. Additionally, all further payments must be financed from either reserves (earned already or from savings) or from profits/surplus that have been generated as a result of operations being completed and customers/consumers being sufficiently satisfied. Hiring or leasing may shift some of the associated costs (although the rates that will need to be paid for use of the resource will reflect this).

Ultimately, the decision about whether to buy, hire or lease will be made with regard to purchase price, usage rates and relative costs of operating the resource. As the next section explains, analysis should be carried out in order to ascertain the value that can be obtained from usage of the resource. If it will be used daily, then it is probable that purchase is sensible (provided that any associated costs are not prohibitive). However, intermittent or irregular use will probably make short-term hiring more appropriate.

---

[29]This can include people (both manual and professional) who are procured from agencies on a short-term basis. However, the experience of construction demonstrates that treating people in a way that suggests their contribution is no more than short-term is not conducive to longer-term improvement.

Analysing costs of resources

As the previous section has made clear, the costs associated with use of a particular resource should be scrutinised carefully. As well as the cost of purchase, there will be costs that are directly related to output (such as fuel), whereby the more it is used, the more it costs to run. Expenditure on the former is normally considered to be a 'fixed' cost which will be incurred regardless of whether there is any production or not. The latter, however, is regarded as being 'variable' and will normally be determined by the rate of use. However, some costs that might be regarded as variable effectively become fixed in that they will be borne by the organisation regardless of output. For example, regular maintenance will normally need to be financed (albeit that there may be an increase if the resource is used intensively). Additionally, in the case of specialised equipment (such as a piece of construction plant), there may be a need for a trained operator whose wages will be paid regardless of whether the resource is used or not.

Operational efficiency is concerned with analysing the most effective way to produce optimal output whilst ensuring that costs are maintained at an acceptable level. At very low rates of output there will be diseconomies because the fixed costs will be high relative to output. Therefore, the objective is to raise output to ensure that the ratio of costs to output reduces significantly. However, there comes a point at which output is so great as to produce diseconomies because the resources become 'stretched' (overworked) and breakdowns become likely. Knowing the realities of production output and their associated cost is a skill that will come with experience (the relationship that can be plotted sometimes being referred to as the 'experience curve'). Being able to make an effective decision about whether to increase the output of existing resources or to procure additional resources will, as well as determining overall efficiency, have a major impact on costs.

6.11   Using integration as a tactic to obtain resources

Integration refers to the situation where an organisation, which has relied on other organisations to produce its materials or components, decides that the uncertainties make it sensible to procure the productive resources – usually by merger or acquisition. This is known as **backward horizontal integration**. Clearly, there are cost implications for this strategy which would need to be outweighed by the benefits that accrue as a result of greater control. Acquiring the productive resources that are possessed by those organisations that an organisation sells to, such as distributors or retailers, is known as **forward vertical integration**. In the case of the former, some retailers have found that gaining control over **their** suppliers makes good business sense. In the case of the latter, some manufacturers believe that gaining control over those in closest contact with end-customers gives them advantage over competitors in terms of understanding market behaviour and future trends.

6.12   Conclusion

The importance of resources cannot be overestimated. All organisations, no matter how powerful, will have limited access to any resource. There is a cost to everything that is employed; even non-profit-making organisations in which there is voluntary input by people will be able to rely on only a limited number who will do this. Governments may be able to raise enormous sums of money to finance projects but even they will have a limit (imposed internally or externally) beyond which they should not go. Therefore, in carrying out strategy, having sufficient and appropriate resources to achieve the overall objectives is critical to success. The real skill among strategic managers is in being able to predict what the future resource base should be. This may mean procuring additional resources or, alternatively, adapting or developing the existing capability. Whether this involves people and/or plant/equipment will be dependent on the context. As previous chapters have described, knowing the environment and appreciating the expectation of customers (and consumers) are vital. Crucially, though, what is most important is that the key resource of an organisation – people – should be central to the future. The next chapter describes how considerations concerning the organisation will fundamentally require analysis of people in terms of their introduction, development and cooperation. To attempt to implement plans in pursuance of strategic objectives without due regard to people will, at the very least, be foolhardy. At worst, it will seriously undermine the potential to successfully attain strategic goals.

Chapter 7

'Organisation[al] matters' – a strategic perspective of the importance of how to manage people

'Never tell people how to do things. Tell them what to do and they will surprise you with their ingenuity.' George Smith Patton Jr III (1885–1945), distinguished and controversial American military commander

'Every company has two organizational structures: The formal one is written on the charts; the other is the everyday relationship of the men and women in the organization.' Harold 'Hal' Sydney Geneen (1910–1997), American businessman and ex-president of the ITT Corporation

7.1 Objectives of this chapter – what is organisation?

When we talk of organisations there are certain assumptions that are implied. The first is that we are considering the most important aspect which creates the very essence of what the organisation stands for: people. It follows that the organisation consists of people who both individually and collectively are committed to the overall aspirations of the strategy as articulated in the mission[1]. The word, organise (a verb), can be thought of as the attempt to create an orderly structure (OED) and the word, organisation (a noun), is one that is defined as being the 'the state of being organised' and 'an organised body, esp. in business, charity etc.' (OED). Therefore, for an organisation to achieve its objectives, as well as possessing appropriate resources which are considered to be sufficient to carry out the tasks required (see Chapter 6), it should be organised and configured in such a way as to do precisely that. This chapter provides an overview of the connection and implications of strategy and organisation and considers how those involved in construction should respond.

It is important that managers should not lose sight of the fact that the most essential component of organisation is people. They, after all, create the sense of what any organisation is. Take them away and you are left with machines and, possibly, systems for doing things. However, as Johnson *et al.* assert, it is people who have the ability to make strategy successful (or not):

> Creating a climate where people strive to achieve success and the motivation of individuals are crucial roles of any manager and are a central part of their involvement in their organisation's strategy (2005, p.448)

---

[1]There may be underlying conflicts in which some may pursue other goals which are consistent with their own personal (or group) goals. Such a situation, involving what is usually referred to as 'politics', can, if unchecked, seriously undermine the organisation.

As this chapter will explain, whilst having a dedicated and committed workforce is considered to be axiomatic in achieving strategic objectives, people management is not without problems. They require leadership and need a culture that is conducive to encouraging them to strive to be the best. Chapter 5 described how so-called 'excellent' or 'world-class' organisations are usually characterised by the skills, knowledge and willingness of all people involved in all aspects of production and processes to ensure that the end result is as good as it possibly can be. The quality guru, W. Edwards Deming (see Chapter 5), stressed the fact that being able to produce excellence requires investment both in people and in the systems in which they operate. Accordingly, providing the right tools and technology is vital. However, too often senior managers believe that investment in new equipment and production systems can be done without equal expenditure on people. As the next section describes, the tradition of considering structure has its origins in thinking about organisation in a way that sees people as incidental.

7.2   Organisation and structure – a consideration of contemporary changes in perception

For a considerable period of time the perceived wisdom was that strategy was the precursor to organisation. The belief was that strategy must be formulated before structure was even considered. Strategic thinker, Alfred Chandler, was especially influential in propagating this paradigm (see *Strategy and Structure* published in 1962). Since the 1960s, however, and in recognition of the turbulence of the environment in which organisations operate, the debate has been whether this assumption is still true. As many commentators argue, if change occurs rapidly – to which strategy must adapt – the particular structure of the organisation should not act as an impediment to a timely response. Therefore, it is suggested, organisational structure and strategy are bound to one another in a way that creates symbiosis: mutual reliance. If this is the case, though, there are implications for all people involved. In particular, managers responsible for formulating strategy will need to be able to provide the sort of leadership and inspiration that will enable people to respond to change (Chapter 9 deals with this topic in detail). Crucially, the people employed by the organisation will need to have the skills and ability to cope with continual change. This means that the traditional approach to organisation, which involves hierarchy based on rigidity and certainty, will be much less appropriate than it was in the past[2].

As the influential writer on organisational change, Rosabeth Moss Kanter, believes, the need to for both individuals and organisations to constantly adapt in order to 'improve performance and to pursue excellence' has created a number of demands that we are expected to comply with, even though they 'seem incompatible and impossible' (1989, p.20). For managers, she suggests that such demands might include the following:

1. Thinking strategically and investing in the future which, whilst requiring risk, must be done without putting the business at risk because of failure.
2. In order to become more customer focused, there is a need to do everything better and faster.
3. Within the organisation it is necessary that managers spend 'more time communicating with employees, serving on teams and launching new projects' (1989).
4. 'Know every detail of your business – but delegate more responsibility to others' (1989).

---

[2]Construction can be viewed as an industry in which organisations have a greater ability to be **contingent** (able to be flexible in response to market changes) than other industries. The use of subcontracting by large contractors as the means by which to deal with increases and decreases in workloads can be seen as an effective way to deal with risk and reduce capital investment.

5. That managers must display 'vision' and be able to demonstrate commitment by 'fanatic' action to achieve implementation (although it must be in a way that is 'flexible [and] responsive' to the prevailing environmental circumstances).

Managers, she accepts, are expected to achieve no less than absolute success and beat all other competitors[3]. As a consequence, organisations (although the word she actually uses is 'corporations') face pressure to achieve what she believes are increasingly impossible targets (1989, p.21):

- The need to get both 'lean' (see Chapter 5) and 'mean' by constant reconfiguration, although it is a 'great place to work', the overriding key objective is efficiency.
- A desire to be seen to be innovative (see Chapter 8) and creative so that products and services are better than others that are offered by competitors.
- Like Peters and Waterman's (1982) advice that success is about doing what you do best, 'sticking to the knitting', it is important that effort is put into the constant search for new opportunities.

In what may seem to be advice that stretches the organisation to 'breaking point', she contends that they should be able to carry out tasks to achieve faster 'execution' although it must 'take more time to deliberately plan for the future'. Moreover, consistent with the 'think big, act small' maxim, she argues that it is important to decentralise and empower to smaller 'business units' to act in ways that are appropriate to their needs (and align with customer expectations), even though they should also 'centralize to capture efficiencies and combine resources in innovative ways'. Managers reading such texts could be forgiven for being somewhat perplexed by the apparent contradictions. The dilemma faced by such managers then (and still now) is in knowing what the secret of successful organisations is – particularly in how they are **organised**.

In contrasting the business environment that existed in the part of the century when Chandler carried out his research that led to his beliefs with that which pertains today, Lynch demonstrates that there are a number of significant influences which are crucial in creating the circumstances within which all organisations must cope. Most especially, he suggests that the following factors are likely to alter the way in which decision-making should occur (2006, p.577):

- Workers are likely to have a much wider 'skill-set' and, normally, will be computer-literate.
- The complexity of work carried out and the inter-relationships involved are many times greater.
- Management of the parties involved requires ability and skills that are very different (not least in the way that internationalisation and multi-culturism has occurred).
- The technology now used creates a fast-moving environment in which people can communicate with each other regardless of time zones or distance.

Inherent in this list is **uncertainty** which makes forward-planning, the key objective of strategy, problematic. If the objectives of the organisation are dynamic, then the dilemma is to answer the question: what is the best way to organise so as to achieve them? Accordingly, the belief that formally imposed structure is the essential precondition to successful strategy has come under scrutiny. Rather, it is proposed, organisations should be designed and operated with the explicit objective of being able to cope with the way that the environment alters. In so doing, they should have the ability to respond appropriately and in a timely way. Such an approach to organisation is based on continual small-step change rather than the assumption of a 'big bang' whereby

---

[3]In the 'zero-sum' game of competitive business, if one organisation is the best, then all others must be seen to be less than the best!

everything is altered in anticipation of change. J.B. Quinn proposed what he termed **logical incrementalism** in 1980 which eschews the belief that there is an ideal structure that can be imposed prior to implementing strategy. As the following diagram shows, logical incrementalism has a number of stages which, although not intended to be prescriptive, suggest that there are certain aspects of change that need to be considered:

Table 7.1 A ten-step approach to implementing incremental change in strategy and potential organisational consequences (based on Quinn, 1980)

Strategic stage	Organisational consequences
1. A belief that change is needed	Use of informal methods and networks to encourage change
2. Clarify what is required, by whom and the range of options that exist	Consultation with all those involved (will be affected); this can be through a variety of mechanisms
3. Use 'change symbols' to signal what is needed	Communicate widely
4. Create appropriate periods during which discussion occurs and people become familiar with implications of the change(s) on their daily routines or practices	Encourage this to happen widely using formal and informal methods
5. Ensure that there is constant clarification so that short-term plans and tactics are aligned with the overall strategy	Involve senior managers and change agents in this so that there is two-way communication and connection between strategists and the doers
6. Keep broadening the basis of support amongst people	Establish committees, project teams, improvement groups and give support and encouragement
7. Consolidate progress	Use regular reports and celebration of success (small gains will create momentum and increasing confidence)
8. Continue to build consensus and re-evaluate strategy to ensure that it adapts to changes in external environment and market	Use a variety of tools and techniques to monitor and analyse external factors; these should be formal and informal
9. Keep the process proactive by looking further into the future (usually three to five years)	As new people enter the organisation, develop their willingness and confidence in the need to constantly adapt and change
10. If required, the organisation goes through major change which will impact on the strategy	Consider change in structure and processes; major alteration in attitudes and behaviour (organisational culture) possibly needed

Clearly, though, there is an assumed desire on the part of managers to try to create stability. Therefore, if the 'new organisation' works in that it succeeds in being able to achieve the desired outcomes, even for a relatively short period, it will remain in place. However, consistent with the model shown, once there is a sense that the conditions are likely to alter, the organisation must consider changing its structure once more. Analogous to a chameleon, the organisation is continually trying to 'blend' into the environment and ensure that its structure is relevant. This is known as achieving 'strategic fit' (see Galbraith and Kazanjian, 1986). The challenge still remains to know what is going to work best in the prevailing circumstances in which an organisation finds itself. Managers, therefore, must consider exactly what it is that constitutes 'organisation' and how this is used to achieve purpose (see below).

7.3 Structure, processes and relationships – virtuous combinations?

Johnson *et al.* believe that because the world is quite so unpredictable and 'fast-moving' the most important objective for an organisation is to harness 'valuable knowledge' (2005, p.396). Congruent with this, they propose that 'formal structures and processes need to be aligned with the informal processes and relationships into coherent *configurations* [their italics]' (2005). The latter, they explain, integrates three key elements that enable an organisation to 'operate'. These are structure, processes and relationships. Accordingly, if the three elements work well they will form a virtuous circle. In terms of what the three elements consist of, each will have particular characteristics that will contribute towards overall effectiveness. Structure is crucial because it is the way that formal roles and responsibilities are assigned and thereafter interconnected. The assumption is that if the structure is 'appropriate' then all processes and relationships will occur effectively. The reality, however, is that processes and relationships can occur despite, and not because of, structure. This poses the question of how much importance should be attached to the organisational structure. As the next section explains, what is important is that every person is clear about the purpose of the organisation and that there is no confusion (or conflict) with their personal role.

7.4 Purpose and organisation

Most organisations are, when judged superficially, simple. They frequently seek to achieve fairly straightforward objectives. Purpose, therefore, is believed to be easy to define in most organisations from fairly rudimentary analysis. Lynch is one commentator who argues that viewing purpose in simplistic and unchallengeable terms is incorrect. Rather, he advises, organisations should carefully consider purpose in a way that incorporates all of the factors that will influence it. With Chaharbaghi[4], he provides a model which is based on a polygon which, he explains, is apposite as it is a 'multi-sided figure with no obvious dominant side' (1999b, p.414). As well as being contextually dependent, each factor is inherently dynamic and its relevant importance is bound to change. Each of the factors shown can be defined as follows:

- **Time dimension** which considers the period over which the purpose is expected to be relevant. In most cases this is likely to be no less than for a period of months and, whilst long-term planning is useful, trying to make predictions for more than five years is notoriously problematic. For construction, whilst individual projects, which are undoubtedly fundamental to success, may be transient, the overall purpose of organisations involved will be much longer. Time, therefore, is a factor that is very much dependent on the particular conditions that create the environment in which the organisation operates.
- **Timing** which considers when intervention and change of purpose are required. Those managers who have a sense for the right time to alter purpose can greatly influence success[5]. Whether this is down to luck or to their innate ability to read market trends before other competitors is questionable.

---

[4] See Chaharbaghi and Lynch (1999b).
[5] A reason why successful senior managers are able to command high salaries.

- **Innovation** can be a major influence on how purpose is created and sustained. An organisation that regards innovation as being central to what it stands for will be likely to be a more dynamic place than one which is conservative. This will lead to an environment in which creative people flourish (see Chapter 8).
- **Value-added dimension** which means that the organisation operates (and cooperates) in order to seek ways to achieve the overall strategic objectives. Making sure that all of the parts are thinking in a way that emphasises the ability to do everything better is crucial. It is this factor that has been a major characteristic of the excellent Japanese organisations such as Toyota (see Chapter 5).
- **Survival dimension** is something that many construction organisations, particularly those that survived the recessions of the 1970s, 1980s or 1990s, can endorse (see Chapter 3). Those organisations that have experienced tough times and survived will have learned useful lessons and an ability to cope in a way that others may yet have to experience. Whilst this dimension may create a financially prudent approach and fixed structure (probably hierarchical which allows for greater control), it may also mean that those with direct experience will be aware of the potential to become complacent.
- **Growth dimension** in which, as in the previous factor, complacency should be viewed as negative. Therefore, the desire to grow the organisation by seeking new opportunities or simply to consolidate markets will have a major impact on the intended purpose.
- **Leadership** (which is dealt with in detail below) is bound to be a major influence on purpose. The leader(s) will provide the sense of direction and values that create the very essence of what the organisation stands for. Leaders, of course, have different levels of willingness to engage with others (see stakeholders below) in terms of developing consensus. In terms of structure, leaders who wish to impose their belief are more likely to prefer a formal approach than those who believe in a shared belief in purpose.
- **Stakeholder dimension** will include all of those who have a vested interest in purpose. This dimension is connected to leadership (above).
- **Values and lifestyle** that people believe in (and espouse) will have a large influence in determining the sort of organisation that results. For example, one in which people are committed to 'fair play' and the stated objectives consistent with what is known as 'Respect for People' will be one that operates in a very different way (and therefore accordingly organised) from one in which there is an accepted belief that any behaviour and treatment of others can be tolerated. Importantly, it is argued, as clients become more inclined to conduct their own affairs on the basis of people-oriented agendas, they too will be expected to ensure that the construction work they procure is operated on similar lines. To do otherwise would be hypocritical.
- **Knowledge** is increasingly seen as the way that value may be added through the ability of people in the organisation to achieve the overall aims by constantly searching for new and innovative ways to produce radical solutions to existing problems. Given the traditional fragmented way in which the industry operates, achieving genuine knowledge acquisition and management is more difficult than in other industries. However, in an increasingly competitive environment, ignoring the wealth of knowledge of the people who routinely carry out day-to-day processes is a mistake. As was described in Chapter 5, this is precisely how those organisations that have achieved pre-eminence, particularly Japanese, ensured that their strategy was dedicated to excellence.

7.5   Setting objectives – the importance of mission

Purpose needs to be articulated in such a way as to ensure there is clarity to both insiders and outsiders. Purpose is primarily to communicate to all stakeholders what the organisation is about (stated intentions or objectives) and how it intends to achieve them (by what means). This is achieved by writing what is known as a **mission statement** which aims to summarise all of the main aims and objectives that are pursued as integral parts of its purpose. Doing this is important in that it forces those managers responsible for strategy to carefully consider whether what **they** believe to be the case actually aligns with what all others think is both true and achievable. Crucially, can they do so in a way that is pithy and, therefore, readily understood by every member, regardless of level or grade or intellect? Importantly, any inconsistencies or lack of realistic intention to achieve the stated aims (through, for instance, under-investment in resources) will be seized upon by critics or sceptics as demonstrating that the overall intention cannot be realised.

Mission statements may be judged according to the following criteria:

- Be reflective of the distinctive features that the organisation currently possesses (or intends to develop);
- Be considered credible;
- Be sufficiently capable of being adapted should the external circumstances alter (which would cause the existing mission to become outdated anyway).

Many large organisations make an explicit link between their purpose and their people in their strategy. The message is intended to be clear: 'we look after our employees because we recognise that they are the key component in producing the level of quality and service that our customers expect.'

7.5.1   Making the connection between mission and objectives

The mission is deliberately broad in both intention and wording; it sets direction. Accordingly, it does not provide the particular targets that will be needed to translate intentions into actions. This requires that the mission is connected to particular objectives which, collectively, will ensure corporate success. Objectives, therefore, provide managers with something that allows them to translate broad visions into more tangible outcomes. Importantly, the provision of definite targets means that there is a datum against which progress can be measured (and any variance identified, analysed and learned from – particularly where corrective action is needed).

Setting objectives is a task that should be done in a coordinated way to avoid the possibility of some objectives being in conflict with others. For example, increasing turnover or profit might require existing employees to be expected to cope with much greater pressure in the short to medium term. Whilst this might not necessarily be something that people in the organisation see as being negative (indeed they may welcome the challenges and opportunities presented, especially if expansion, resulting in new posts and potential promotions too, is part of the plan), if the organisation states that a better work–life balance and reduced stress are objectives, there will be a danger of inconsistency. Time does, however, have a bearing on how conflict is dealt with. Short-term issues of inconsistency may be explained as being a necessary way of successfully achieving longer-term objectives. An example might be where cost reductions on the basis of job losses are required to restore the organisation to longer-term surplus or profitability. The example of motorcycle manufacturer, Harley Davidson, was a case in point. Without job losses it would have gone bankrupt. Having reduced cost and implemented vigorous quality management techniques to deal with the fact that Japanese imports were far superior in terms of performance, the organisation was able to ask many of those who had lost their jobs to come back when production could be increased following rapidly improved sales. Like many organisations that have successfully implemented quality improvement techniques, Harley Davidson discovered that short-term pain and investment reap longer-term benefits.

Like the criteria against which the mission is judged, objectives must also be seen to be SMART:

Specific – Objectives should specify what they want to achieve.
Measurable – You should be able to measure whether you are meeting the objectives or not.
Achievable – Are the objectives you set achievable and attainable?
Realistic – Can you realistically achieve the objectives with the resources you have?
Timely – Can you achieve the objectives in the time that is either available or can be realistically allocated?

Nonetheless, objectives must also be seen to 'stretch' people by causing everyone involved in all processes to consider how, by doing things either differently or innovatively, greater value can be obtained; this is the way that competitive advantage is achieved. As already explained, an organisation will have an embedded style of doing things (its culture) that creates an environment in which taking risk and dealing with challenge are perceived to be normal (as opposed to one where innate conservatism militates against trying to alter established routines and patterns – and makes change extremely difficult to implement or maintain).

7.6   Organisational configuration

The key consideration for managers is in deciding what structure will be most likely to deliver the corporate strategy through attainment of objectives (and therefore the overall mission). Knowing what works best in terms of the configuration of departments and, of course, people is often seen as the key skill of successful managers. Lack of success in any context is often viewed as a failure of management to get the best from the people under their 'control'. The reality is frequently that there is no secret to success and that what may work perfectly well for a manager in one setting may be much less effective in another. Indeed, the corporate world is replete with examples of senior managers who have been appointed to 'shake the organisation up' and, having implemented widespread changes in structure, have then witnessed the (presumably) unintended consequences of chaos and confusion. In searching for a reason, they will inevitably find that there are many factors that contribute to the inability of people to dedicate themselves to the achievement of corporate goals.

Failure to recognise that people are the main 'component' of an organisation is all too frequently suggested as the reason why managers fail. In their eagerness to implement what they believe will create success, such managers often ignore those managers who are responsible for implementing policies and procedures that will ensure that people involved are consulted. Such managers, very frequently referred to as 'human resource management' (see below), should be seen as having a crucial role in considering the most effective combinations that will create the ideal structure from which the strategic vision can be realised.

Lynch provides nine factors that he believes are 'primary determinants' of organisational configuration (2006, p.582). These are:

1. Age of the organisation – whereby as it grows older, it tends to become more formal.
2. Size of the organisation – whereby as it grows larger, there is a belief that increased bureaucracy and hierarchy are required to ensure that overall aims can be achieved.
3. The environment – (see also Chapter 4) which will create the influence that suggests the type of structure and combination of resources that are believed to be most appropriate for the circumstances. In a turbulent environment it is probably best to have a structure that is capable of responding quickly, most likely where there is flexibility in terms of resources (people, see below, being most important).
4. Centralisation/decentralisation – this is one of the ongoing debates of management and organisation: to keep all of the main functions in one place (such as head office) or to dissipate such resources around the organisation. Construction, of course, is an industry which,

because it is project-based, will have much less centralisation than might be the case else-where. As Lynch advises, much depends on the need for responsiveness and 'the need for a local service' (2006).

5. The type of work to be undertaken – which in the case of construction is both project-based and to a very large extent still an industry that relies on people. As such, organisa-tional structures that tend to exist are those based on the ability of people to cooperate at any level where decisions and activities take place. As such, on every large project it is not exceptional to have members of the strategic team based on site so that they are close to 'the action'.

6. Technical content – like the previous factor, the type of work is very important to creating the influences that determine structure. In carrying out work that is repetitive and low-skill, the organisation can be based on simplicity and routine. However, where operations and activities involve greater complexity and lack of repetition, the organisation should be created around the knowledge and ability of those with the expertise and confidence to know what works best and gives customer satisfaction.

7. Different tasks in different parts of the organisation – in which the attempt to impose 'one approach fits all' is eschewed. The needs and aspirations of the people who carry out particu-lar functions vary and, accordingly, so does they way that they operate. This requires struc-tures that work for the people involved. Attempting to impose standardised ways of operating will frequently create dysfunction and inefficiencies that undermine the desires of people who are essential.

8. Culture – this is linked to the softer (social) aspects of organisation and, as a result, will strongly influence the organisational configuration that works best. The difficulty is that people are not 'consistent and inter-changeable components' that can be fitted together in such a way as to give guaranteed results. Therefore, knowing the individual preferences and styles of the individuals involved will be very useful in determining the structures that are likely to work best.

9. Leadership – is based on the particular approach of the individuals who have most influence and power within the organisation. The amount of influence that leaders 'enjoy' will allow them to decide what they consider to be the structure and combination of resources that works best.

In his book, *The Structuring of Organisations* (1979), Henry Mintzberg proposed that there are four main characteristics of environment that strongly influence the type of organisational structure that is believed to work best. These are:

- **Rate of change** which can range from static to dynamic – an organisation which has to deal with greater change will need to be flexible and have a looser structure which means that people should be able to cope; this requires training and education to build their confidence and ability.

- **Degree of complexity** which can range from simple to complex – organisations should have structures that cope with complexity in an appropriate way; formal structures and procedures are likely to work only in situations of stability and simplicity, otherwise decentralisation and keeping operations and processes flexible and 'close to the customer' will be likely to be best.

- **Diversity of customers/consumers** which can range from simple (singular) to complex variety and will require the organisation to organise and reorganise in such a way as to respond.

- **The competitiveness of the environment** which can range from being very static and benign to extreme. The contemporary environment is one in which cooperation is encouraged. However, construction is an industry in which competition is normally prevalent and, especially when the economy becomes 'difficult', hostility is more likely to be the means by which to survive. As such, the organisation will probably become more centralised having departments whose objective is to explore ways of dealing with the 'opposition'.

Mintzberg used these characteristics in order to develop six organisational configurations that incorporate four key aspects of consideration (Environmental Analysis **EA**; Resource Analysis **RA**; Essential Element of the Organisation **EEO**; Major Coordinating Device **MCD**):

1. **Entrepreneurial organisation** in which EA assumes a simplicity and a dynamic environment, RA consists of relatively small units doing whatever is needed, EEO is probably the owner or boss, and MCD is hands-on involvement.

An example would be a newly created contracting organisation that carries out local work and aspires to develop rapidly.

2. **Machine organisation** in which EA assumes that there is potential for growth but in a consistent way, RA is based on tasks being carried out which are regularised using defined processes and procedures, EEO is what is called a 'techno-structure' (one in which the emphasis is solely on technical aspects of production), and MCD is on standardisation of work.

An example of such an organisation would be a specialist manufacturer of components to the construction industry.

3. **Professional organisation** in which EA assumes stability although inherent complexity will exclude those with the requisite knowledge, RA is by managers who possess such knowledge, EEO is based on developing a coordinated team whose MCD is dedicated to standardisation of skills.

An example would be a professional practice of surveyors or engineers.

4. **Divisionalised structure** in which EA is diversity, RA is based on a structure that is regulated and routinised (having carried out the functions many times before), EEO is the cadre of middle managers who operate the systems, and MCD is focused on standardisation of outputs.

An example would be a contractor producing a consistent product such as a type of building where variation is minimised – for instance, prefabricated units for housing or retailing.

5. **Innovative organisation** (also referred to by some as 'adhocracy') in which EA assumes both complexity and dynamics that make predictability very difficult, RA requires those with ability to work in a way that is without prescribed procedures (they must be capable of thinking 'on their feet'), EEO is one which relies on the willingness of staff to support each other, and MCD is based on continual adaptation to the particular circumstances.

An example would be a design practice which carries out novel work for clients who believe that their building should suggest that their objectives are unique.

6. **Missionary organisation** in which the objectives are benevolence and altruism and therefore, EA assumes simplicity and regularity, RA will be based on people having a belief in the objectives of the organisation (identifying with the culture), EEO is ideological, and MCD is about acceptance of norms.

An example of such an organisation would be a cooperative or a self-build group.

Whilst such a classification is useful from a theoretical perspective, there are many critics who believe that any attempt to derive typical structures is fallacious; the variables and the context are just too unpredictable. Moreover, some organisations are successful by being a hybrid. Therefore, it is frequently a case of 'what works, is what works!'. As the next chapter will explain, any organisation, which is not attempting to reconsider what it does (and how it does it by reconfiguring), is in danger of become staid and too obsessed by the past. As a consequence, it may discover that both its strategy and organisation are too outdated to deal with the challenges of contemporary markets. Indeed, as Lynch explains so amply, 'It is difficult to specify clear and unambiguous rules to translate strategy into organisational structures and people processes' (2006, p.585). The most important thing, therefore, is to build the organisation around the people who put into operation the processes that are intended to achieve the overall objectives.

Contemporary management theory recognises that overbearing procedures and formalised hierarchical structures tend to achieve nothing other than the worst aspects of bureaucracy. In Burnes' analysis of organisational change (2004), he refers to the 'culture-excellence' approach in which organisations are capable of achieving extremely high levels of quality and service. As would be suggested, this makes a very clear link between the concept of culture and the achievement of excellence (see Chapter 5 in which the ability of organisations to be judged as 'world-class' was described). Organisational culture is explored in a subsequent section. The next section of this chapter considers the different types of configuration of organisation that commonly exist.

7.6.1   Types of organisational configuration

As suggested already, the way that organisations are structured has altered over the past century to recognise changes that have occurred in society, technology and business relationships. Congruent with the previous section, it is useful to consider what the 'standard' types of organisational structure are and how they may be applied to implement strategy. These are:

- Functional;
- Multi-divisional;
- Corporate;
- Matrix;
- Innovative.

Functional

The first of these is one in which all of the major activities that are carried out in the organisation (the 'functions') are grouped together, usually in departments under the management of someone who has expertise in that area. An example would be to have all of the quantity surveyors or buyers working together. This allows collective expertise and resources dedicated to achieving the objectives of that department in what is believed to be the most effective way. Such grouping has the advantage of simplicity and allowing sole focus on the particular function.

However, those who work in functional departments have a tendency to become obsessed with the functional objectives only. The need to appreciate the objectives of other functions and to engage in cooperation so as to achieve the overall organisational goals may lead to conflict or rivalry. As such, the organisation needs stewardship that is sufficiently able and willing to coordinate (and control) the various functional departments.

This configuration is appropriate in all types of organisation as the basis of carrying out day-to-day tasks. Accordingly, it becomes the 'building block' of larger organisations, like the next to be considered: multi-divisional.

Multi-divisional

As organisations become larger, the smaller departments that may have served well in carrying out functions may be expected to carry out a greater quantity of work. This will require expansion that, after initially working well, may be likely to create diseconomies and complexities in day-to-day management. If the organisation becomes sufficiently large, it may need to reorganise itself around particular products, clients or on a geographical basis, depending on the circumstances, for example, major contractors that operate on a national basis.

If reorganisation does happen it will be sensible to subdivide the functional departments into smaller units that serve the needs of each division. Whilst this will spread the expertise of each function around the organisation, it will make overall coordination more challenging. An alternative might be that functions in different divisions are autonomous. However, this may result in duplication of activities and, quite frequently, the same task being carried out in inconsistent ways (the differences becoming apparent only when there is an attempt to standardise procedures such as when quality assurance systems are implemented). The challenge to senior management is how much control they think should be exercised from the centre or whether allowing divisions to manage their own affairs is more likely to lead to attainment of strategic objectives.

Corporate

Increasingly, organisations may have interests in a number of very diverse fields of operation. As such, there may be different businesses that operate entirely separately from one another. For example, a main contractor might have separate businesses for property development or plant hire. Whilst there is a possibility of 'cross-over', they would be expected to operate in a way that protects their own interest. So, in the case of the property company, it would be perfectly acceptable to elicit bids for construction work from more than just the contracting company. The advantage of organising in this way is that each of the companies operates in an autonomous way that allows expansion and contraction into and out of markets with less risk in comparison to that of internal expansion within one company. In construction there has been a rapid rise in the number of joint ventures, alliances and partnerships that are overseen by what is often referred to as a 'holding company'.

Whilst construction has not tended to move outside its traditional area of expertise, some larger companies have developed interests in sectors that would not be seen as having any logical connection. For example, many have either formed or bought companies that are intended to provide services such as facilities management. The difficulty for managers at the centre is in ensuring that, whilst all companies can be managed effectively and without conflict, there is a willingness to allow managers to have sufficient authority to make decisions on their own behalf, interference usually being unwelcome.

Matrix

Matrix organisations are appropriate where there is a desire to bring together various functions in order to achieve dedicated objectives such as a project for a particular purpose or client. As such, the matrix organisation is allowed to decide on its own strategy and how best to implement it. It is crucial to have people working in the matrix who subscribe to the objectives as set by the controlling managers and who are willing to forgo any long-standing loyalty to their functional departments. The emphasis is on action and consensus in order to make effective and timely decisions rather than on creating systems and bureaucracy which may be too slow. The last organisation configuration takes this desire one step further.

Innovative

As the title suggests, this sort of organisation is dedicated to the constant search for opportunities to carry out tasks or processes in new ways that are creative and novel. The emphasis is to bring together people who believe that such a culture is one in which they feel comfortable and that they can make suggestions in an open way without fear of ridicule if their solution or alternative is radical. This sort of arrangement is best suited to 'thinking the unthinkable' or 'outside the traditional ways'. As such, there will be a need for cooperative working through teams. It is very likely that the organisation will have a high dependence on social norms and interdependency rather than on formal structures. Innovation (and technology) is dealt with in greater detail in Chapter 8.

The form that organisations take and the way that they operate are dependent on one component that has significance beyond all other things: people. This component should be considered from two perspectives: those who take key decisions (leaders); and those who implement processes intended to produce action. Whilst some may argue that the latter are influenced by what their leaders say and do, others counter that great leaders emerge because of their willingness to listen to what their followers tell them. The following sections consider leadership and people.

7.7   Leadership

There are many definitions of what leadership is but they will usually contain the essence of someone being able to influence followers to do things that they might not otherwise feel capable of doing. Whilst a leader may resort to the use of force or threats, it is more likely that an effective leader will possess powers of persuasion that enable them to convince their followers that what is expected of them is right (is moral and compatible with their personal sense of values) but still consistent with the organisational goals and, therefore, any success will be good both for them individually and for everyone collectively.

Unsurprisingly, leaders are widely accepted as having great influence on how their organisations operate (what people do) and on the form that they take (configuration). Success in organisations, therefore, has much to do with whoever is leading them. To quote the seminal management writer, Charles Handy, the search for the answer to what constitutes leadership has become similar to trying to find the Holy Grail. Nonetheless, leadership does have certain features that are common.

7.7.1   *Vision*

One of the key characteristics of a leader is someone who has vision. They can see things in a way that others can not and, most especially, are able to identify opportunities and possibilities that will enable the organisation (all who follow) to prosper and succeed. The difficulty for a leader is in articulating their vision in such a way as to persuade others; particularly if what they believe to be possible is radical or a major departure from what is believed to be standard practice. Empirical observation of construction will teach you that there are a great many people who are 'characters'.

Closer analysis will demonstrate that these people are able to convince others to do things that are often perceived to be extremely difficult (or need to be carried out under less than ideal conditions). The history of construction is replete with examples of engineers who carried out schemes that, even today, are truly amazing. This, in turn, leads to the next characteristic of what creates great leaders: the ability to communicate their ideas and beliefs in such a way as to convince others. If they cannot communicate, their vision can never be realised and, at best, they remain a fantasist!

7.7.2   Communication

The ability to send a message to others in such a way as to ensure understanding between sender and receiver is crucial. Much depends on language (encoding) and the medium for transmission. However, the measure of success of the ability of the leader to communicate will be to judge the effect that the message has on others through their actions. The willingness of followers to achieve things, which are either radical or significantly beyond what would normally be expected, is dependent on their desire to listen to the message in the first place.

Cultural identification plays a significant role in this process. If a leader is out of step with what followers believe the value system to be, they will be intolerant of any message that does not align with what they see as important. As a consequence, the leader will have to accompany their message with implied or explicit threats to elicit action; this is something that will not be effective and will merely cause resentment. As a subsequent section describes, organisational culture is vital in getting people to work together effectively to achieve corporate goals.

7.7.3   Other characteristics

There are a number of views as to what creates a leader. For example, many commentators believe that possessing self-confidence is essential. This is part of what is known as trait theory. This theory assumes that there are particular characteristics that any leader must have. In addition to the trait of self-assurance in their own judgement, this theory cites a leader's intelligence as enabling them to spot opportunities that others cannot. Trait theory also suggests that a leader should have the ability to inspire by their charismatic personality. Whilst trait theory is believed to be a sensible way of considering leadership, extensive research shows that there is no agreement on what characteristics create the perfect leader. Perhaps the biggest challenge facing any person trying to lead an organisation is dealing with change. Having a particular set of characteristics may mean that they can provide leadership only when the conditions are favourable to that combination. Given that environments are unpredictable, and that conditions may change very rapidly, trait theory is subject to criticism. Therefore, other theories provide alternatives to the trait approach.

7.7.4   Style and contingency

Empirical research into leadership does demonstrate that a leader's style and, in particular, their ability to both recognise and adapt to the context and the prevailing circumstances is extremely useful. In the former, the leader may adopt a relaxed approach if the circumstances are favourable and the people are sufficiently skilled to know what is required. As such, they would adopt a democratic style. However, should circumstances alter and people are unsure of what they must do, a more authoritarian style would probably be more appropriate (especially if the organisation's very survival is at stake). An organisation in crisis because its strategy is found to be 'wanting' will be more likely to benefit from being led by someone with willingness to take immediate decisions rather than by someone who prevaricates – like Churchill's influence in Britain during the Second World War. The difficulty for an individual may be in altering their style sufficiently quickly to deal with changing conditions. Recognition of how to best deal with particular circumstances considers contingency.

Contingency is based on the assumption that a leader should be able to adapt to the situation and influences affecting the organisation. Moreover, they should be able to work with the people to formulate the best way of dealing with the new circumstances. Whilst there may be an argument that there are two extremes of contingent leader, one who leads through rapid and dramatic change and another who leads by creating stability, there is no reason why the same person cannot do both. A great deal depends on the organisational requirements, the expectations of the leader and the pressures on people to respond. A workable compromise that may be the most appropriate is what is known as the 'best-fit analytical approach'. This assumes that an organisation must find the best way of agreeing on how it can respond to change. It certainly eschews the belief that changing the leader is a sensible way to cope. They will be expected to understand the extent of change that is necessary and to be able to work with people to produce the requisite alteration to their own practices with a willingness to accept new processes and procedures. As such, they should be able to engender trust, enthusiasm and commitment; something that seminal leadership writers, Warren Bennis and Burt Nanus, advocate in their book, *Leaders: Strategies for taking charge* (1997). As they recommend, in order for a leader to be successful, they should achieve the following to convince followers:

- Be able to demonstrate through their actions that they have integrity and should be trusted with respect to their strategic vision;
- Be willing to engage with people to elicit their support and cooperation and, most especially, to utilise the knowledge and expertise that exists within the organisation (but because it is tacit, it will need to be 'surfaced');
- Show 'passion and commitment' for the purpose that has emerged through the process of consultation and decision-making.

Before consideration of the most vital component of organisation – people – it is worth remembering that successful leaders rarely emerge in a planned way or that people read a particular text containing theoretical models of which there are many. Their skills will be in their willingness to recognise that their efforts will always be judged as a result of the efforts of others and that such motivation does not occur by accident. Effective human resource management is a consequence of foresight and planning and this will happen only by senior managers (leaders) being in place.

7.8   Getting the best from people – the 'secret' of really successful organisations

In Chapter 5, the fundamental principles of becoming an excellent (and potentially 'world-class') organisation were described. As should be clear, such organisations are able to achieve success because of the commitment that is dedicated to the most important resource that any organisation possesses: its people. Whilst many believe that technology – especially using information technology – has enabled organisations to regularise systems for dealing with customers, people are still crucial. People still carry out many of the day-to-day tasks that are required. If change is necessary to the strategy, which, in turn, causes alterations in processes and practices that are used to carry out these tasks, the very people who use them should be actively encouraged to support them. The term that is applied to the management of people in organisations is normally referred to as 'Human Resource Management'. Many definitions exist but all have as their essence the desire to proactively achieve the most effective input of people's skills and abilities to maximise the organisational outputs (and therefore achieve the corporate goals). The definition provided by Torrington *et al.* demonstrates precisely the desire to produce coalescence between what the organisation wants to achieve and the personal objectives of people:

It is when employer and employee – or business [objective] and supplier of skills – accept that mutuality and reciprocal dependence that human resource management [becomes] productive of business success (2005, p.14)

Guest believes that there are three key aspects that impact to produce effective HRM (2000, p.2):

1. Ensuring and enhancing the competence of employees;
2. Achieving high levels of motivation and commitment in employees;
3. Actively creating systems of work that enable every person to contribute towards the overall objectives.

By so doing, he contends, organisations can attain higher levels of quality and productivity than if aspects of HRM are treated as unimportant or insignificant. HRM, therefore, is a fundamental part of strategic management.

7.8.1   Theoretical considerations of HRM as a strategic function

There are three distinct approaches to strategic HRM:

- Universalist;
- Fit or contingency;
- Resource-based.

Universalist

This approach assumes that there are certain key objectives in HRM that are intended to produce the best (most efficient) organisational outcomes. Guest (1989) proposes four such objectives that should form the basis of the aspiration of producing the best possible outcomes for the organisation:

- Strategic integration – in which HRM is considered to be part of everyday practice and intrinsic to all managers' roles;
- Commitment – in which every person believes in the goals of the organisation and will dedicate themselves to attaining them;
- Flexibility – whereby the organisation will adapt its structures and processes in pursuit of the overall goals;
- Quality – becomes the overriding objective for the organisation through employee involvement in the design and use of processes and procedures.

In this approach, employees are encouraged to play an extremely active role through the operation of teams. Nonetheless, some suggest that this approach is too prescriptive and simplistic (Pursell, 1991 and Whipp, 1992). Most especially, what concerns these commentators is the contradiction between what people want and the desire of the organisation to do whatever is necessary to achieve short- to medium-term objectives. In a development of the universalist approach, the Harvard model was proposed by Beer *et al.* (1984) as a way to deal with some of the apparent dilemmas.

Table 7.2 Connection between business strategy, employee role behaviour and HRM policies (based on Schuler and Jackson, 1987)

Strategy	Employee Role Behaviour	HRM Policies
1. *Innovation*	Creative behaviour	Encourage jobs that develop cooperation and coordination
	Long-term focus	Encourage people to develop solutions to problems based on strategic objectives
	Cooperation and inter-dependence	Jobs that allow people to develop their expertise and skill base
2. *Quality management and excellence*	Regular involvement and concern with 'quality issues'	Facilitate improvement initiatives
3. *Cost reduction*	Concern with reducing waste and improvement in efficiency of processes	Training in understanding of cost structure and consequences of ineffective processes

The next approach is one that more readily tries to assimilate the desire to do what works best in the circumstances.

Fit or contingency

This considers HRM from the external and internal perspectives. The former is concerned with what the external environment requires in order to survive and prosper. The latter is intended to achieve the best configuration of resources that, of course, can be achieved to deal with the former. In essence, both aspects are a question of trial and error to derive the best combination: if it doesn't work, change it! See models produced by Fombrun *et al.* (1984).

A very important part of this approach is to develop policies for selection, appraisal, development and reward that work in harmony to create the ideal environment for employee performance. Work by Schuler and Jackson (1987) used previous work by Porter (1980) to consider what the key type of behaviour would be for a particular strategy. This approach incorporates aspects of organisational culture. As many suggest, if you want to alter behaviour, then you must get people to reconsider what their value systems are and find how it is possible to alter beliefs that are widely held, 'the way things are done around here.'

Resource-based approach

In this approach, the emphasis is on developing the skills, knowledge, attitudes and competencies of the people that you employ. To do so, it is argued, will create the sense of excellence that is inherent in all individuals and which can be harnessed by proactive and innovative management (see Barney, 1991). Because people are different, every organisation can create a unique set of resources that cannot be replicated elsewhere. Moreover, the desire to give people the opportunity to develop themselves is crucial to developing a workforce that is capable of providing the level of analysis and 'intellectual capital' that will be able to deal with the challenges of the future (Briggs and Keogh, 1999). As such, this belief in the inherent capability of resources to provide the basis of strategic improvement and excellence is one that requires four criteria:

1. Value – in which every individual has a particular contribution to make; the objective is to appreciate what each person potentially has to offer and work with them to develop;

2. Rarity – which considers the ability of the person and, in particular, the organisation's need for skills to be possessed by individuals which, depending on the job that must be carried out, may be very limited;
3. Imitability – which is based on how difficult it is for others to identify (and therefore attempt to replicate) the skills and abilities possessed by people in an organisation that produces excellence;
4. Non-substitutability – which is based on the fact that some people possess knowledge and understanding of processes, technology and the way that the organisation can produce excellence that is literally irreplaceable.

The advantage that this approach has over the other two, it is suggested, is that strategy is viewed as being evolutionary and, therefore, the development of people should also be viewed as being part of the ongoing adaptation that is required to deal with change. HRM is an integral and symbiotic part of the strategic process. As organisations are increasingly recognising, their people are as vital a part of the capital basis as the more tangible assets, albeit much harder to quantify. However, as Chapter 5 described, the use of benchmarking techniques (such as the EFQM Excellence Model) can provide an extremely valuable qualitative measure of the value that people in the organisation provide in the attainment of objectives.

Lynch provides a useful list of what should be considered as part of the audit to determine the basis of HRM (2006, p.244):

People
● Leadership (because it is key to people);
● Employee skills and their current contribution to the organisation;
● Number of people currently employed and relative turnover;
● The structure that exists;
● Reward systems and their effectiveness;
● Current motivation and morale;
● Policies for recruitment, selection, induction, training and development;
● Contribution of support departments (including HRM);
● Future development.

As a consequence, a HRM strategy should be developed which is entirely in harmony with the overall organisational aims and objectives. It should be dedicated to ensuring that people's involvement in all aspects of operations is directed towards achieving the corporate goals through their efforts.

7.9   The importance of organisational culture

Organisational culture has become a focus for many commentators who suggest that it is the key to success. Definitions of organisational culture, of which there are literally hundreds[6], tend to be characterised by a common theme which includes shared values and beliefs which causes all 'members' to act in a coordinated way. It is important to remember that culture's provenance is derived from the work of anthropologists who did not consider 'people' to be employees. Brown provides fifteen such definitions which range from Jaques' classical explanation from the 1950s (see below) to the more contemporary which tend to emphasise the importance of people believing in the purpose (mission) of the organisation.

---

[6]Brown (1995) cites a study by Kroeber and Kluckhohn carried out in 1952 which discovered 164 definitions.

> The culture of an [organisation] is its customary and traditional way of thinking and of doing things, which is shared to a greater or lesser degree by all its members, and which new members must learn, and at least partially accept, in order to be accepted ... [it] covers a wide range of behaviour: the methods of production; job skills and technical knowledge; attitudes towards discipline and punishment; the customs and habits of managerial behaviour; the objectives of the concern; its ways of doing business; the methods of payment; the values placed on different types of work ... and the less conscious conventions and taboos (Jaques 1952, p.251)

Johnson *et al.* use Edgar Schein's belief that it is 'the basic *assumptions and beliefs* [their italics] that are shared by members of an organisation, that operate unconsciously and define in a basic taken-for-granted fashion an organisation's view of itself and its environment' (Schein, 1997 p.6). Culture, they contend, is important because it 'contributes to how groups of people respond and behave in relation to issues they face' (2005, p.199). As they go on to state, it has 'important influences on the development and change of organisational strategy' (2005). Culture is therefore crucial because having influence over it will mean that managers can achieve objectives through people's willingness to believe in what is required rather than having to be forced to do so through threat and imposition of procedures.

Understanding what motivates people and causes them to act in a coordinated way is obviously attractive to managers. As managers might ponder with respect to the people in their organisation, if culture allows us to engender in them the values we believe in, this will be an extremely valuable 'tool'. The belief that there was a potential connection between an organisation's culture and its success was given potency by the seminal, if somewhat simplistic and flawed, book, *In Search of Excellence*[7], written by Tom Peters and Robert Waterman. They believed that they had established a very definite link between culture and success:

> Without exception, the dominance and coherence of culture proved to be an essential quality of the excellent companies [they studied 62]. Moreover, the stronger the culture and the more it was directed toward the marketplace, the less need was there for policy manuals, organization charts, or detailed procedures and rules. In these companies, people way down the line know what they are supposed to do in most situations because the handful of guiding principles is crystal clear. (1982, pp.75–6)

Given that *In Search of Excellence* has achieved sales in excess of six million, there certainly appears to be a strong desire by managers to learn what becoming excellent means and, in particular, to implement a similar approach to organisational culture. Peters and Waterman believed that in their analysis of the excellent companies they were able to identify eight consistent attributes:

1. A bias for action;
2. Close to the customer;
3. Autonomy and entrepreneurship;
4. Productivity through people;
5. Hands-on, value-driven;
6. Stick to the knitting;
7. Simple form, lean staff;
8. Simultaneous loose-tight properties.

---

[7]Its full title is *In Search of Excellence: Lessons from America's Best-Run Companies.*

Whilst it is notable that within two years of publication of *In Search of Excellence* half of the excellent companies that Peters and Waterman analysed had ceased to exist, its continuing sales testify to the belief that Peters and Waterman's book offered practical guidance for all organisations. In particular they presented what is known as the McKinsey S Framework that they developed whilst working for the management consultancy of the same name. The seven Ss are:

- Strategy;
- Structure;
- Systems;
- Staff;
- Style;
- Shared values;
- Skills.

Importantly, strategy, structure and systems are regarded as being 'hard' Ss. These tangible elements of organisation are usually believed to be the things that managers need to concentrate their effort towards in order to achieve success. Peters and Waterman, however, stressed that it is the other four Ss (staff, style, shared values, and skills), believed to be 'soft', which are the key elements that contribute to organisational success. They argued that the apparent obsession with the 'hard' elements was merely the result of the traditional (and highly rational) view of organisations. Peters and Waterman believed that this approach results in many of the strategic problems that beset organisations which don't enjoy the success that they might otherwise have. Such problems include the difficulty of understanding the complexity that is inevitable in analysing any organisation (what Peters and Waterman describe as 'wrong-headed analysis'). They also suggest that blind faith in rationality leads to what they call 'paralysis through analysis' in which it is more important to understand than to actually achieve anything. This may be in combination with 'irrational rationality' which occurs when managers believe that they can, with sufficient data, knowledge and expertise, find the correct answer to the organisation's problems.

Peters and Waterman's view of the apparent ease and simplicity with which culture can be managed in order to achieve success has been criticised (see Wilson, 1992). However, the concerns voiced about *In Search of Excellence* did not deter others from providing advice to managers as to how to improve the capability of their organisations (see Handy, 1984 and Kanter, 1989). As Burnes asserts, the combination of these 'perspectives' has created the basis of what he calls the 'culture-excellence' approach. Whilst there are particular differences in the message that each of the three commentators provides, the common link is the need for organisations to learn what Japanese manufacturing had achieved and, in particular, the absolute importance of getting people to understand and support changes intended to improve competitiveness. As Chapter 5 explained, Japanese manufacturing had demonstrated the ability of organisations to achieve remarkable results by concentrating on improvement of processes and engaging workers in such initiatives. Mastering organisational culture, it seemed, would provide the elixir that would increase the chances of success. The question that still needed to be answered was: what does managing culture really involve?

7.9.1   What is organisational culture?

Organisational culture is believed to consist of four aspects or 'layers' (see Johnson *et al.*, 2008, p.194). These are often shown as concentric circles (in order of central outwards):

- Taken-for-granted assumptions;
- Behaviours;

- Beliefs;
- Values.

In the centre are what are known as **taken-for-granted assumptions** which create a paradigm for behaviour and action. They are important in that they provide the core of what the organisation really stands for, and in order for strategy to be more likely to be successful, managers should be prepared to accept such assumptions. Any attempt to impose assumptions that people do not truly believe in will be resisted and, as a consequence, the strategy is unlikely to be successful. Surrounding the taken-for-granted assumptions are **behaviours** that people exhibit. Because behaviour can be seen within and without the organisation it enables people to quickly identify routines or ways of doing things that are particular to that organisation.

Around behaviours are **beliefs** which according to Johnson *et al.* are important in that they 'can typically be discerned in how people talk about issues that the organisation faces' (2008, p.195). These can be very powerful in providing the 'common thread' that unites people in what they consider to be important. If a person enters an organisation that has collective beliefs that are conflicting with their own they will experience potential difficulties. Reconciliation may require them to radically alter their beliefs. Alternatively, they may try and persuade those around them to change (with the attendant problems of dispute). Otherwise, they may simply leave. The most explicit demonstration of an organisation's culture is the circle that surrounds beliefs: **values**. These are usually formally recorded in statements used to describe the organisation's mission or objectives. The values that are espoused in this way should be aligned to the layers within. If not, there is a danger that the organisation will be accused of purporting to stand for things that its people do not really believe in. For example, if it wishes to be known for taking a strong ethical stance, ideally every person should believe in the necessity of doing this. Any deviation will be spotted and used against it as evidence of lack of sincerity (which is never good!).

Another way to consider organisational culture was presented by Schein (1985) in which its particular aspects can be considered in terms of depth (or superficiality). Accordingly, the deepest are the basic assumptions which are equivalent to the taken-for-granted assumptions described above. The most shallow are what are known as 'artefacts' which, according to Brown, 'take the form of stories, myths, jokes, metaphors, rites, rituals and ceremonies, heroes and symbols' (1993, p.9). Between artefacts and assumptions are 'beliefs, values and attitudes'. Brown warns that the apparent simplicity and elegance of this model are misleading. Because of the 'complexity', 'uncertainties' and 'ambiguities', aspects of organisational culture are rarely as easy to identify and 'deal with' as such models would suggest. Though it is possible to talk about an organisation having its own 'culture' – using a method of analysis based on the schema described below – because of the inherent irrationality and variability of people (and especially their views), it may vary. Nonetheless, understanding the elements that influence an organisation's culture is valuable. Even though culture is a phenomenon that cannot be objectively measured (though some may argue to the contrary), knowing what these elements are and how they combine to create influence should be something that all managers involved in strategy are cognisant of.

Carrying out analysis of an organisation's culture

Lynch suggests that there are four elements to organisational culture (2006, p.246):

1. Environmental analysis;
2. Cultural factors that are specific to the organisation;
3. Analysis of the strategic implications;
4. Its cultural style which is based on the following:

a. Power
b. Role
c. Task
d. Personal

7.9.2   Environmental analysis

According to Lynch, there are a number of factors that will exert particular influences on the organisation. These are the people it employs and the nature of the industry that creates the sense of what is important (or not) and which gives it 'essence'. In terms of people, he considers the following to be useful indicators of what sort of culture will exist in the organisation (2006):

- The age profile of members – young people usually having different opinions and views from those who are older. Conversely, if the members are older and, potentially, due to leave (retirement), there will likely be rapid changes.
- The socio-economic group(s) from which members come will have impact on the values that are embedded in people. If all employees are very similar in background (and upbringing), they will frequently possess similar outlooks and opinions which will 'feed into' the culture.
- Balance between male and female which is much less significant than in the past but still potentially of influence.
- Ethnicity, language and religion which in a multicultural country are things that should be acknowledged and celebrated.
- Government policy which, as Lynch suggests, will determine 'education, training, social welfare provisions, health and pensions provision […and] people development inside the organisation' (2006).

As Chapter 3 described in detail, the socio-historical development of construction has created strong influences and reasons why it tends to operate in the way that it does (and which some commentators criticise as militating against change and improvement). Accordingly, the general culture of the industry will mean that there is a strong tendency for each organisation to comply with what is the norm. However, there is nothing to stop an organisation from being differentiated by its willingness to be radical. Indeed, the imperative to be different from the opposition may be a very worthwhile strategic decision – especially if it breaks a negative cycle such as poor treatment of people (see explanation of the 'Respect for People' initiative in Chapter 3).

7.9.3   Cultural factors that are specific to the organisation

There are many things that influence an organisation and upon which its culture will be contingent:

- Its history and ownership whereby a young organisation is likely to be different from those that have been in existence for years or decades (and, in some cases, centuries, i.e. the English royal family).
- Size in which small organisations will tend to be very different from those that are large and possibly based on a centralised system of management and control.
- Technology will possibly change the way that things are done and the values that people possess. In the last twenty years the development of information technology and personal computers has had a large impact on the way that certain types of organisation operate.

Construction, still being very labour intensive (certainly on site), has arguably been less affected than, for example, retailing, especially by the Internet.

Lynch makes the point that the organisation's leaders will strongly influence the beliefs and values that people tend to possess. In an industry such as construction in which there are many small organisations led by one person, there will be a tendency for them to be 'populated' by people willing to share the same views and attitudes as that person. As organisations become larger, the influence of leaders will become less immediate (or direct), albeit that they will tend to surround themselves with those who share their beliefs (which, in turn, they will do to those around them). It takes someone very secure in their position who will recruit those willing to challenge them and, possibly, alter the 'value system' (usually happening only after a crisis, takeover or death).

Lynch presents the **cultural web** which is based on Johnson (1992) as a means by which to explore the particular factors that 'characterise some aspects' of the culture of an organisation (2006, p.247). Many of these factors are resonant with what was described earlier in this section:

- Symbols;
- Power structures;
- Organisational structure;
- Control systems;
- Routines and rituals;
- Stories.

When an organisation considers its culture (or outsiders carry out analysis), it should be possible to identify each factor in the above list and provide an overview of what it is really like. Symbols are things such as the level of equality that might exist between all levels of staff (both manual and non-manual). The sort of accommodation that is used, particularly that which is for central functions, might present a very clear sense of the symbolism that the organisation wishes to be seen as having. Anyone who has been to New York will be struck by the grandiose buildings that major commercial organisations constructed as a way of demonstrating their corporate power and dominance over their competitors[8].

The rest of the list will be less explicit to outsiders but, with some judicious investigation, it will be possible to appreciate the way that an organisation is structured and, more especially, who is most important and how power (authority) is devolved. In any organisation there will be control systems which equate to the amount of bureaucracy that is in existence.

Normally, the larger an organisation the more likely it will be that there are rules and procedures to provide guidance to people in carrying out their day-to-day duties. This will impact on the way that routines and rituals operate. Importantly, Lynch makes the point that using the cultural web can be valuable in distinguishing between what is done '*officially* [and] *unofficially* [his italics]' (2006, p.248). Chapter 9 describes how to create strategic change. It is crucial to remember that if the strategy does not start from a point of recognising what the reality is, and addressing problems or difficulties, any change is unlikely to succeed without being imposed. Such an approach is almost guaranteed to be resisted and, as a consequence, not willingly accepted. The corollary is that the culture is not one that exists without the involvement of managers (possibly using threats or sanctions). This is most certainly not the way that those examples of excellent organisations described in Chapter 5 believe that organisational culture should occur.

---

[8]The Chrysler building is an example of an organisation whose current market position is much less healthy (to say the least) than when this was built in the 1930s.

7.9.4   *The cultural style of the organisation*

Harrison (1972) and Handy (1993) believe that there are four main types of organisational style each of which will have particular consequences for an organisation's culture:

1. Power;
2. Role;
3. Task;
4. Personal.

Handy used Greek gods to typify the four types. Accordingly, a **power** culture in which there is dominance by either one person (often the founder) or a small coterie is resonant with Zeus (an omnipotent power). The beliefs that exist (and which determine values so crucial to culture) are heavily influenced by those with most power. Any change in these key people will normally have a rapid and profound impact on people. A **role** culture is one that will tend to be reliant on the use of rational analysis and formal systems of administration (procedures and processes). People are expected to abide by what has been decided as being the most efficient and/or sensible way of carrying out the operations that will lead to corporate objectives; deviation is dealt with by sanctions. In such a bureaucratic organisation those who have the ability to write the 'rules' can effectively control behaviour (although not necessarily people's values). Handy suggests that role culture is typified by Apollo, the god of reason.

In a **task** culture the emphasis is on getting things done. The value system that exists is one where ability to achieve is likely to be celebrated. People will be influenced to believe that if action is successful, they are justified. There is, of course, a danger in this. The need to abide by formal rules or procedures will not be seen as important and there is a risk of failure. Advocates of a task culture counter that the need to use a bureaucratic approach (role) will tend to make the organisational culture moribund and, certainly, averse to risk. In Handy's use of Greek gods, such a culture is considered to be similar to Athena (though he has since suggested there may be no particular God). The last type of culture, **personal**, is one in which the strategy is operated to serve the interests of individuals. The relevant god is Dionysus. In such a culture the organisation is not important to these people; they may operate competitively but also cooperatively. As Lynch suggests, 'work is undertaken in teams that are flexible and tackle identified [particular] issues' (2006, p.249). An example in construction would be a design practice in which each person is nominally a partner but all might engage in competition.

Lynch provides a number of guidelines for analysis of organisational culture which include the following (2006, p.251):

1. What is the age of the organisation and the degree of turbulence that exists?
2. Where is ownership (or stewardship) located and how much power do these people possess?
3. What is the nature of the organisational configuration and how does decision-making take place?
4. What are the metrics that are used to evaluate performance and/or success?
5. How are people (especially leaders) judged?
6. Can changes be made easily to respond to changing environmental conditions?
7. In what way are people controlled (influenced)?
8. Do individuals work cooperatively using teams?

The tests for strategic relevance

According to Lynch, these are fivefold and 'might include' the following (2006):

1. Risk;
2. Rewards;

3. Change;
4. Cost reduction;
5. Competitive advantage.

7.10   Conclusion

Having a good organisation is a desirable objective. Having a great organisation, manifested by people who truly believe in what is being done (and the means by which goals are achieved), is even better. As Chapter 5 explained, what are known as excellent organisations, sometimes also referred to as being 'world class', are characterised by having people with dedication and willingness to search for continuous improvement and innovation. Cursory analysis of how these organisations have been able to create the environment (culture) in which people feel motivated and supported so as to provide such input shows that this is no accident. It takes great effort on the part of managers who lead by example and are willing to invest time and money in training, education and the inculcation of a climate in which effort by every person is acknowledged, publicly recognised and celebrated. Such organisations also exhibit an explicit willingness to embrace new methods of working and innovative technology. Additionally, people are able to contemplate change in their ways of working and to alter their behaviour so that the organisation can capitalise on opportunities that present themselves. Chapter 8 deals with knowledge, innovation and technology. Chapter 9 considers the basis of organisational change and how people can be encouraged to support it.

Chapter 8
Knowledge, innovation and technology – the 'keys' to the future

'Imagination is more important than knowledge. For knowledge is limited, whereas imagination embraces the entire world, stimulating progress, giving birth to evolution.' Albert Einstein, theoretical physicist

8.1 Objectives of this chapter

In attempting to create competitive advantage an organisation is actively considering ways to differentiate itself from others. In order to do this it seeks to find ways to alter what it does in terms of its end product(s) or service(s), and the way that processes are used in its production. Organisations that have demonstrated their ability to give customers what they want will have been able to develop a knowledge base that understands what the organisation is capable of (or might be able to achieve in the future) and how such capability can be used to satisfy need. However, the collective behaviour of markets (what all of us consume) is not static and alters as we are provided with more choice that, frequently, consists of products or services that are revolutionary. As history demonstrates, such revolutions are not always predictable. The Industrial Revolution was made possible by the introduction of methods of production using power (initially water but subsequently steam and electricity). Those who invented these sources of power would not have envisaged the innovations that came from the ability to produce in large quantities, that is, using what is now referred to as the system of mass production. What is important is to recognise that the innovations that stemmed from the development of more efficient and consistent power sources were driven by inventors' intellectual curiosity (the evolution of knowledge).

Whilst the search for invention may be based on a need (so-called necessity), it equally may come from chance discovery (serendipity). Importantly, whilst invention might be the catalyst for innovation, it is the foresight of those who can see its wider potential that creates new or radically altered markets. As such, they can harness the newly developed knowledge base to create opportunities that will enable them (or their organisation) to provide products or services that are attractive to potential consumers. Knowing what these consumers need (or want) is the key to being successful. Accordingly, there is a logical connection between developing capability that is based on having the best possible knowledge available and the constant search for innovation by which consumers can be offered something better or, potentially, radically different or completely novel. Offering novelty is, of course, fraught with risk. Offering radical products or services will potentially allow the organisation to be the first to do so (a 'prospector' – see Miles and Snow, 1978). By so doing, a premium may be charged. Naturally, should the product or service be successful, others will seek to offer alternatives that imitate by reducing costs of production or using cheaper components. However, if the product or service is not successful, then

165

the invested costs will have been lost. Worse, others may use the experience to develop a better alternative.

There is no guaranteed formula for success. What is clear to those who have studied the rise and fall of organisations, most especially in the commercial sector, is that potential consumers and customers are less likely to have loyalty to those organisations that simply give them what they have always offered. Being able to develop knowledge within the organisation so as to find ways to innovate is crucial. A strategy based on this desire will potentially result in increased business and higher profits for the organisation by offering the customer the same or lower prices for goods that perform better. Importantly, as the previous chapter explained, there is an organisational benefit. People employed in processes will have more input and say in how tasks are carried out. The American quality advisor, Dr Deming, who assisted Japanese manufacturers, stressed that strategic improvement in every aspect of what goes on can only be achieved by the active involvement of people. The traditional belief that imposition of management can reduce costs and increase efficiency is one that he eschewed in favour of encouraging everyone to collaborate in the desire to innovate for excellence.

This chapter will consider what is involved in using innovation to enhance and develop knowledge and technology in order to provide the basis of implementing a strategy that, at the very least, enables the organisation to remain competitive and, better still, to be potentially superior. As will be explained, construction's reputation is one that is perceived by many to be culturally unable to innovate in the way that other industries have demonstrated they are capable of (the Egan report being especially critical). Consequently, it is an industry that needs to show that it is willing and able to emulate the examples that are so frequently cited from exemplar industries.

8.2   Knowing and doing – linking them together to ensure appropriate action

Organisations don't tend to do things by accident. Even if the original strategy is altered, it is usually by purposeful intention so as to take advantage of the circumstances that present themselves. Ultimately, the objective is to do things that you can achieve as well as the opposition and, if possible, better. The key is in **knowing** exactly what the capabilities of the organisation are and how these can be utilised to achieve outputs by **doing**. Much will depend on the degree of technological innovation that exists in combination with people (who possess particular skills – albeit that they can be trained).

The history of production demonstrates that those who have the ability to adapt technological processes to produce more efficiently will usually have an early advantage over others (even though in the long term it will be copied and 'bettered' by others). So, early industrialists, such as Henry Ford, used techniques of mass production advocated by Taylor to produce cars sufficiently cheaply to be purchased by a much wider market. Since then, the know-how to produce (do) has developed so as to be appropriate to both the environment and the market. Traditionally, if the objective was to widen markets, those engaged in supply sought to reduce cost. If the desire was to increase quality, the assumption was that costs must increase. It is now accepted that decrease in cost and increase in quality are not incompatible. Indeed, those that cannot achieve both simultaneously will lose their competitive position.

Consequently, the degree of investment in technological innovation and human input that exists in different industries is influenced by the desire to either protect or increase market share. The experience of manufacturing, certainly in the west, has been one in which investment in technology and innovation has been matched by the belief that people are an integral element of the production process. Construction organisations, especially those engaged in the productive part of the cycle (contractors and subcontractors), have traditionally sought ways to reduce investment in both technology and people. This has ultimately reduced overall capability, the established knowledge base and, crucially, the collective ability to produce efficiently, cheaply and to standards that are perceived to match those found elsewhere (especially in world-class industries).

According to Lynch, there are three factors that impact on the decisions that are made with respect to investment in technology and people (2006, p. 311):

1. the technological environment that exists;
2. 'discontinuities' in technology;
3. the desire to reduce cost.

In the case of the first factor (the technological environment that exists), there is an assumed constant evolution of technological development and innovation. However, this steady evolution will from time to time be radically altered by invention or a shift in considering how processes are carried out. For instance, the development of elevators in America meant that occupants of a building could be moved more quickly than when using stairs. This stimulated a demand for buildings that could be constructed much higher than before which, in the case of the early so-called 'skyscrapers', necessitated the use of steel and concrete using cranes, formwork and, to cope, newly developed skills by operatives and trades. The spin-offs that this development created were utilised widely throughout the industry in order to search for more efficient means of construction. More recently, the ability to pump concrete made its use more effective than when it had to be transported in hoppers by crane.

The second factor ('discontinuities' in technology) refers to the fact that the sort of changes referred to in the previous factor will be dealt with differently by organisations that have particular ways of coping. So, for example, those that refuse to embrace a change such as the introduction of more technologically oriented ways of production in construction – preferring to remain with one that pursues a labour-intensive approach – might see their market reduced (unless there still remains a large enough client base who prefer to use them at the prices they charge). If the change is particularly rapid and the organisation cannot (or will not) change sufficiently quickly, their very survival might become an issue if they cannot compete and cover their costs. The important thing about discontinuity is to be able to react sufficiently quickly to change. Strebel (1992) contends that there are two phases that occur after such change has become apparent:

- development phase;
- consolidation phase.

In the first phase, the emphasis is on attempting to make the product or service better by using alternative technology and labour inputs, changing relationships that exist with external suppliers or contractors or by innovation (see below). The important thing about this phase is that, if customers see one product or service as being superior to all others, it will set a standard that others are expected to emulate. Should this occur, the general standard achieved across the sector or industry will have consequentially been raised. Importantly, there will be an assumed increase in costs that may or may not be passed on to the customer (although a novel product or service can normally be sold at a premium). However, the next phase is one in which the desire by competitors is to produce more cheaply and, as a result, 'steal' market share from those who cannot reduce prices quickly enough. Following development, consolidation normally occurs when competitors seek to increase market share by imitation and adaptation of their production processes in order to reduce cost (and therefore price). Knowing where the industry is in terms of these phases is important so that they can dedicate their efforts to either catching up (development) or reducing cost (consolidation).

The last of the three factors that Lynch identifies (the desire to reduce cost) is one that, in the last three decades, almost every industry has learned to cope with and with varying degrees of success. For many other industries the cause has largely been due to the increased levels of competition that have emerged from the Far East, such as Japan, Malaysia, Thailand and Indonesia and, more latterly, from China. What these countries have been successful in achieving is a combination of significantly more efficient systems of production and, to some extent (particularly China), lower wage costs. Few commentators in the 1970s were able to recognise that British customers, who had a tradition of buying goods produced in this country out of a sense of patriotism, would switch their allegiance when offered cheaper alternatives that performed better. As such, and as is shown by the analysis carried out by researchers interested in how high-quality production systems operate, customers are increasingly aware of the concept of value (see below) which enables them to purchase more and better products (or service) for the same or less money.

8.3   The context of construction

Construction might claim that it is not the same as other industrial sectors, most particularly those that produce homogenised goods under one roof. As described in the introduction to this book, it is an industry that is used as both a 'regulator' and 'accelerator' to the economy. Therefore, whilst the end product may consist of 'standardised' components, these largely tend to come from home-produced suppliers. Traditionally, construction as a 'product' is large, certainly in comparison to high-value items such as cars or computers. Shipping completed buildings or bulk components like bricks or cement would be uneconomical when they can be locally sourced. However, it is important to remember that this was an argument used by manufacturers in Britain who thought that foreign producers would be unable to compete without having to set their prices much higher to account for the need to transport. As history has shown, such assumptions proved to be fallacious.

Even though the production facilities for bulk components such as bricks or concrete tend to be highly automated (and extremely efficient), the recent downturn in construction has caused many such suppliers to severely curtail their output. Should activity continue to decline, many such suppliers may cease to trade and, as a consequence, the industry will have lost not only productive capability but also technical expertise and skill. This is important because once expertise is lost (even temporarily), it is difficult to replace, especially in the short term. Traditionally, the industry relied on a 'hand-made' approach which was labour intensive and utilised a robust merchant network that had learned to cope with the 'waxing and waning' that was a consequence of the peaks and troughs of British construction. But as Chapter 3 explained, the prolonged periods of reduced construction activity in the 1970s and 1980s led to the loss of many skilled workers. The utilisation of industrialised methods that was advocated by the authors of *Rethinking Construction* (Construction Task Force, 1998) has meant that, even though there was a severe loss of knowledge and expertise, construction has been able to maintain output.

Understanding the basis of the industry's knowledge is therefore important. Equally, the industry must continue to strive to develop innovative capability. Construction has a long tradition of coping with change, especially when new technology is applied (see Chapter 9). Recent history suggests that the rate at which general change takes place in technology, innovation and technical capability is unlikely to slow down. Construction, therefore, will be affected by such changes. Accordingly, any organisation should be capable of developing strategy that both acknowledges and embraces such change. Not to do so will severely undermine its competitive ability.

8.4   Defining knowledge

It was the philosopher, Sir Francis Bacon, in his essay, *Meditationes Sacrae* (published in 1597), who is credited with the origins of the now commonly used phrase, 'knowledge is power'[1]. In

---

[1] The Latin being *'Scientia est potentia'*.

strategic terms, possessing knowledge that is unique or superior to that of other competitors is an undoubted advantage. For instance, an organisation may have particular ability in the way that it sources materials or components (through its 'supply chain'). This can be with respect to the technological aspects of production, as Japanese manufacturers demonstrated. However, it is important to remember that an organisation is a social construction that consists of those people who are its members. Therefore, it follows that the collective knowledge that the people possess can be extremely powerful in creating ability to achieve objectives and creating value for customers. This is precisely what the principles of TQM, which was developed in America, explained as a way of dealing with the threat that appeared in the form of goods from Japan that performed and were built to much higher standards. As the Toyota system of production showed, significant advantage existed within the very organisation if managers were able to find ways of getting people to use their knowledge and expertise. Whilst car manufacturing may be regarded as specialised, Toyota's methods of developing understanding of how it can match its productive capability to customer tastes and alterations in buying behaviour are important and have been used in other sectors, including construction, to improve their productive capability (see Chapter 5). Steven Spear in his seminal *Harvard Business Review* article, 'Learning to lead at Toyota', describes how the research work that he and Kent Bowen carried out allowed them to identify what they call 'the DNA of Toyota' (2004, p.1). As he contends, what they identified were key principles that enabled Toyota to continually innovate to develop improvements in the capability of all workers. Spear makes clear his belief that the strategic desire that Toyota has for using 'a system of nested experiments through which operations are constantly improved' has served it well in terms of 'reliability, flexibility, safety, and efficiency and [significantly] market share and profitability' (2004, p.1). There are, he asserts, four key lessons (2004, pp.5–8):

1. There is no substitute for being directly and intimately involved in observing what goes on in carrying out standard operations.
2. Any changes to operations or routines should be considered to be 'structured as experiments'.
3. Workers and managers should experiment as frequently as possible.
4. Managers should coach not fix.

Given the strategic importance that knowledge has for organisations, it is somewhat surprising that it was not until 1964 that its significance was considered as a phenomenon that had importance for managers. In his book, *Managing for Results*, the seminal writer and commentator, Peter Drucker, suggested that it was crucial (1964):

> Knowledge is the business fully as much as the customer is business. Physical goods or services are only the vehicle for the exchange of customer purchasing – power against business knowledge. (cited in Lynch, 2006, p.378)

Beyond providing a general statement, though, this does not provide a useful indication of what kind of knowledge is most important from a strategic perspective and, more especially, how it should best be used as an organisational asset. Given the somewhat nebulous nature of knowledge and the uncertainty of how it will actually be applied (for good or otherwise), it is perhaps unsurprising that definitions and guidance as to how strategic knowledge should be used are elusive. However, Davenport and Prusack make a very good attempt at doing precisely that. As they explain, it is entirely dependent on the people who have it and the situation that they find themselves in:

> Knowledge is a fluid mix of framed experience, values, contextual information and expert insight that provides a framework for evaluating and

incorporating new experiences and information. It originates and is applied in the minds of the knowers [*sic*]. (1998, p.5)

They stress that, whilst knowledge may be codified and archived, the aspects of organisational culture that exist are highly influential in determining how it is used to suit particular circumstances:

In organisations, it often becomes embedded not in documents or repositories but also in [...] routines, processes, practices and norms. (1998, p.5)

Organisational culture and how it really affects the actions of people, particularly where it is deeply embedded, are notoriously difficult either to identify or to record in the documents that would be routinely used to provide guidance to new members. The advice that you merely 'go and ask people what really goes on' is helpfully suggested as the best way to discover existing knowledge. But like any craftsperson who knows trade secrets that make them superior to others, organisations usually take great care to control the sharing of the most powerful and commercially sensitive knowledge. The way that highly successful organisations use knowledge in developing long-term relationships in order to provide customers with added value is something that competitors will seek to discover using techniques such as, for example, benchmarking. The difficulty in attempting to discover the 'secrets' of such success is frequently a challenge that defies the assumption that it is possible to accurately measure and transpose know-how in a way that resembles the transfer of technology. As the next section describes, the belief of advocates of knowledge-management systems is that such innate information should be recorded in a standard method of operating or on an Intranet. If such information is easily accessible, it can provide a valuable source of knowledge that will assist those who consult it subsequently.

8.4.1   Appreciating the difference in types of knowledge

Organisational knowledge may be obvious in the sense that it can be easily defined. For example, the type of technology that is used in production may be very clear to others and, therefore, easily imitated. Such knowledge is referred to as **explicit** in that is unambiguous, can be documented and does not require interpretation. However, the way that people interact with the technology or systems used, which others may also possess, may produce results that are superior to other competitors. Clearly, the ability to do this will give competitive advantage and remain a secret that will be known only to those inside the organisation. It is usually referred to as **tacit** knowledge in that it will not be documented, will remain ambiguous and will largely remain in the minds of key people[2]. Nonaka and Takeuchi, in their book, *The Knowledge-Creating Company* (1995), identified the need to appreciate the importance of understanding the difference between explicit and tacit knowledge. As they explained, explicit knowledge is important if it can be used effectively to produce products or services in a way that others find difficult to imitate quickly, and particularly if patents exist to make this difficult. However, Nonaka and Takeuchi stressed the importance of an organisation's ability to develop its expertise based on knowledge that is 'fuzzy' and much less easy to understand than that which is explicit. Indeed, they suggest that, by placing importance on members of the organisation as the basis of its knowledge, an organisation will create the potential for embracing change and, in particular, creativity and innovation (see below) which will present opportunities for a 'knowledge-jump'. A key objective of creating

---

[2]Why key personnel, such as managers with intimate knowledge and experience of how to produce organisational success, are frequently offered such large pay increases in order to tempt them to join and, of course, divulge all of the secrets they know.

such 'jumps' is in identifying the existing state of current knowledge and, of course, being able to compare this to competitors. This requires a process that involves audit and, subsequently, knowledge management.

8.4.2   The importance of knowledge audit and management

Skandia, a Swedish insurance company, was the first commercial organisation to recognise the need to understand and identify the state of knowledge as the means by which to create advantage based on what is called 'intellectual capital'. As such, the collective ability that people possess becomes something that, whilst not being tangible like real estate, has value nonetheless. As the managers of Skandia recognised, the way that accounting systems tend to operate fails to make any serious attempt to include the tacit knowledge that people are able to use to make their organisation operate efficiently. As they explain, the knowledge (intellectual capital) that each person has can impact on an organisation's operations in that:

> ... future earnings capacity from a deeper, broader and more human per-
> spective than that described [in its financial reports]. It comprises employ-
> ees as well as customers, business relations, organisational structures and
> the power of renewal (Edvinsson, 1997)

Most importantly, Skandia contend that the intellectual capital that exists will allow people to be much more capable of judging the way that the organisation should respond to future trends and alterations in the environment or market. As such, there is congruence with the concept of what is known as the 'learning organisation' (see below). As Deming stressed to managers, people will create excellence only if they are supported by their managers, and by the implementation of systems and techniques that enable them to carry out their tasks in the most efficient and effective way. Such systems must be designed to enable rather than disable. These systems, techniques and supporting mechanisms are entirely complementary to the human element and are referred to as 'structural capital' and will include customer capital and organisational capital. The former is concerned with providing the means by which the knowledge used is implemented in the achievement of identifying and satisfying (although Deming argued that delight is needed) consumer expectations.

8.5   Learning as a way of organisational life

Some of the expressions that have 'currency' in organisational terms are **the learning organisation** in which there is a constant desire by members to improve their ability and knowledge (this is dependent on a culture that facilitates such – see below). The other is the belief in **knowledge management** in which there is an attempt to use and exploit the intellectual capital that exists within every individual.

8.5.1   The knowledge creation process

The learning organisation has been defined as something that:

> is skilled at creating, acquiring and transferring knowledge, and modifying
> its behaviour to reflect new knowledge and insights (Garvin, 1993)

It is essentially an approach to managing the organisation that encourages its members to engage in continuous learning so as to be willing to create change both in themselves and in the processes they use.

Sadler provides the following diagram that indicates the various processes and influences that are involved in a learning organisation:

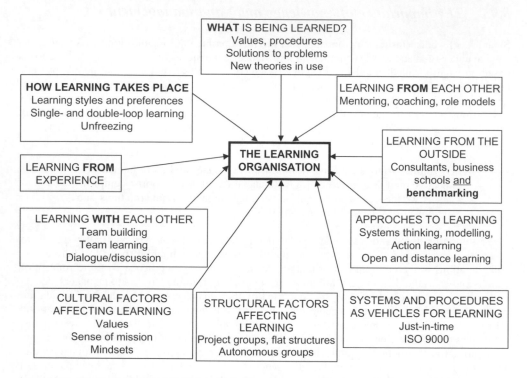

Figure 8.1   The processes and influences that 'create' learning organisations (based on Sadler, 1995, p.126).

Peter Senge, one of the most well-known proponents of organisational learning, contends that there are five 'disciplines' that support the methods being used. These are:

1. **Systems thinking**   which according to Senge *et al.* is the 'forces and relationships that shape the behaviour of systems' (1994, p.6). As Deming advised, it is the duty of management to understand how an organisation's system operates and the limits to its growth and development.
2. **Personal mastery**   whereby individuals are encouraged to have enquiring minds so as to continually seek better ways of doing things. As Crainer describes, people in a learning organisation are expected to 'live life from a creative rather than a reactive [perspective]' (1996, p.238).
3. **Mental models**   are the way that commonly held assumptions are challenged. As such, people are encouraged to develop new models that allow them to consider radical ways of carrying out day-to-day activities.
4. **Shared vision**   is, according to Senge, vital. As Sadler asserts, in any organisation where there is a shared vision among its members, 'people are capable of outstanding achievements' (1995, p.128).

5. **Team learning.** This, of course, follows on from what had been described previously in this chapter – that teamwork is a vital component of improvement. Indeed, as Senge *et al.* contend, learning by groups will always be greater than the 'sum of individual members' talents' (1994).

Argyris (1993), another well-known proponent of learning, is similar to Sadler in that he believes that learning takes place at two levels:

1. single-loop learning;
2. double-loop learning.

As he explains, in an organisation that provides customers with a product or service that is predictable, it is appropriate to provide instructions to employees that are mandatory and must be learned by rote. An example of this is the McDonalds food chain. Employees are expected to provide food that, despite some variation (for particular regional reasons), has a consistent taste throughout the world. However, as Argyris explains, double-loop learning occurs where employees are allowed to question the accepted method of doing things. The emphasis, therefore, is to encourage employees to continuously seek alternative methods regardless of how long the existing methods have existed.

In addition, Argyris draws attention to what he calls espoused theories and theories-in-use that managers use. The former, he explains, are what managers describe 'should' go on in the organisation. The latter, he explains, are what people who carry out day-to-day operations describe as 'actually' happening. According to Argyris, any deviance between espoused theories and theories-in-use frequently results from managers being out of touch with what goes on. Therefore, he suggests, managers should be prepared to realign their espoused theories to match reality. Any change is based on consensus as to what is required, not on an artificial view of what managers believe may happen. This reinforces a point that Sadler makes concerning the need for honesty in achieving an organisation where the desire to engage in learning becomes really embedded.

What is learned?
    values
    procedures
Solutions to problems
    new 'theories-in-use'
Learning **from** each other
Mentoring
    coaching
    role models
Learning **from** outside
Use of benchmarking
    consultants
    business schools
Approaches **to** learning
Systems thinking
    action learning
    open and distance learning
Systems and procedures as **vehicles** for learning
ISO 9000
    internal logic
    established ways of doing things here (culture)
Structural factors **affecting** learning

Project groups
> organisational make-up – flat/hierarchical etc.
> use of autonomy as a means to empower

Cultural factors **affecting** learning

Values
> mission (strategy)
> mindsets
> history
> dominance by leaders (founders?)

Learning **with** each other

Teamwork
> dialogue and discussion
> quality circles
> suggestion schemes
> forums for debate

Learning **from** experience

Our history
> what worked in the past
> confidence

Education

How does learning take place?

Learning styles

Single-loop versus double-loop
> encouraging personal preferences
> unfreezing current practices

How to change personal preferences   People need to be encouraged to adopt styles of learning that more readily suit their desire to create different approaches to carrying out day-to-day tasks. This involves moving away from single-loop learning (where people unquestioningly follow instructions) to double-loop learning where there is an acceptance that it is legitimate to consider different ways of doing things.

8.5.2   Characteristics of a learning organisation

1. Constant desire by senior management to consider alternatives;
2. Managers become coaches or facilitators rather than controllers;
3. Feedback from all levels is normal;
4. Freedom in choosing style of learning;
5. Analysis of critical incidents to consider better methods of doing things in the future;
6. The use of blame and search for scapegoats is no longer acceptable;
7. Activities are carried out in a way that stimulates mutual cooperation and trust;
8. What goes on is fun! People are no longer inhibited.

Chris Agyris who is one of the best-known advocates of learning believes that a part of change is for managers to consider their 'espoused theories', what they suggest goes on, and 'theories-in-use', which are the beliefs and assumptions that they draw on to make decisions and judgements. The key to success is ensuring that people identify gaps between the two and attempt to consider how they might be reduced.

8.5.3   *Knowledge transfer processes – Nonaka and Takeuchi's model*

In considering knowledge, it is important to remember the fact that, even though its existence may be identified, it is necessary to carefully consider how it might be appropriately applied to best effect. As we know, from our own experiences, information about what goes on, who does what, and how they do it, will be a process of understanding the way that people use knowledge. As has already been described, knowledge is either tacit or explicit and, therefore, transference or conversion can only be with one or other of these. Usually, tacit knowledge can be transferred to others and remain tacit because it has been achieved verbally. Some tacit knowledge may be seen as being sufficiently important enough to need to be converted into an explicit form that can be accessed by others through, for example, procedures or via the organisational Intranet. The existence of explicit knowledge can be converted into another explicit form, such as written procedures into a pocket-sized handbook or prompt cards that can be easily carried around. Alternatively, explicit knowledge can become tacit by the fact that individuals might be encouraged to interpret written procedures or guidance manuals into action that they believe is best suited to the circumstances. This would be especially useful in an environment of rapid change or where novelty in the situation is common. Nonaka and Takeuchi present a matrix which indicates their belief in how the four 'modes' of knowledge conversion they have identified takes place:

	To tacit knowledge	To explicit knowledge
From tacit knowledge	**Socialisation**	**Externalisation**
From explicit knowledge	**Internalisation**	**Combination**

Figure 8.2   'Four modes of knowledge conversion' from *The Knowledge-Creating Company: How Japanese Companies Create the Dynamics of Innovation* by Ikujiro Nonaka and Hirotaka Takeuchi (1995), Figure 3.2 , p.62, by permission of Oxford University Press, Inc.

As should be obvious, the most difficult form of knowledge to attempt to 'manage' is tacit. It is often the most important because it can hold insightful clues as to how things really happen and, potentially, where change may best occur. Therefore, attempts to try and capture tacit knowledge, despite being fraught with problems, are nonetheless worthwhile. Accordingly, any advice that can be provided to assist in the process by which deep-seated tacit knowledge can be externalised is useful. Davenport and Prusak do precisely this by recommending six ways in which the 'creation' of knowledge may occur (1998):

1. **Acquisition** – 'borrowing' from others (usually referred to as benchmarking – see Chapter 5 for a full definition).
2. **Rental** – by buying in those people who have the requisite knowledge that will create difference and advantage, i.e. consultants/academics.
3. **Dedicated resources** – by setting up a project group or improvement team whose specific objective is to achieve particular outcomes. Importantly, these people will need absolute support during the process of developing solutions and more especially during the implementation of proposals.
4. **Fusion** – merging departments or bringing together those who possess particular skill sets which, it is anticipated, when combined, will create synergy and new ideas (they will innovate – see below).
5. **Adaptation** – the need to do things because of external forces, such as a dramatic shift in consumer behaviour which creates a need to develop understanding quickly to ensure survival (initially) and change in order to cope with new markets.
6. **Networks** – these exist in all domains for mutual sharing of ideas. Clearly, the more that an organisation can encourage people to engage in networks, the more likely it is that they will be able to elicit useful knowledge which may be used to create advantage.

The key to knowledge creation and transfer is that, once innovative ideas have been identified and tested (and found to be potentially useful), they should be shared as widely and as quickly as the organisation's systems of communication will allow. Any delay risks others recognising the same factors and implementing them faster. Leonard-Barton (1995) proposed a model that summarises how knowledge is 'created and diffused'. She believes in an organisation's need to engage in solving problems that undermine their ability to achieve strategy. In the future, however, the organisation can experiment to attempt to find new ideas or innovations that might have potential. Understandably, if such ideas don't have potential, they should be abandoned as soon as possible. This is not always easy as much time and expense will have been invested and all that may be needed are persistence and good marketing. However, sticking with an idea that will not work risks not just ridicule but financial disaster if too much money is dedicated to this development. As she also believes, knowledge is acquired internally through discussion and group development whereas externally it is imported.

8.6   What is innovation?

Innovation is concerned with considering ways of generating ideas that will allow the organisation to create new products or services or, alternatively, to improve those that already exist. As such, it may be based on technological development that alters the processes used in production – such as to increase efficiency – or the actual composition of the product. Importantly, though, innovation can be used to improve the people's ability to carry out their tasks more effectively or, as the Japanese demonstrated, to give them the confidence and techniques by which they are constantly searching for alternatives that will create further innovation .By so doing, the culture of the organisation becomes one in which creativity and constant enquiry is the 'way things are done around here'. As such, innovation should be viewed as a vital component of the organisation's HRM strategy (see previous chapter). As a subsequent section will explain, creating a knowledge-oriented organisation can be crucial.

According to Lynch, innovation can enable an organisation to achieve 'three priceless assets' which will be vital to attaining its corporate strategy (2006, p.396):

1. Growth in terms of markets and access to potential customers;
2. To be seen to be different (and better) from others who provide products or services that are viewed as comparable or direct substitutes;
3. The ability to leapfrog major competitors.

The real difficulty, of course, is in knowing what effect the innovation(s) will have in terms of either producing more effectively or efficiently, or in achieving enhanced customer satisfaction. It will be assumed that any change made will be intended to make things better, but if it does not, then it should be immediately abandoned or reversed or, at the very least, reviewed and modified.

Baker (1992) has suggested that there are three main 'drivers' of innovation (although the third is a combination of the first two):

- **Analysis of customer needs** – so-called market 'pull';
- **The development of new technology** – what is usually called technology 'push';
- **The combination of pull and push** – referred to as 'disruptive innovation'.

8.6.1   Pull

If an organisation has been able to identify opportunities in the market that are not currently being served (either because they have not been considered important enough or possibly because there was not sufficient expertise until now), this creates potential that might be served. Using innovation to identify such opportunities is known as 'market pull'. It is a very useful way of achieving strategy that enables the organisation to derive competitive advantage over others until they provide acceptable alternatives. Clearly, there is a need for the organisation to invest in analysis to understand that current market so that 'gaps' or deficiencies can be identified.

A potential example of pull in the context of construction would be the ability of the contractor to utilise technology (such as modularisation) either to complete work much faster than others, or to produce a standard of finish and performance that is superior to that capable of by competitors. As such, the desire is to emulate what the Japanese car manufacturers were able to achieve by the reorganisation of production systems and supply-chain management in the 1980s.

8.6.2   Push

If pull is about identifying opportunities for developing new innovations in existing markets, push is about taking what may exist elsewhere and applying it in a new context. As such, this type of innovation is based on the desire to diffuse (to spread) ideas beyond the location or technological application where they originated. The development of radical strategies is frequently based on taking simple ideas but applying them in a way that others did not perceive as being relevant. For example, the use of a production system that has been applied elsewhere might be used to provide a more effective flow that suits a client's need. This is what lean construction (which is based on the Toyota production system) claims is possible.

8.6.3   Disruptive innovation

This is based on the objective of using technology that might already exist but in a way that allows the production of goods and services to be provided to customers so that they are available as a cheaper alternative. For instance, the use of prefabricated components, produced by using cheaper labour in the Far East, might allow a contractor to offer the client a building that is much cheaper than that which would be possible by using local sources (as has been shown by auto-manufacturing). Where organisations are able to utilise existing systems, technology or people, and where there are ways that enable them to cut costs, this will potentially allow them to 'disrupt' the existing markets that established providers currently serve. The obvious difficulty of this strategy is that if you can do it, others will do the same and, as a consequence, there is no longer a strategic advantage.

According to Baker (1992), innovation is a process by which new ideas will be taken up slowly at first. This makes the introduction of new innovations fraught with danger. If there has been great investment in the idea but little interest, perhaps because the initial selling price is high (to reflect its novelty or cachet value), a commercial organisation may risk losses that seriously undermine its reputation. However, if the innovation becomes much more commonplace, there will be opportunities for those who are able to enter the market with alternatives that are not only cheaper but which also did not require the investment that the prototype needed. Therefore, according to Baker, being the next best-placed 'follower' is potentially a less risky strategy than being the first into the market with a new innovation. Mansfield (1992) concurs with this view and, following extensive research, suggests that those who were second into a new market, after the innovative company, are able to imitate the product or service in two-thirds of the cost and time.

Clearly, knowing when to innovate and when to imitate is an extremely important decision that can, if made correctly, deliver significant advantage. According to Utterback (1996), innovation has three phases that will assist an organisation in knowing what is the best decision to make:

- **Fluid** – in which some organisations (possibly only one) attempt something radical in a haphazard way which means that there may be duplication. At this stage it may be difficult to know which of the new innovations will become successful; this will become apparent in a subsequent phase.
- **Transitional** – in which demand has increased and potentially more competitors will have entered the market (choice increases and, usually, costs fall). Those with the most innovative products (often coupled with reduced costs) are best placed to survive.
- **Specific** – at this point it is likely that innovation will have slowed and any new alterations will be minor. It is also likely that the market will be dominated by a small number of large organisations that are capable of producing in significant numbers at lower cost than those smaller organisations that entered earlier.

8.6.4   How to achieve innovation?

Because innovation is something that is so potentially valuable to an organisation, it makes sense to be able to encourage people to constantly search for opportunities to implement it. The problem is that people are usually so obsessed with carrying out their day-to-day duties that they have little time to spend innovating. Moreover, there is also a tendency for a culture to evolve in which risk-taking, normally a corollary of innovation, is frowned upon and, should the outcome be unfavourable, that blame be assigned to those involved. The bigger an organisation becomes, the greater this likelihood becomes. This is something that the seminal writer on innovation, Professor James Quinn (1985, p.73), identified and he suggested that every organisation should try and become entrepreneurial in nature by 'acting small'. In particular, he advised that innovation should foster a climate of 'controlled chaos' in which the organisation is willing to depart from preconceived beliefs that follow particular rules or procedures and this will create innovation. In particular, he believed, the following conditions should exist:

- **Atmosphere and vision** – where there is a clear indication that innovation is welcomed by senior managers and those involved are provided with total support (both morally and in terms of resources).

- **Small, flat organisational configuration** – in which rules are minimal and the emphasis is on being able to adapt to change and respond quickly to new opportunities.
- **Small teams of innovators** – where people are encouraged to cooperate in teams based on creating new solutions that create opportunities.
- **Competitive selection criteria** – whereby generation of novel ideas is seen as being similar to a game but where the losers are equally valued as the winners (because second-best solutions may still be worthy of consideration in the future and nothing should be done to deter people from wishing to become involved in the future).
- **Interactive learning** – which, of course, is consistent with the explanations presented earlier.

Quinn's beliefs have much to recommend them, especially given the emphasis they place on people as being the originators of innovation. Whilst there is an acceptance that there is no ideal set of circumstances that create innovation, there are studies that stress the need for people within the organisation to be prepared and able to cope with the climate that engenders the proposal of new ideas. Nonaka (1990) stressed the need for a willingness to accept that chance will play a part and that the ability to respond in a timely and proactive way is important. In their influential book, *Competing for the Future*, Hamel and Prahalad (1994) assert that there are seven guidelines that should be used in an organisation to facilitate innovation:

1. Question the present strategy and market definitions in the hope that something new may emerge.
2. Consider the purpose served by current products or services because the accepted wisdom may be out of date.
3. Explore external timing and market opportunities and be constantly on the 'look-out' for opportunity.
4. Seek out competitors' weaknesses because, by so doing, they can be exploited (and they would do the same).
5. Deliver new and better value for money because this is now the only way to attract new customers and retain existing customers.
6. Search far and wide using, for example, techniques such as benchmarking.
7. Seek to challenge conventional wisdom by allowing people to ask difficult questions (even though senior managers may resent being expected to justify their decisions or actions).

These guidelines are a common-sense approach that any organisation would adapt to deal with its consideration of what it does and how it might do it better (both in terms of processes used and final outcomes). So, constantly re-examining where the organisation is 'positioned' is sensible. New ideas may emerge from the process that might otherwise have been overlooked or ignored until a formal review was carried out at a predetermined date. Tom Peters strongly advocates that a key component of success in terms of innovation is to populate the organisation with 'curious people' (Peters and Waterman, 1982). As a result, they will be likely to think creatively and outside of the current systems that will have been designed to regulate outputs that were appropriate for a different period. Most importantly, though, they will be more inclined to comply with the last two of Hamel and Prahalad's guidelines. Finally, Tidd, Bessant and Pavitt (2005) believe that there are four 'clusters of behaviour' that are crucial to innovation management.

As they advise, in the strategy 'domain' it is important that the organisation is willing to learn from past experience and have an ability to analyse which, in turn, depends on its position vis-à-vis its products/services, processes and the embedment of technological innovation. With respect to implementation, they advise that the organisation is able to move from consideration of innovation to achievement through systematic decision-making based on solution of problems:

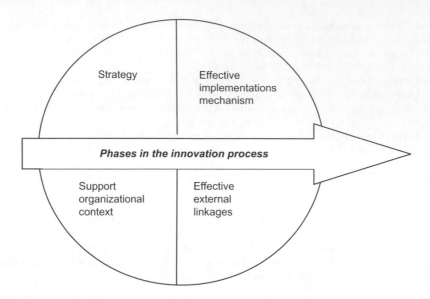

Figure 8.3   The four key 'clusters of behaviour' in innovation management (reproduced from Tidd, Bessant and Pavitt, 2005, p.560).

reactive (existent) and proactive (future). Skill and dedication are essential. Moreover, the management of change will be a key attribute (see Chapter 9). Effective external linkages require close interaction and understanding of what the organisation does and the impact on external parties such as customers and suppliers. As they recommend, partnership through 'open innovation' (Tidd *et al.*, 2005, p.561) is increasingly a way to share learning and development that is both collaborative and mutually beneficial. The last domain, 'supporting organisational context', draws on the importance of what was explained in Chapter 7 and the desire to create a culture in which learning is the norm. As well as stressing the need for appropriate structures and systems for communication, reward and employee development, they emphasise the need for every person to actively support the management of innovation.

8.7   The importance of technology

Technology is the method by which almost everything we experience can be changed. The world we all experience every day is largely dependent on, and frequently influenced by, the technology that currently exists. Our worlds have become utterly reliant on the systems that enable the infrastructure to operate (consider what happens when power is turned off). It allows communications and transport to occur. Importantly, for so-called 'civilised society', it is the basis on which commercial enterprise and public services can be facilitated.

That technology, a word that alters depending on the context in which it is used, should be so crucial is not surprising. Since humanity has existed, it has attempted to harness whatever resources are at hand to create and improve the environment to exist in safety and comfort (for example, canals, railways and sewers). However, remembering the basis on which the origins of strategy developed reminds us that technology has also been used to dominate and subjugate others through the use of arms. Technology, therefore, can be used for good and bad. The Industrial Revolution may have been the origin of the world we currently know, but some realised that its effects would irrevocably alter their world by mechanising tasks they carried out individually: the Luddites. As these people feared, it gave power to those who had the capability to create organisation on the basis of combining technology and people. But as the Luddites had to accept, no matter how grudgingly, the impact of the changes that the Industrial Revolution wrought were unstoppable. In the same way, contemporary organisations are forced to acknowledge that the continuous evolution of knowledge and innovation will cause technology to alter, which, depending on the circumstances and product/services, requires a strategy of change.

8.7.1   *Understanding and utilising technology*

Like the early factory owners, managers who take major strategic decisions need to be aware of the technology that currently exists, what they currently possess (and how it compares to 'state-of-the-art') and, crucially, what developments can be reasonably contemplated in the short to medium term[3]. The key objective is to be able to utilise technology in such a way as to create sustainable competitive advantage. Importantly, as Johnson *et al.* contend, whilst technology itself is important, if it is equally easy for all competitors to acquire, there will be no advantage (2005, p.478). Rather, they believe, the important factor is how it is exploited. Using a matrix that was originally proposed by Tidd *et al.*, (2001), Johnson *et al.* suggest that there are many 'mechanisms' that may be used to derive advantage:

- novelty in the product or service;
- novelty in the process;
- complexity;
- legal protection of the intellectual property;
- robust design;
- rewriting the rules.

So, in the first three of these mechanisms, the objective is to do things that others have not yet considered or mastered. The problem will be that once it becomes explicit what has been achieved, others will imitate the features that the product or service has or re-engineer the processes by which production takes place. However, as Japanese manufacturers of cars and electronic products have demonstrated, even though others may imitate, being able to match superior levels of quality or service at lower costs may be more difficult to achieve. Using legal protection is very useful if there are features or components of the product or service that can be 'copyrighted' or patented. Others will, of course, attempt to provide alternatives that are just about legal. The use of a licence-based system may be a way of allowing others to use the technological developments that have been created in a controlled way; the advantage is that those who purchase them do not have to spend a great deal of time or money in developing their own alternatives. 'Robust design' and 'rewriting the rules' are the desire to achieve something that is so utilitarian or unique that it provides the basis for long-term use (albeit that variations may be introduced). In the case of the former, there are examples of formwork or scaffolding systems that some contractors have introduced which work so well that others cannot find ways to better them. The latter is viewed as a revolution that, once it has been introduced, sets the standard (benchmark) that others will want to emulate.

According to Tidd *et al.* (2001), there are five different types of context in which technological development has particular consequences. These are known as 'technological paths' (or

---

[3]In the longer term technology may shift so dramatically that it is impossible to predict. For instance, could any manager in the 1980s have envisaged the way in which their organisation would have to adapt to a world so reliant on the information and communication systems using the Internet?

trajectories) and will be dependent on factors such as size of the market, the type of product or service and the state of innovation and knowledge. These are:

1. **Supplier-dominated developments** – in which an organisation that is ostensibly outside the production chain makes a breakthrough that allows some, although not all, of those who produce for customers or consumers to achieve advantage. For example, some proprietary systems are so expensive and complex to use (and will require particular expertise and resources) that they will only allow certain firms to take advantage of them.
2. **Scale-intensive developments** – like the previous factor, some technological innovations or developments will require large 'runs' or outputs that will only be economically viable to those organisations that have sufficiently large markets.
3. **Information-intensive developments** – when the use of information technology can be employed to radically alter the ability of some organisations to serve customer or consumer needs and expectations.
4. **Science-based developments** – in some sectors there is constant development in terms of research and development in order to create new products or breakthroughs for those organisations that can create a product or service that others have not yet achieved; this is especially so in sectors such as pharmaceuticals and engineering.
5. **Specialised-supplier organisations** – in which some organisations can, because of their understanding of customer needs, develop particularly effective and innovative solutions that are suitable for wider application. Quite frequently these organisations can be quite small; their expertise is something that allows them to create distinctiveness with respect to technological expertise.

8.7.2   *Connecting technologically strategic issues and strategy*

Because technology is so important to what organisations seek to achieve and how they do so, there is bound to be an impact on all considerations of strategy. Johnson *et al.* believe that there are three key aspects of technology that should be linked to strategy (2005, p.480):

● How technology can be incorporated into the competitive situation by analysis of 'forces' and through transmission in all processes used;
● The way that technology can be incorporated so that it underpins and enhances the overall capability of the organisation;
● Consideration of how technology can be organised to ensure improvement.

Each of these is dealt with in more detail below.

8.7.3   *Technology and the competitive situation*

Using Porter's Five Forces model as inspiration, they consider how each might be affected by technology. Therefore, 'barriers to entry' may become easier or harder depending on the costs and complexity of the new technology. The use of personal computers has undoubtedly allowed small organisations to compete with larger ones, for example, in publishing. In construction, the use of low-cost, hand-held tools has enabled small organisations, frequently one-person concerns, to carry out work as effectively as their larger counterparts. However, if the technology is so expensive as to be prohibitive to all but the very large, the barriers inevitably increase. As has been explained in a previous chapter, the construction sector has a long tradition of not investing large amounts of money in equipment; it prefers to hire on a job-by-job basis.

Substitution can be facilitated by new technology allowing consumers to derive equal satisfaction from the introduction of new products of services. The widespread use of UPVC since the 1980s has largely replaced the use of timber for use in components such as doors and windows (to name but two). The increased use of prefabrication has equally reduced the need to produce many components *in situ* and this has reduced the requirement for skilled labour.

The **relative power of suppliers or buyers** can be altered by technology in that its influence can alter the dynamics of markets or the relative power that is 'enjoyed' by some. In the case of some materials routinely used in construction, and because of the high capital costs of start-up, there is a very high degree of power. An example might be in gypsum-based products. However, as international competition becomes commonplace, it is more likely that cheaper alternatives will be available. Finally, **competitive rivalry** can be altered by technological development. If one major organisation introduces new techniques that others find difficult to imitate, the relative competitive positions will be altered.

In addition to the above, Johnson *et al.* stress the importance of an organisation understanding the nature and pace at which new technology enters the environment in which it operates (2005, p.481). As they contend, there will be both supply and demand side issues that will guide any considerations. The former are concerned with what others are doing and how their thinking is guided. So, for instance, the organisation will look at things such as the rate at which change is occurring (see also subsequent chapter). It will also look at factors such as the amount of experimentation and research and development that are currently being undertaken within the industry or in institutions outside (such as academia). Judging the findings and predictions can assist in making better guesses as to what the future developments will be. In the case of the latter, demand side issues, the focus is on consideration of what customers and consumers believe technology will achieve for them. Clearly, much depends on what the current state of awareness is and how quickly the new technology is being adopted. As has been demonstrated in the past, technology is often slow in terms of 'take-up' (especially if it is expensive) but once its value is better understood and becomes more cheaply available, the rate of use increases dramatically. Customers and consumers no longer need to be convinced of the virtues; they demand that they 'must' have it! This is usually referred to as the 'tipping point'.

8.7.4   Technology and strategic capability

Much depends on the so-called 'core competences' that an organisation possesses. These provide the organisation with its particular capabilities; the more distinctive and unique they are, the better. Clearly, if technology is used to improve or enhance them, their capability is likely to be increased. The important thing is to develop or acquire technology that is going to underpin and support core competences. As Tidd *et al.* assert, there are three ways in which technology can be developed or acquired:

1.   within the organisation ('in house');
2.   alliances;
3.   by acquiring it from others using licences.

Clearly, of these, the first has the attraction of doing for itself that will mean that there is likely to be advantage over others. If the level of advancement is far greater than that of competitors, there will be distinctiveness for a sufficient time to build on success. The use of any means by which to protect the developments will be extremely valuable. However, there is likely to be considerable cost and risk involved which may count for little if the advancement is easily imitated. The second is a method by which both cost and risk may be reduced. The problem is that any gains will also be diminished by the need to share them. The third is a very conservative way of acquiring technology with speed and low risk.

8.7.5   *Organising technology to achieve advantage*

This may be seen as a two-stage process whereby stage one considers what the organisation currently has, and stage two implements those developments or initiatives that are most likely to produce beneficial change. As Lynch explains, like marketing (using the Boston Portfolio Matrix), technology can be classified similarly.

The real challenge, of course, is in how to know what will work and what will be less effective (and potentially disastrous from an investment point of view). The need to be at least as good as others cannot be over-emphasised. As Chapter 5 explained, the use of constant monitoring and 'benchmarking' is an effective way to ensure that an organisation can achieve this. Once a new innovation or development has become known, then the main objective will be to implement and organise its introduction as quickly and effectively as possible.

8.8   Conclusion

The real difficulty for analysts is in knowing what is possible (limited only by imagination) and whether it is possible to produce (consider cost – although if sufficient demand, not usually an impediment), and whether potential customers will be persuaded that the products/services are needed. Some of the things we now take for granted have been the result of genuine need. However, just as many innovations have been the consequence of unintended discovery; chance can be the arbiter of success.

The key is to assemble the best possible mix of people (where the new ideas will always come from) with technology to create an innovative culture. It is not by chance that the likes of Microsoft provide extremely good working conditions and perks to assist in stimulating people. The argument of many is that what is crucial is a knowledge-based workforce, encouraged to be creative.

Chapter 9

Change – the only constant in strategy

'It is not the strongest of the species that survive, nor the most intelligent, but the one most responsive to change.' Charles Darwin

9.1 Objectives of this chapter

Change and its 'management' have become a dominant theme in literature dedicated to informing managers how to cope with a world in which there seem to be continual alteration, evolution and, quite frequently, revolution. As previous chapters have explained, history is replete with examples of shifts in taste and behaviour that have been stimulated by, for example, new developments in technology or culture. The important thing for organisations is to be able to anticipate these shifts and, more especially, to be able to cope with the ramifications. This requires decisions to be made about what the organisation does and how the objectives (including outputs) are achieved. In dealing with change, there will be consequences, most particularly for the people. They, after all, will have to use the new technology to operate systems, produce goods and services and, if required, acquire new skills or adapt to alternative patterns of work.

The fact that change occurs is, of course, not new. However, the focus that management literature has placed on it is, in particular, on what is involved and how it can be managed in such a way as to create success. The last word is, in strategic terms, crucial. If the managers in an organisation wish to be successful they will usually look elsewhere to discover how others have achieved it. The fact that Japanese firms in the 1980s were demonstrating their ability to produce effectively, efficiently and to extremely high levels of consistency provided the inspiration to learn how they did it and, naturally, to attempt to change in order to enjoy similar success. What has followed, in the wake of the desire to learn the secrets of Japanese success, has been the production of a large number of texts that are dedicated to explaining to managers how they should implement change in their organisations. Change, as far as strategic management is concerned, seems to have become a constant theme.

Lynch makes the important point that there is a distinction to be made between organisational change that, he states, 'happens in every organisation', and strategic change that, he believes, 'can be managed' (2006, p.752). Organisational change is something that is ongoing and, it is suggested by Lynch, a phenomenon that occurs either slowly because of normal events that happen over a period of time, or very fast because of an event that creates consequences that need to be dealt with immediately. Change that occurs slowly will often be achieved by deliberate intervention and will be 'managed'. However, change that occurs quickly is often in response to an event such as a crisis like the loss of a major customer; it will create many consequences, some of which may be unanticipated (and unwelcome). Therefore, it is likely that change that occurs quickly will involve much less proactive management than if it occurs slowly. The important point, though, is that change – whatever its origins – will have an impact on strategy. Therefore, it is necessary to consider aspects of strategic change and how these can best be managed. Prior to doing this, it is useful to consider the theoretical basis on which change management is based.

9.2   The theoretical basis of change management[1]

Burnes makes the very valid point about change management (which can equally apply to strategy and general management) that it is not a subject that has 'rigid and clearly-defined boundaries' (2004, p.261). Rather, he explains, whilst it is a part of management that draws inspiration and knowledge from a wide variety of disciplines, much of what is accepted as part of 'established theory' came from thinkers aligned largely to psychology. In an attempt to provide an understanding of the theories that have helped to create the subject of change management, he suggests that there are three 'schools of thought' on which development has occurred:

- the individual perspective;
- the group dynamics;
- open systems.

9.2.1   The individual perspective

This perspective is one that suggests that if an organisation is to change, it is essential that the very constituent component – people – must be prepared to alter themselves as individuals. Whereas the other two schools of thought are based on the assumption that it is organisation through groups of alliances that is important, the individual perspective, as would be suggested from the label, instead concentrates on consideration of ways to encourage and motivate each person to want to associate themselves with change. This can occur through the normal extrinsic motivators such as money – support the change and the organisation will become more successful which will allow us to pay you more. Alternatively, intrinsic motivators may be important in that every person will want to be part of an organisation that aspires to become known for its ability to become successful through radical innovation or technology.

Within the individual perspective there are two groups: 'behaviourists' and 'Gestalt-Field psychologists'. Behaviourists are interested in the ways that a person can be influenced by the environment they experience. Accordingly, if you want to change a person's behaviour, it is necessary to alter the environment. The early behavioural theorists believed that individuals who experience gratification would be more likely to repeat what earned it in the first place, rather as a child would do. This view is, of course, based on the simplistic assumption that people can be easily manipulated by external rewards or stimuli. People, we all know, are frequently hard to predict and do not always respond to external influences in the way that behavioural theory would suggest. Gestalt-Field psychologists are more willing to accept that there are greater complexities involved in causing people's behaviour to change. As Burnes puts it:

> For Gestalt-Field theorists, learning is a process of gaining or changing insights, outlooks, expectation or thought patterns. [… they consider] not only a person's actions and the responses these elicit, but also the interpretation the individual places on these (2004, p.262)

---

[1] This section has greatly benefited from Bernard Burnes' seminal book, *Managing Change* (2004).

So, whereas the behaviourists believe it is simply about external influences, the Gestalt-Field psychologists concentrate on assisting each individual to appreciate the situation they are currently in, their own role and, significantly, their importance in creating change.

9.2.2   The group dynamics perspective

The work of Kurt Lewin (see below) was extremely important to the early development of this perspective. His belief was that people in organisations tend to work in groups and, therefore, change must focus on how to cause these to alter. Therefore, in order to create change, the group should be influenced in ways that alter the dynamics (forces) that each person is subject to, such as the norms and expectations.

The emphasis is on understanding the way that groups currently operate (their dynamics) through norms, roles and values. Using teams and teamworking have become an essential element of how advocates of group dynamics believe change can occur. Indeed, a major part of management thinking, so-called 'Organisational Dynamics', is based on the importance of precisely this. French and Bell in their book, *Organizational Dynamics*, stress this point:

> ... [in OD] teambuilding activities, the goals of which are improved and increased effectiveness of various teams within the organization (1984, p.128)

9.2.3   The open systems perspective

Systems theory attempts to understand organisations as being composed of various subsystems each of which operates in conjunction with each other. The more that the subsystems cooperate and rely on each other, the more effective the organisation becomes. The analogy of the human system is apposite. We know that our health depends on what we eat, how we exercise and, of course, the degree of exercise we take that ensures we have effective lungs, heart rhythms and state of mind (which is improved by exercise).

The open systems perspective is based on the belief that for change to be effective, those attempting to inculcate it must consider all the parts of the organisation (the subsystems). Importantly, though, and like us as individuals, these parts are subject to influence from the external environment. For example, an organisation may be very effective in all respects until a sudden event causes a radical shift in the patterns of working in a certain department, such as a sudden lack of skilled professionals (for example, quantity surveyors). This may cause consequences that impact on many other departments and create problems throughout the system (organisation) as a whole.

In carrying out analysis consistent with an open systems perspective, it is necessary to consider the four principal organisational subsystems (see Miller, 1967):

1. its goals and values;
2. technical;
3. psychosocial;
4. managerial.

9.3   Strategic change in organisations – deriving an understanding of what is involved

Johnson *et al.* (2005) believe that the important aspect of change is that whatever its origins, it should ensure that strategy is altered in such a way as to ensure that operations and everyday aspects are in conjunction with each other, an aspiration that may not be entirely straightforward (see subsequent section):

> This emphasises the importance not only of translating strategic change into detailed resource plans, key tasks and the way that the organisation is managed through control processes, but also of how change is communicated through everyday aspects of the organisation (2005, p.504)

Lynch provides an alternative definition of what strategic change involves and, using Schein (1990), he argues that it requires a fundamental alteration to the way that people do things (in terms of actions, beliefs and attitudes):

> Strategic change is the *proactive management of change* [his italics] to achieve clearly identified strategic objectives [... it] involves the implementation of new strategies that involve substantive change *beyond normal routines* (Lynch, 2006, p.753)

A common aspect of these definitions is the implied need to consult and engage with people. As such, strategic change is often explained in terms of the effects that it has on the organisational culture. Given the importance that the word, culture, has assumed in management literature in the aftermath of the 'quality revolution' that followed the introduction of Japanese products into British markets, this is not surprising. Perhaps the most influential text to appear was Peters and Waterman's *In Search of Excellence* (1982) in which they argued that a 'dominant culture', in which the beliefs of the people in the strategic objectives of the organisation, was a common feature of the so-called 'excellent organisations' they studied. The message was clear: ensure that people are in agreement with what is being attempted and their attitudes and approaches will change in accordance with the desired intentions of senior managers. This raises the key question: if people are crucial to strategic change, how can their support be elicited? Tichy, in his book, *Managing Strategic Change* (1983), believes that crucial 'pressure points' exist in this process (and require people's involvement in successful implementation):

1. formal organisation structure and processes;
2. people;
3. tasks;
4. informal organisation structure and processes.

The first of these refers to the way that authority is vested within the organisation that is provided in the official version (usually by the senior management). It concerns the way that relationships are believed to occur and how decision-making occurs in terms of the designated roles and positions. If this is compared to the last pressure point, a contrast can be made in terms of how unofficial relationships and social relationships are formed and sustained outside of the version that is presented by management. In achieving a match between action and strategy, it is important that as much effort is dedicated towards the informal (or as sometimes called, 'shadow') organisation as the formal one, which will, inevitably, be the main focus of activity and influence.

People, as a 'pressure point', consists of leadership, motivation, culture and the sociological aspects that create the very essence of the organisation. Tasks, however, consists of the way that work is carried out and the processes that are used to achieve outputs. Finally, informal organisation requires those implementing change to acknowledge the existence of those groups of people who have influence over processes (in terms of collective behaviour) and, therefore, can be key to receiving support. Being able to access these 'networks' may, of course, not always be straightforward due to the fact that their very existence is deliberately intended to be outside the formal hierarchy (and somewhat illicit!). Ultimately, of course, it is imperative that management, in their attempt to achieve strategic change, engage with people in all aspects of change so that any alteration is consistent with the desired outcomes.

9.4   Types of strategic change

There are a number of ways in which change may occur. According to Balogun and Hope Bailey (2004), there are four types of strategic change which are based on two axes:

- Analysis of the nature of change (which can be either 'incremental' or 'big bang' in orientation);
- Consideration of the scope of change (which involves either 'realignment' or 'transformation').

The four types are:

1. adaptation;
2. reconstruction;
3. evolution;
4. revolution.

Adaptation is based on an approach that is both incremental and aimed at creating realignment. It is the sort of strategic change that most organisations will attempt to implement. It is relatively slow and usually based on predictable assumptions. The problem that occurs for an organisation is that their plans can be made irrelevant by events in the environment that were not anticipated. This requires change that must be achieved very quickly to respond: what is known as **reconstruction**. For example, there could be an economic crisis which means that savings must be made, potentially resulting in job cuts, reduction in wages or elimination of bonuses. Therefore, reconstruction may involve people having to accept changes which cause their roles or their departments to be restructured. Reconstruction requires managers to be able to communicate the sense of impending crisis and convince all those who will be affected of the urgency of very rapid (and potentially painful) strategic change.

If the managers wish to transform the organisation, they can do this by one of two approaches: **evolution** or **revolution**. As is explicitly suggested, the first attempts to create the transformation by a series of gradual steps intended to ensure that the organisation is different (albeit that the change is intended). The other transformation, revolution, is more likely to be caused by a catalyst that is outside the immediate influence (and, probably, contemplation) of managers. Revolutionary strategic change will usually be rapid and lead to consequences that will be extremely difficult to predict. As such, all that managers can do is assist the people affected, including themselves, to deal with the consequences by implementing whatever alterations are needed to stave off the immediate crisis (short term), ensure survival (medium term) and attempt to regroup and prosper (long term).

9.5   The causes of change

There are particular models that exist which purport to suggest what the main causes of strategic change are. Tichy (1983) believes that there are four influences that create change in organisations. These are:

1. **The environment –** with respect to the economy and general conditions in which the organisation trades (or exists);
2. **Business relationships –** considering how others in the overall supply chain can create greater synergy and opportunity to provide services or products that ensure value;
3. **Technology –** which was considered in the previous chapter and will have impact on work routines and processes;
4. **People –** because as society alters and the skill base and educational aspirations shift, there will be consequences for organisations.

1. All of these have appeared as influences in previous chapters. The key, of course, is that the organisation is able to anticipate how each one will impact on their ability to attain strategic objectives. Kanter *et al.* (1992)[2] believe that there are three influences for strategic change:environment – see Tichy (above);
2. life-cycle differences;
3. political changes that happen within the organisation.

The second refers to alterations that occur in the greater supply chain (or within the larger organisation) and which necessitate change to respond to increased expectations or demands. The third considers how the process of a shift in power may occur to reflect altered circumstances, such as when production is required to have a greater realisation of the feedback that the sales department have elicited. Political aspects are a part of the process that is the hardest to manage. The way that individuals or, to be more accurate, particular groups of people react to whatever changes are proposed will potentially create dynamics that may be hard to predict and certainly difficult to control. As subsequent sections describe, imposed change, regardless of the cause or rationale, will be less likely to be welcomed than that which has emerged from those who have a vested interest in its implementation.

9.6   How to manage change

There are two dominant theoretical perspectives on strategic change, which might be seen as diametrically opposite to one another. These, prescriptive and emergent, are based on assumptions about the processes by which strategic objectives are determined and implemented. So, for instance, a prescriptive approach to change is one that assumes it is a process by which deliberate outcomes can be decided on and a definite method be selected as the means by which they can be attained; it is referred to by some as being 'planned'. Such an approach would be most likely to be appropriate where there is a potential crisis that necessitates very rapid or immediate imposition of change(s) to deal with anticipated consequences that, without action, might prove extremely detrimental. Crucially, a prescriptive approach assumes that managers are able to believe that the future environment in which the organisation will operate is likely to be stable and predictable. Moreover, people are, with sufficient encouragement and convincing (usually by means of training and briefing), expected to actively support the proposed changes.

An emergent approach to change, however, is one that eschews the belief that the future is either stable or predictable. Because of the uncertainty and turbulence that typifies the contemporary environment in which organisations operate, it is proposed as a way of 'coping' by continual adaptation and experimentation. Models proposed as part of the emergent approach are

---

[2]This publication is the source of a model of emergent change that is examined later in the chapter, what is known as their 'ten commandments'.

consistent with a belief that engendering constant learning is an entirely appropriate way for people in the organisation to behave. As Lynch stresses, the choice of which approach is believed to be most appropriate will be influenced by the circumstances ('context') that exist at the time (2006, p.763).

9.6.1   *Prescriptive models of change – the influence of Kurt Lewin*

Probably the most widely recognised advocate of prescriptive approaches to change management is Kurt Lewin. Whilst his work was seminal in many areas such as social conflict, disadvantage leadership and group dynamics, his work on change management has allowed him to achieve a significant place in management theory. Like many other notable theorists, such as Dr Deming and Dr Juran who are credited with assisting post-Second-World-War Japanese industry to achieve pre-eminence in terms of quality and service, Lewin's experience during the conflict was hugely influential, particularly as a German-born Jew. His desire was that what the Nazi regime did through authoritarianism should never be allowed to occur again. Rather, he argued, democracy was important in every instance of human endeavour and this should include organisational settings.

Lewin recognised that organisations frequently engage in processes that are deliberately intended to cause change so as to produce particular outcomes: a phenomenon he termed 'planned change'. Achieving change, he believed, was through the willingness of people to cooperate in groups. Accordingly, his work on change management can be considered as consisting of four 'themes' which despite usually being thought of as being separate are, Burnes suggests, better thought of as a 'unified whole with each element supporting and reinforcing the others' (2004, p.270). These themes are field theory, group dynamics (see above), action research, and the three-step model.

9.6.2   *Force field analysis*

Understanding the influences that cause people to adopt certain behaviour was crucial according to Lewin. Most importantly, he argued, an individual's willingness to behave in a certain way was the product of the way in which the group operated. This influence or environment was what he called the 'field', a term he defined as being 'a totality of coexisting facts which are conceived of as mutually interdependent' (1946, p.240). So, he argued, if change was to occur, it was necessary to understand all of the forces that exist within the field. These forces may push a group in one way or another. As well as this, even though there may be forces acting on the group, they might oppose the change by resistance that, theorists of physics would purport, is another force. According to Burnes, Lewin believed that if it was possible to 'identify, plot and establish the potency of these forces' then it would be entirely feasible to 'understand why individuals, groups and organisations act as they do', and to determine how much force needs be applied to derive the particular consequence in terms of creating change.

So, similar to applying sufficient force to create an action, such as pushing over a heavy weight (like a bookcase), the force acting on the group has to be sufficiently strong as to overcome resistance. Once the force had been applied, and the outcomes found to be satisfactory (as successful as expected), the situation would return to a state of rest; Lewin suggested that it would be in 'quasi-stationary equilibrium'. The problem of considering organisations and people in this way, of course, is that there is an implicit assumption that the magnitude of such 'forces' that act on or exist within the group (as resistance) can be accurately defined (as would happen in experiments in physics).

9.6.3   Group dynamics and their influence on 'Action research' and the 'Three-step model' of change

Lewin's work on how groups interact and cooperate led to his identification of the importance of appreciating the way that groups can be altered, by understanding the particular dynamics that exist among the members. As Schein (1990) believed, this involves consideration of norms, roles, interactions and socialisation. It was such work that led to Lewin's proposal of the importance of what he called 'Action research' and the 'Three-step model'. The first emanated from work he was carrying out in the immediate aftermath of the war and was based on the belief that groups will change if they feel there is a need to do things differently: what he called 'felt need'. To assist in the process, he argued, people should be encouraged to reflect on and gain new insights about their environment and how objectives might be attained. In essence, 'action research' requires people to experiment with alternatives to explore their worlds. Given his belief in Gestalt psychology, the emphasis is on the group being able to consider the totality of all of their experiences in the given situation. By doing this, he asserted, they will learn about their behaviour and relative desires to achieve particular objectives. Accordingly, he believed, the end result should be that all of those involved are able to develop a shared consensus from their shared learning based on actions.

Lewin was concerned that change must be viewed as something that has intended permanence. If this is not an objective, then change may be something that is short-lived and the group will be tempted to return to previous ways of operating (and any potential gains lost). Thus, in order to create a sense of a permanent objective, Lewin devised his 'Three-step model' that suggests that organisational change occurs in three stages and is analogous to ice:

1. **Unfreezing**   in which the senior managers decide that the current way of working, structures and value system are no longer appropriate. Accordingly, strategic decisions are taken to make alterations that dismantle ('unfreeze') the organisation in readiness for the new way of working.
2. **Forming**   which after some consultation with people to garner potential solutions for change that matches future expectations, a predetermined configuration and culture is selected and implemented using a change-management programme.
3. **Freezing**   by which the preferred solution to change is embedded by the use of reinforcement through 'soft' methods such as team-building, and 'hard' techniques such as new structures and procedures intended to elicit intended action.

9.6.4   Kanter, Stein and Jick's three-stage model

As many have remarked about Lewin's model, it is extremely simplistic and takes no account of the people who will be affected. An alternative prescriptive model is that proposed by Kanter *et al.* (1992). Like Lewin's model, this is similar in that there are three 'forms' (which might be viewed as stages or phases). Importantly, these three forms involve three categories of people who will be crucial in facilitating the required changes. The three forms are:

●   changing the identity of the organisation;
●   coordination and transition issues as the organisation alters;
●   managing the political consequences.

The first form will be based on certain beliefs about why the organisation should change and the form that it should take. The speed at which this change takes place will be influenced by the imperatives that have been imposed by external events such as a change in markets or tastes that customers possess. The second phase is concerned with the management of the various phases through which the organisation will move on the journey towards the intended 'destination' (the intended outcome of the change programme). As described above, the fact that organisations are social entities will create the potential for alliances and dynamics based on beliefs among key individuals and groups as to the benefits that they may enjoy or consequences they may suffer as a result of the intended change. The three key categories of people that Kanter *et al.* believe are crucial to the implementation of change are:

- change strategists;
- change implementers;
- change recipients.

As should be apparent from the labels used, these people equate to the three forms. Accordingly, strategists are those who are responsible for the decision to implement the changes. They will usually be senior managers and will look to the next category, implementers, to carry out the intended changes. Such people, often referred to as 'change agents', will be given the task of ensuring that the strategic intentions (changes) are successfully implemented. As such, they will be expected to deal effectively with any resistance. Their mandate will allow them to threaten sanctions or punishment on those who are concerned about the effects and who are, possibly, actively resisting and encouraging others to do so. Recipients are those on whom the changes will be imposed and who will be expected to support the intended consequences. Typically, they will be informed through sessions intended to communicate to them the importance of success and the need to embrace new ways of working and thinking. Unwillingness to do so, they will be told, will jeopardise not only their future but possibly the organisation as a whole. Such arguments frequently adopt the moral argument that runs, 'you are either with us or against us'!

9.6.5   Other prescriptive models of change

Burnes makes the point that there has been much debate about how change management should take place and, in particular, the pace and direction that is believed to be required (2004, p.282). Because of the inherent uncertainties that exist in the environment, change has become a phenomenon that every organisation will have to cope with at some time. The main issue is one of how frequent and to what degree does the change need to be. The three models that exist are:

- incremental;
- punctuated equilibrium of transformation;
- continuous transformation.

The first model is based on the belief that change occurs in short bursts of activity to deal with particular problems or issues in certain parts of the organisation. The debate among commentators is how frequent these 'bursts' need to be and how intense the intended change is. The second model, punctuated equilibrium of transformation, is based on the belief that organisations have a tendency towards stability (equilibrium) that will be punctuated by temporary periods of change (which may be rapid and revolutionary). Unlike the incremental model, this suggests that once change has been successfully implemented, the organisation will seek to return to stability. The final model, continuous transformation, is one that assumes change is something that is a constant phenomenon of the contemporary world. Therefore, organisations should be prepared to constantly adapt (transform) to the prevailing influences.

9.7   A change for the better? Challenging the assumptions of the planned approach

The belief that change can be a simple and linear process is one that came under scrutiny in the period from the late 1970s onwards. Indeed, the assumption that change has various stages or phases that can be effectively managed and the outcomes predicted was challenged by those who believed that such assumptions were no longer appropriate in a world undergoing rapid transformation and challenges. Experience of increased competitiveness from the Far East (Japan) was, according to critics of the planned approach, evidence that alternative perspectives were required. Burnes contends that those who aligned themselves with the argument for 'culture-excellence' were the most 'vociferous' in their condemnation of organisations that were not only bureaucratic, inflexible and lacking in ways to innovate, but were unable to respond adequately to the ability of those who showed that they could give customers what they wanted (2004, p.290). Rather, it was argued, organisations must alter their processes and their members willingly look for new ways to carry them out so as to provide excellent products and service. Planned change was, according to critics, certainly not the way that organisations should attempt to achieve alternative methods of achieving outcomes. An alternative approach was necessary, one that eschewed the belief that managers, who had a deliberate intention about what the outcomes should be, can impose change on people. Such an approach, one that relies on greater involvement by organisational members, has been proposed and is known as being 'emergent'.

An emergent approach is so-called because, as Burnes explains, it advocates seeing 'change as emerging from the day-to-day actions and decisions of members of the organisation' (2004, p.291). Weick explains it as follows:

> Emergent change consists of ongoing accommodations, adaptations, and alternations that produce fundamental change without *a priori* intentions to do so [my italics]. Emergent change occurs when people reaccomplish routines and when they deal with contingencies, breakdowns, and opportunities in everyday work. (2000, p.237)

Furthermore, it is added, such change is usually seen as being insignificant because 'small alterations are lumped together as noise in otherwise inertia' (Weick, 2000, p.237). Change, therefore, is one that is based on processes that are much less predictable and straightforward than the rational pre-planned approaches would suggest. It is argued that change tends to be a process that does not always end in the anticipated consequences; rather, it is an emergent approach in which, essentially, what occurs will be for the best. Therefore, the process and what goes on are given primacy; hence, it is widely associated with those who are known as processualists, the most widely recognised of whom is Andrew Pettigrew whose belief is explained by Burnes as follows:

> For Pettigrew, change cuts across functions, spans hierarchical divisions, and has no neat starting or finishing point; instead it is a 'complex analytical, political, and cultural process of challenging and changing the core beliefs, structure and strategy of the firm' [he quotes Pettigrew, 1987, p.650] (Burnes, 2004, p.292)

Processualists are also aware that creating change in organisations involves recognition of the politics and power that exists and which, they stress, must be dealt with. In this sense, at least, there is resonance with planned approaches that acknowledge the fact that there are group dynamics in any social situation. Carnell (2003) believes that change is something that every manager must be involved in (certainly in terms of facilitation) and requires four competences in order to be successful:

1. decision making;
2. coalition building;
3. achieving action;
4. maintaining momentum and effort.

Accordingly, an emergent approach is used in order to address the constant changes that occur in the market and general environment. As Pettigrew and Whipp (1993b) advocate, the only way that managers and those they work with can effectively deal with the complexity and uncertainty of the contemporary environment is to ensure that the key objective is constant learning. Therefore, whilst the role of managers is certainly lesser than would be expected in a planned approach, they are encouraged to actively engage with people to assist them in understanding their importance in developing change that best achieves the overall strategy.

In attempting to create change, it may be necessary to avoid using methods that attempt to impose it on people. If the approach is to make people change, then it is likely to be resisted which, of course, makes success much harder to achieve. As suggested above, this may be an option if time is extremely limited, such as in a crisis, when the proposed changes are considered absolutely essential. Resistance, it may be argued, may potentially put the organisation at severe risk. However, not all change necessarily requires that the organisation has to react to potential crisis or the need to impose solutions quickly. Therefore, an alternative to the prescriptive approach may be used, one that affords the organisation the ability to encourage people to question what the end objectives are or indeed may be? Moreover, because the future is uncertain anyway, making definitive decisions on what the organisation should attempt to achieve frequently turns out to be inaccurate. Therefore, if change is to occur, it should be managed in such a way as to prepare people to constantly adapt. People should also be encouraged to question what they currently do and how **they** believe they might do things differently in order to attain the corporate objectives. Such an approach emphasises the need for people to become intimately involved in the change process and, ideally, to take ownership. This is the basis of what is known as an emergent approach to change.

Burnes makes the point that whilst there are many proponents of the need to develop alternative approaches to using methods that might be regarded as being planned, there is no precise agreement about what any model should involve (2004, p.298). Given the uncertainty and 'fuzziness' that are implicit in an emergent approach, this is probably not surprising. However, whilst there is no precise definition about what an emergent model should contain, there is, Burnes believes, broad agreement about the existence of 'five features of organisations that either promote or obstruct success' (2004, p.298). These features are:

1. organisational structures;
2. organisational culture;
3. organisational learning;
4. managerial behaviour;
5. power and politics.

9.7.1   Organisational structures

In dealing with issues of power and relationships that will need to be addressed, it is believed that any effort to create change must acknowledge the structures that currently exist. But as advocates of emergent change argued, the types of structure that usually existed and which tended to be hierarchical were not appropriate to foster the sense of creativity and dynamism that they stressed was needed to deal with the increased levels of competition and innovation. Rather, organisations should become more able to inculcate improvisation and constant communication to facilitate improvement in customer services and products (Brown and Eisenhardt, 1997). According to Burnes:

A common aspect of these new structures is the move to create customer-centred organisations with structures that reflect, and are responsive to, different markets rather than different functions. (2004, p.299)

Importantly, he adds, the emphasis is on ensuring 'effective horizontal processes' in which the belief is that 'everyone is someone else's customer' (2004, p.299) – a crucial aspect of the implementation of TQM. Therefore, as a consequence, organisations are advised that structures should be flattened in order that the connection between management taking strategic decisions and those carrying out day-to-day processes should be closer. Allied to this, those lower down the organisation who traditionally might have been excluded from decision-making should be included and given greater autonomy over their work. Kotter makes the very persuasive argument that when an organisation does this and encourages more delegation, '[it] is in a far superior position to manoeuvre than one with a big, change-resistant lump in the middle' (1996, p.169).

9.7.2   Organisational culture

This concept was dealt with in Chapter 7. Given the importance that people play in any change process, it is regarded as being critical that everything that can be done to facilitate a change in attitudes and behaviour should indeed be tried. However, achieving cultural change is far from easy. Many organisations that have attempted to align the extant culture to a new belief system have found that the reality of ensuring success is that it is extremely difficult. As Wilson stresses:

> ... [achieving] change in an organization simply by attempting to change its culture assumes an unwarranted linear connection between something called organizational culture and performance. Not only is this concept of organizational culture multi-faceted, it is not always clear precisely how culture and change are related, if at all, and if so, in which direction. (1992, p.91)

Nonetheless, as has previously been explained, if the change is seen to be important and worth pursuing, all those involved must be willing to actively engage in altering their own practices and behaviour. Indeed, what they do must become the norm, not simply something they do because it is expected. This is where the new culture becomes embedded. Leaders and so-called 'change agents' (see below) become a crucial element in helping others to consider what they do and accept that alternatives may exist and, of course, potentially result in improvement. Allied to this, they encourage people to engage in organisational learning which is considered in the next section.

9.7.3   Organisational learning

Getting people to question what they currently do and to seek alternative ways of carrying out their day-to-day tasks are entirely consistent with the principle of a learning organisation (see also Chapter 7). Fundamentally, it is based on the ability of people to search for aspects of current practice that they believe are capable of improvement. What stemmed from the analysis of those organisations that have been able to achieve excellence in terms of their products and/or services (mainly Japanese) was the fact that they encouraged their people to become active learners. Importantly, people must be aware of their ability to create opportunities for improvement in every aspect of what goes on and that this can have a major influence in terms of strategic position. This will mean that they should be aware of all aspects of the organisation's performance and financial position.

As Pettigrew and Whipp contend, organisational learning (which they describe as 'collective') should be done in such a way as to ensure that people are given every encouragement to look beyond what they do simply to achieve day-to-day tasks but to consider longer-term aspirations (1993a, p.18). As advocates of learning assert, learning is a two-way process in that individuals benefit by being able to develop themselves and the organisation progresses by being able to employ dedicated people who wish to be associated with success. As Clarke (1994) and Nadler (1993) both stress, this is a major 'pay-off' for learning organisations; people who engage in self-development and increase their confidence will be more willing to take risks in the search for opportunities that may lead to innovation and, therefore, competitive advantage (as Japanese manufacturers have so amply demonstrated).

9.7.4   *Managerial behaviour*

All of the preceding sections have emphasised the belief that managers are crucial to creating the conditions for change. Implicit in this, though, is that managers are able and willing to reflect on themselves and be prepared to alter their own behaviour. Instead of being in control, managers should be willing to support and encourage others. As Burnes puts it:

> Instead of directing change from the top, managers are expected to operate as facilitators and coaches who, through their ability to span hierarchical, functional and organisational boundaries, can bring together and empower teams and groups to identify the need for, and achieve, change (2004, p.302)

Changes in managerial behaviour will necessitate their willingness to accept that their role may be scrutinised and subject to constructive criticism. If they expect others to be willing to agree that accepted ways of doing things must alter and, as a consequence, deal with uncertainty and ambiguity, then they too must do so. Additionally, they must be willing to engage in 'open and free' communications in which everyone is able to ask questions, seek information and solve problems. This will develop their skills and ability to think strategically:

> … by encouraging a collective pooling of knowledge and information […] a better understanding of the pressures and possibilities of change can be achieved, which should enable managers to improve the quality of strategic decisions (Burnes, 2004, p.303)

Another skill which managers will need to develop, as proponents of emergent change advise, is sensitivity in dealing with resistance that may occur amongst people who believe that their ways of working and processes are going to alter in a way that affects them negatively. Whereas in the past, managers may have dealt with resistance by the threat of punishment and sanctions to 'bring people into line', such an approach is inappropriate with emergent change, its advocates assert. Rather, they suggest, managers should allow people to express their concerns in a climate that most certainly does not cause them to fear retribution.

9.7.5 Power and politics

Dealing with the various groups and key individuals who can wield influence is a part of implementing organisational change. So, unless those who seek to engender change are willing to recognise and confront the power and politics that inevitably exist in any organisation, the likelihood of overall success will be reduced. Accordingly, eliciting support and developing coalitions from key players and groups with influence are an important part of the process of emergent change. Senior (2002) who developed the work of Nadler (1988) proposes four steps that those involved in organisational change should take in order to effectively manage 'political dynamics':

1. Ensure or develop the support of key power groups;
2. Use leader behaviour to generate support for the change;
3. Use language and rituals/symbols that demonstrate the level of support the change has at every level – most especially amongst senior managers who will tend to have most influence;
4. Build in stability by using power to ensure that some aspects remain the same (especially those things that people believe are important).

9.8 Emergent models of change

9.8.1 Pettigrew and Whipp's Five Factors Model

The model proposed by Pettigrew and Whipp (1991) was developed on the basis of their research into four sectors of industry in the UK. Importantly, their research suggested that change was certainly not like the models proposed by those who advocate prescriptive models in that there are distinct phases and a predetermined outcome. Rather, they believed, having carried out research into organisations that have successfully achieved change, it is a messy and uncertain process in which there is a need for people to constantly attempt to do things, observe the effects and, should they work, try the same thing again. If, however, the outcomes are not as intended, there may be some very valuable learning which will create better informed initiatives in the future. In short, people are encouraged to experiment in an environment that allows mistakes and most certainly does not castigate those who are willing to try something different. This model, described below, proposes that there are five factors involved in change that are all interlinked. These factors are:

- coherence;
- leading change;
- linking strategic and operational change;
- human resources as assets and liabilities;
- environmental assessment.

Coherence

The first factor attempts to find ways to ensure that all of the other factors are combined in a way so that they operate in harmony with each other. As such, the change process is being carried out in a way that recognises the complexity of doing many things simultaneously and in a coordinated fashion. There are particular mechanisms that can be used to provide ways to underpin the process:

- consistency;
- consonance;
- competitive advantage;
- feasibility.

Consistency is about trying to coordinate efforts and to ensure that conflict is avoided.

Consonance considers how the process is carried out so as to be appropriate to the environmental conditions that exist.

Competitive advantage is the key objective of coherence; that is why the change process is being implemented.

Feasibility considers what all of the options are and how the most appropriate and potentially effective one is selected. Clearly, those options that are least likely to be feasible will be rejected.

Leading change

The ability to provide effective and inspirational leadership that elicits support from followers (those in the parts of the organisation where change is most desired) is important. Much will depend on what is happening, where it is occurring and why. Leaders will frequently emerge to suit the particular period in which the change is happening. They may be senior managers but, as the emergent approach fully acknowledges (and encourages), leaders may equally be located at lower levels including at operations.

Linking strategic and operational change

This may seem obvious, but it is important to ensure that whatever desires are the basis of strategy can realistically be translated into meaningful action. If not, the process is academic. Clearly should the strategy changes have emerged from those who work at operational level, they will be far more inclined to ensure that success occurs than if the decisions had been taken remotely and without reference to those likely to be affected.

Human resources as assets and liabilities

Chapters 6 and 7 stressed the absolute importance of people as a resource that the organisation utilises. Being able to encourage people to alter their attitudes and behaviour is a vital aspect of change. Improving their skills and ability to cope with the future will increase their worth and, as a corollary, the potential worth of the organisation.

Environmental assessment

As Chapter 4 explained, this should be an ongoing and essential part of 'strategic orientation'. The ability to use tools and techniques to analyse change and sense potential shifts in markets, consumer behaviour or other influences that will have an impact on the organisation is essential.

9.8.2   *Carnell's model of political skills*

Carnell (2003) developed a model that suggests that there are three types of skill that a manager should be willing to acknowledge in achieving the change process:

- The ability to use resources;
- The aptitude to understand and manage processes;
- The capacity to engage in various activities that will involve 'politics' and require careful and considered discussion/persuasion.

9.8.3   Kanter, Stein and Jick's ten commandments for managing change

Kanter *et al.*, in their book, *The Challenge of Organizational Change*, propose a ten-step approach ('commandments') which should be used to manage organisational change (1992, pp.382–3):

1. Analyse what goes on and what needs to be altered;
2. Create a vision so that people believe in a sense of purpose that is shared;
3. Make a break with the old ways;
4. Create a sense that change is urgently needed – it cannot wait;
5. Provide support for those who lead;
6. Put in place the resources required to demonstrate support;
7. Develop a plan that will implement appropriate action;
8. Ensure that the organisational structures are 'enabling';
9. Communicate with everyone and with honesty;
10. Continually reinforce the message that change is necessary and celebrate the successes.

   In their analysis of change, Kanter *et al.* suggest that there are two ways of attempting to achieve the intended outcomes: 'bold strokes' and 'long marches'. The first is usually tried in order to implement a change that has immediacy in terms of impact, such as the belief that new technology is needed to alter production so that it is more efficient. As such, it may create changes in attitudes but not in the short term and possibly in ways that cannot be anticipated (or effectively managed). Whilst it perhaps has relevance as a planned approach at the outset, the longer-term effects may be viewed as being emergent. On the other hand, 'long marches', which is based on a desire to try smaller-scale initiatives on a semi-regular basis (or very regularly if the environment is suitable), would certainly assume that people are consulted and intimately involved in planning and implementation of the initiatives. Indeed, whilst senior management would be most likely to make major decisions that lead to the 'bold strokes', they should create a climate in which people at operational or departmental level see it as part of their normal routine to suggest the small-scale changes.

9.8.4   Kotter's Eight-Step Model

John Kotter (1996) believed that too many organisations that attempt to change fail in their objective because of not having prepared properly in a number of key areas: he believed there were eight. The result is, of course, disappointment that the outcomes are not as intended, costs are greater than anticipated, and time and effort are wasted. Therefore, he advised, there were certain things that should be done in order to increase the chances that the change might be more successful:

1. establish a sense of urgency;
2. create a guiding coalition;
3. develop a vision and strategy;

4. communicate the change vision;
5. empower broad-based action;
6. generate short-term wins;
7. consolidate gains and produce more change;
8. anchor new approaches in the culture.

At first glance, this list has remarkable resemblance to that provided by Kanter *et al.*'s 'ten commandments'. This is unsurprising as there was broad agreement about the way to achieve change. However, the major difference was emphasis in implementation. Kotter stressed the need to see all eight steps as a process that requires each to be implemented in a way that allows sufficient time and effort (and resources) so that the next one has a sound basis on which progress can be made.

9.8.5   *Balogun and Hope Bailey's contextual framework*

In contemplating strategic change, there are, according to Balogun and Hope Bailey, many contextual factors that may have an impact on the success of the change being initiated (2004). These are:

1. **Time**   How quickly is the change required?
2. **Scope**   How much change is believed to be needed?
3. **Preservation**   What parts of the organisation (or culture) should remain intact?
4. **Diversity**   How consistent (or not) are people?
5. **Capability**   Are the managers able to carry out what is required or will training be needed?
6. **Capacity**   Does the organisation have sufficient resources?
7. **Readiness**   Are people prepared for the change that is needed?
8. **Power**   How much influence does the leader(s) possess to influence people?

The magnitude and effects of each of these factors will, as the title of Balogun and Hope Bailey's framework indicates, be dependent on the context in which they occur: the organisation. Therefore, it is incumbent on those involved in the change initiative to appreciate the exact existence and impact of these factors. As would be expected, the advice given to those involved is to think carefully about where the organisation starts from, in terms of context and, of course, culture, and how much can be done in consideration of the particular factors that exist.

9.9   The role of managers in change

The importance of managers in either initiating or leading change is clear from the explanations of the models above. Whilst the process may be 'bottom-up', it is probable to assume that managers will play their part in facilitation and resource distribution/allocation[3]. As such, there are a number of styles that may be adopted by managers to produce change. The most useful analysis of these styles has been produced by Dunphy and Stace (1993) in which they suggest there are five that range from those which are emergent to those that are resonant with planned change:

1. education and communication;
2. collaboration;

---

[3]Even the example of the Semco organisation, in which Ricardo Semler (1993) claims that there are no managers, still requires those who oversee strategic processes.

3. intervention;
4. direction;
5. coercion/edict.

9.9.1 Education and communication

This will require people to be informed in groups as to what is going on and should overcome fear and assist in dealing with rumours. The problem is that the process is very time-consuming and may allow others to pursue their own 'agendas'. This is best suited to incremental change where the 'timeline' is sufficiently long.

9.9.2 Collaboration

This style is based on the desire to achieve consensus through every person being involved in collective decision-making (although use of groups and networks is useful). Increasing a sense of ownership and giving people a belief that they have control over the process are extremely valuable. Like education and communication, this style can be potentially time-consuming and is best suited to incremental change which is taking place over a long 'timeline'.

9.9.3 Intervention

This is based on the desire to engender change through key individuals who, because of their credibility (and authority), will be able to assist others by guidance in what is needed and how to achieve desired outcomes. As such, those being encouraged to change may feel that there is support for their efforts. However, others may feel that they are being manipulated to do things that are simply going to benefit their managers. This style is best suited where there is a need to create change which may be incremental but could also be seen to be transformational, but not on the basis of a perceived emergency or crisis.

9.9.4 Direction

This style of change is based on management being willing to engage with people in a way that uses formal imposition of authority to tell them exactly what is needed, when and by what means. Its main virtue is that it will not be as time-consuming as the previous styles but, of course, it risks causing alienation and resistance. It is most suited to circumstances where the objective is to engender transformational change quickly, for example, where there is a need for the organisation to respond to dramatic alteration in markets or shifts in technology.

9.9.5 Coercion/edict

There can be little doubt as to how this style is implemented. There is absolute imposition of the change and those responsible will have no reticence about using their power to ensure compliance. This is because the change is likely to have been caused by a crisis that needs immediate and radical action. As such, there will be little doubt by people about the nature of the change required. Nonetheless, there may potentially be a great deal of stress and emotional trauma for those affected, especially if there are major consequences in terms of job losses or changes in work patterns and traditions.

Importantly, sensitivity and willingness to deal with the circumstances that exist (context again) will be crucial. Therefore, it is necessary that managers consider the willingness of people to engage with the change initiative. Those that are most willing require least coercion and, in all likelihood, simply require support and guidance through education and collaboration. The personalities of those involved – both managers and people – will influence style. There is also a view that managers may be well advised to adapt their styles as the process develops and people become more confident and, it is anticipated, more supportive, with styles ranging from coercion at the outset to participative at the conclusion.

9.10   Strategic leadership approaches

Farkas and Wetlaufer (1996) have analysed approaches that leaders adopt during change which suggest that particular outcomes need them to appreciate the nature of the task (and objectives). In turn, this requires them to alter their behaviour and adapt their approach. They contend that there are a number of key areas that leaders must address when implementing change:

- strategy;
- human assets;
- expertise;
- control.

They explain that there is a range of approaches that can be adopted and that these may be influenced by position within the organisational structure that, in turn, may be affected by position and power. The particular style that is adopted (see Dunphy and Stace above) is likely to be highly influenced by the approach of the senior managers (leaders). Accordingly, each area requires a leader's focus which alters in terms of their skill base (expertise). Importantly, this recognises that whilst senior managers may have different expectations and goals from middle managers and that they too may have differences from operational managers, all can be engaged in change (and be doing so in coincidence). Moreover, an organisation should develop what are known as **change agents** at every level. These are the individuals who will actively support and 'champion' the importance of the need for change amongst colleagues and peers. As such, they will not be seeking to impose but to persuade. Those doing so at operational level may be crucial in dealing effectively with potential resistance.

Based on research carried out by Floyd and Wooldridge (1996), Johnson *et al.* stress the importance of middle managers in providing the ability of an organisation to engage in change, especially in attempting to reduce the impediment of 'hierarchy' and where it is large and/or complex (2005, p.521). Crucially, they assert, middle managers act as a bridge between those above (strategists) and those below (supervisors and operatives). Additionally, they believe that there are a number of roles that middle managers may fulfil (2005, p.521):

1. The 'systematic role of *implementation and control*' [their italics].
2. To be 'translators' in which they understand what those above them wish to achieve strategically by implementing and engendering change, but to know how it will be best received and supported by those who will be affected.
3. As change unfolds and develops, that they should be willing to continually 'reinterpret' and 'adjust' in order to ensure that people who will be expected to be willing to support the changes are fully informed and appreciate what is required.

4. That they provide a 'relevance bridge' which allows them to report back to senior managers what is actually occurring and, potentially, can get them to adapt the strategic-change management plans to be more appropriate. It is important to remember that change often fails because those who initiate it (senior managers) fail to acknowledge the practical issues or difficulties that people have in understanding why it is needed; forcing people to engage is probably the worst way to proceed (see below)!
5. In addition to 4., to be willing to act as 'advisors' as to what will really work.

9.11 Changing the way things are really done

In Chapter 7, corporate culture was described. What is important is that if you want to get people to change their attitudes and behaviours, it is necessary that **they** believe in the need to do things differently. If they don't, the only way to alter practices will be to force them, which is precisely not what culture change involves. Organisations are not unlike people; they develop particular ways of doing things that evolve over time and are based on beliefs that certain **routines** are 'the way things are done around here' (see Deal and Kennedy, 1982). The problem may be that established routines militate against the proposed changes. The challenge, therefore, is in getting people to consider the routines they carry out, why they do it in whatever way they do and, of course, how alteration might be in the best interests not only of the organisation, but also of them.

For example, many organisations have had to become more willing to be 'customer-focused'. This often requires changing working patterns so that customers can contact people within the organisation outside of 'normal' working hours. This might not be popular but if competitors are able to get their people to change their routines and you do not, there may be a potential loss of business.

On the basis of the five 'styles' of managing strategic change which Dunphy and Stace propose (see section 9.9 above), Johnson *et al.* believe it is crucial for managers to carefully consider how each of these applies, most especially with respect to the 'context' (2005, p.516). They draw attention to the fact that each style will have 'appropriateness' depending on the following factors (2005, p.517):

The state of 'readiness' of people will influence selection;

'Time and scope' in terms of the urgency of change and, therefore, whether it is necessary to attempt to be inclusive/participative or to impose a solution;

The hierarchy of relationships and the organisational structure – a flatter structure tending to be democratic and more likely to embrace collaboration (as opposed to one in which power is vested in senior managers who are remote).

Johnson *et al.* also draw attention to the need for managers to recognise and manage both 'power' and 'political processes' which, they argue, can be usefully combined to effect change (2005, p.531). As they maintain, 'manipulation of *organisational resources*' in a way that gains support from powerful groups together with adaptation of the 'organisational subsystems' can allow managers to achieve change in three ways:

1. Building a power base which requires key resources (influence) to be identified and garnered;
2. Overcoming resistance by which people may either be convinced or (should it be considered necessary, especially where there is urgency or crisis) managed;
3. Achieving compliance by the use of imposition.

Power and politics in organisations are nothing new; Florentine court advisor, Nicola Machiavelli, wrote about such matters in the sixteenth century and, as Johnson *et al.* believe, his observations and recommendations can still be considered relevant to contemporary managers contemplating change in their organisations.

9.12   Communicating change

Inherent in what has been described in preceding sections is the need for those propagating change to explain it to those who will be affected by its consequences, especially in terms of altering routines or culture. This requires communication. This is a word that is often used in a way that assumes it is straightforward and unproblematic. The reality is that communication in any respect is something that we can all work to improve. With respect to managing change, it is crucial, particularly if people are to be convinced of the reason why it is needed and why their cooperation and support are vital to success.

9.13   Potential problems with change

Harris and Ogbonna (2002) suggest that there are many things that can go wrong with change because of the failure of those responsible to appreciate the real difficulties that will occur. These include:

- making organisational change constant;
- erosion;
- reinvention;
- expecting too much too soon;
- inattention to symbols;
- not making intentions clear;
- forcing it through.

9.14   Conclusion

Tom Peters, in a video presentation he provides to accompany his book, *The Tom Peters Seminar: Crazy Times Call for Crazy Organizations* (1993), stresses his belief that the only certainty that managers can have is the constancy of change. Over fifteen years later this is still true, probably even more so now than it was then. Construction, like all other sectors, has been subject to constant exhortation to change what it does and how it does it. The recommendations contained in *Rethinking Construction* (1998), as well as the 'descendants' it has spawned, have spurred the industry through all of its constituent parts to examine traditional practices and, where improvement can be made, alter. For many organisations this has meant that there has been a need to engage in change. Moreover, many organisations have reflected on existing strategy and found it wanting. Accordingly, managers are required to consider alternative strategies and, once an optimal choice has been decided on, to implement it. Chapters 10 and 11 respectively explain how strategic options are considered and, having made a choice, what implementation involves (including dealing with difficulties that may present themselves).

Chapter 10
Considering the development of strategic options

'You will either step forward into growth or you will step back into safety'
Abraham Maslow

10.1 Objectives of this chapter

In considering strategy, it is necessary for an organisation to make some key decisions about what it wants to do (its mission) and how it intends to achieve it (tactics). This requires those involved to consider what the potential options are and, of course, to select the one most likely to produce success. Effectively, this means that there are two things to be done. Firstly, it is important to look at what existing and potential customers and consumers will be likely to want. Secondly, those responsible for resources (usually management) must consider the best way to utilise what currently exists (or develop alternatives). This will need some method to evaluate the choices that may exist (or be likely to present themselves).

10.2 What to do?

According to Wickham, the need to creatively generate ideas is crucial to the process by which strategy selection can take place (2000, p.201). This, of course, is important in ensuring that the corporate objectives are attained. However, in order to do this, someone will need to take the key decisions as to what needs to be done. As Wickham advises though, 'selection must take account of the [organisation's] resources and capabilities and the external competitive situation in which it operates' (2000, p.201).

Some of the ways that options can be used are considered below:

10.2.1 Considering options

Using the Ansoff matrix

- Igor Ansoff (1987) developed this as a method that attempts to consider the various options that might exist for any organisation given the particular condition that exists in the 'market' environment. In this matrix there are two axes:increase in technical innovation of the product or service;
- increase in the number of customers.

The matrix is created by considering each axis to be either existing or new. Accordingly, four options are presented:

1. market penetration (although there may be the possibility of withdrawal altogether);
2. product development;
3. market development;
4. diversification.

Market penetration

So, as described in a previous chapter, the organisation might consider one of the three generic strategies that Porter recommended (either to reduce cost or differentiate or focus on certain 'niches' that others cannot access). The objective is to attract additional custom either by attracting existing customers from competitors or by growing the market by encouraging people into the market. Therefore, it might be possible to 'steal' customers by making the product or service cheaper or better. The former may be 'doubly good' in that it takes away customers from competitors and also attracts new customers who are now able to afford to make the purchases.

As the market share grows larger, it becomes a greater challenge to continue to penetrate. There will be a point at which the market may become saturated and continued effort aimed at increasing penetration will be pointless (or certainly not very cost-effective). Organisations with a very high market share will need to consider how to retain existing customers, especially if there is a prospect of new competitors entering. An understanding of the dynamics of the market and appreciation of the behaviour of those who make the purchasing decision will influence the tactics that should be employed, that is, to reduce costs or improve quality to increase the likelihood of repeat business.

As suggested, an alternative strategy may be one that considers the possibility of withdrawing. In some situations this may be an entirely sensible thing to do, particularly if the costs of remaining in operation outweigh any benefits (such as the level of profit). Even public-sector organisations that would have been able to continue to cope with year-on-year deficit, because of considerations of the impact on communities, have been forced to withdraw services to reduce costs.

Product development

If the product (or service) can be significantly developed, then it is possible to implement a strategy that allows the organisation to secure its existing market share. This will be especially useful if there is threat from others who are attempting to increase their market share by, for example, lowering their costs. Being able to offer something that is innovative or revolutionary may be very attractive to customers. This scenario is particularly important in an environment that is characterised by constant technological innovation as has been the case with microelectronics and computers in the past ten years.

Market development

Being able to attract new customers for the existing product or service is an extremely useful way to develop strategy. This might be done by changing the advertising and marketing to attract custom from those who may have been unaware of what is on offer. Additionally, it may be a matter of using new networks or franchise arrangements to be able to gain access to these new consumers. Alternatively, simply looking beyond the traditional geographical region of operation may provide the new consumers.

Diversification

Movement into different markets on the basis of product and/or service innovation and development will provide benefit in terms of potential growth. This can be either by moving into related markets in which there will be some connection ('linkages') or by moving into entirely unrelated markets. In the former, there is the advantage of having existing knowledge on which the strategy can be built. Furthermore, moving into entirely new markets increases the risk of failure.

10.2.2   Going beyond Ansoff – the 'expansion matrix'

This is proposed by Lynch (2006, p.466) as a way of considering various methods by which an organisation might expand its operations in ways that include what was suggested in Ansoff's matrix but, significantly, goes beyond it. So, for instance, it suggests that an organisation in a domestic market develop internally and consider how it might expand its activities externally (by merger, acquisition, joint venture, alliances and franchises). Further, he believes that the choices, as to the methods by which to expand inside or outside the country of normal operation, are based on careful analysis of all aspects of the potential markets.

When international options are considered, many organisations now operate on a truly global basis. For major manufacturers this provides significant advantage in terms of being able to produce in very large numbers (advantage of economy of scale). Provided that there is sufficient homogeneity, this strategy works. However, such organisations frequently discover that an ability to respond to local tastes, customs, and regulations or laws is a vital part of ensuring 'fit' to local markets. A key to success is local knowledge. This requires the willingness to use methods that incorporate the expertise of those who possess insights and understanding. Even so, there will be significant risks involved, most especially if the countries in which this expansion is attempted are considered to have risks such as unstable economic or political systems.

The pros and cons of expansion

In the previous section there is an implicit assumption that expansion is a beneficial thing. Undoubtedly, increasing the size of the market by growing the market is going to produce advantage. However, it will also bring problems. There will be a need for investment from sources either within or outside of the organisation. Capital does not come free; there is a cost in terms of the opportunity that has been lost by using it for investment. At the very least the owners will want a return (dividend) to make up for the potential loss they might otherwise have enjoyed (interest). Additionally, they may want a higher rate of return on their investment to allow for any risks they believe are possible in this venture.

10.2.3   To cut, differ or get into a niche – using Porter

Porter (1985) argued that there are only three strategies that can be pursued by any organisation in order to achieve competitive advantage (see Chapter 4). These are:

1. cost leadership;
2. differentiation;
3. focus.

The first of these is the desire to create advantage by being the cheapest. The organisation will attempt to reduce costs by whatever methods are available: producing more efficiently by introducing different methods or technology; lowering wage costs (by either paying less or moving production); searching for radical innovations such as standardisation; or rationalising the supply chain. The second strategy is to ensure that products and/or services are perceived to be significantly different from those offered by competitors. If the organisation can achieve this, it may be possible to develop a reputation that means a higher price can be charged (which may be justified if there is new technology of innovative methods involved). Differentiation is not the same as focus which is about developing a strategy that is based on the organisation wanting to target its efforts on specific markets or groups of customers (frequently called 'niches'). In addition to these three strategies, Porter offered the advice that an organisation should avoid trying to achieve a compromise that means they become 'stuck in the middle'. Some commentators assert that, whatever Porter may believe, some markets are so dynamic or difficult to predict that this is possibly the best way of surviving (see Kay, 1993).

10.3   The importance of resources in making choice

Using resources effectively can, according to Hamel and Prahalad, be a very effective way to achieve both 'stretch' and 'leverage' which, they believe, will assist an organisation in attempting to gain advantage (1994). There are ten ways by which this may be achieved:

1. Convergence – which is the gap between what the organisation wants to achieve and the capability that exists within the resource base. There is a belief that the gap should be 'right' (too little complacency is a danger, too much disillusionment will become apparent in those who think that they are expected to achieve the impossible).
2. Focus – where managers ensure that every effort is made to use resources in a way that delivers real advantage.
3. Extraction – by which every person uses opportunities to explore the world in which the organisation operates and 'extracts' knowledge that can be used to competitive advantage. The encouragement of a culture of learning is a vital aspect of this concept.
4. Borrowing – in which knowledge and additional resources can be gleaned from others outside the organisation. Benchmarking can be a way of achieving this and will certainly provide information of where potential deficiencies or opportunities might exist.
5. Blending – having a selection of resources from which to choose is valuable but combining them ('blending') in the most appropriate configuration is what will create the basis of success. This requires groups to come together to discuss how they believe such combinations can be made and linkages made which will enhance the value of each resource.
6. Balancing – this follows the logic of blending in that every part of the organisation works together in an effort to create value (see the value chain [below]). Balance is dedicated to making sure that potential weakness, particularly the lack of connection between parts of the organisation, is identified early and, of course, remedied.
7. Recycling – which considers how knowledge and expertise that may exist in different parts of the organisation (or subsidiaries) are identified and applied to create similar advantage. The use of internal benchmarking would assist in doing this.
8. Co-option – like borrowing, this concept attempts to exploit resources that can be garnered from outside the organisation from those with whom the organisation has a regular relationship (suppliers, subcontractors, those who provide professional services). Clearly, there should be mutual advantage and a willingness to engage in reciprocity.
9. Shielding – by which the benefits enjoyed from the use of resources are protected by whatever means can be employed: patents, branding or the unique way in which processes are implemented. The harder it is for competitors to imitate, the more effective the shield will become.
10. Recovery – by which resources are used and reused as many times as possible. Expertise and knowledge gained in one area should be reapplied in a way that attempts to ensure efficiency gains are enjoyed elsewhere.

10.4   Using the value chain to consider resources

In considering which strategy is likely to be most appropriate, it is essential to remember that resources (considered in Chapter 6) are key to decisions. As was explained, resources and how they are used are the basis of developing 'value' which, it is hoped, will provide competitive advantage. Importantly, there is a distinction that can be made between activities that are either 'upstream' or 'downstream'. Upstream activities are those that can include aspects of the overall process such as procurement of important materials, design and actual production. Downstream activities are those that will allow the product to be differentiated through advertising, branding and reputation based on service and excellence.

Lynch provides a list of resource options that tend to be 'associated' with upstream and downstream activities (2006, p.475):

Upstream
- The use of standardisation;
- Increased technology and investment to lower the existing costs of production;
- Innovations and developments in both the actual product and allied service that ensure superior performance is enjoyed by customers;
- Attempting to sell or provide a consistent product and/or service to as wide a range of potential consumers as possible.

Downstream
- Variation to suit segments that have been identified;
- Research to develop the product and service;
- Constant innovation;
- The search for value-added is constant and never-ending.

10.5   Andrews and SWOT

Undoubtedly, SWOT remains the most recognised tool of analysis that organisations use in order to identify how they will achieve intended outcomes. The connection between resources and activities and attainment of the corporate mission was something that was considered by Andrews (1971). Accordingly, it is the S (strengths) and W (weaknesses) that will be considered most directly with respect to resources, although they provide the basis of dealing with the O (opportunities) and T (threats).

The important thing to remember about SWOT is that whilst its advantage is simplicity, it can be used to explore almost every aspect of what the organisation does and how it does it. Importantly though, much can be generated in the process of consideration of the issues; ensuring consistency and integration is vital.

10.6   The use of a resource-based view

Using resources in a way that ensures differentiation from competitors is extremely beneficial. Most especially, according to Kay (1993), they can be tested for what is referred to as 'distinctiveness' from others by considering them against the following concepts:

- architecture;
- reputation;
- innovation.

The first is the network of relationships that exist within and outside the organisation that allows the best possible combination of resources (see Hamel and Prahalad, 1994). The second refers to the way that the organisation is perceived by its customers: the higher the reputation, the better. Reputation is something that is very hard to develop (and requires constant dedication to resources) but is easily lost by careless decisions or inappropriate application of resources. Finally, innovation (see Chapter 8) is something that the organisation should be willing to constantly develop and encourage every person to be willing to dedicate effort towards achieving improvement in the tasks they routinely carry out and the processes they use.

Organisations that can combine resources in a way that enhances all of these concepts will be in a much stronger position than those that do not. As Japanese manufacturers of cars and micro-electronics have demonstrated, using these concepts and the strategy to implement improvement can provide the basis of sustainable competitive advantage.

10.7   The importance of core competences

Core competences, a concept developed by Hamel and Prahalad (1994), are the attributes that an organisation possesses which can be used to provide the ability to deliver particular product(s) and/ or service(s). In effect, they will be based on both the technology that exists to carry out 'production' and the skill and experience of the people employed. Accordingly, some organisations can develop a set of competences that provide its very essence in terms of what it does and how well. Grant states that the objective is 'capabilities' that are relative to other firms (2002, p.145).

Competitive advantage assumes that the organisation will wish to focus its effort on those things that it does well and will seek to exploit those competences that give greatest advantage of all over potential rivals. As Lynch states, 'core competences form the basis of core products [and services] which, in turn, form the basis of the business area of the company' (2006, p.25). Clearly, therefore, in terms of analysing the potential options for strategy, all aspects relating to the competences of the organisation should be thoroughly explored. Like SWOT, whilst this may seem straightforward, this can be quite a complicated task. In particular, it needs a systematic analysis that considers what happens in each of the main areas of activity (functions) and through-out all of the interrelated aspects that flow across the phases of 'production' (this can use the value chain).

According to Grant, the functional aspects of an organisation are those that relate to the principal areas such as (2002, p.147):

Corporate
Including all aspects of financial management and control, strategic management, innovation (strategic), acquisition management.

Management information
Which considers all aspects of the systems that are intended to provide management with information and data by which effective decisions can be made.

Research and development
Of all aspects of the organisation in order to create improvement.

'Production'[1]
Considering ways to create efficiency and to implement continuous improvement in all processes.

Product [service] design
To develop capability and respond to perceived customer needs.

---

[1] Grant uses the word 'manufacturing'.

Marketing
Considering brand management and promotion and analysing the environment for trends in consumer behaviour and taste.
Sales and distribution
To ensure effective sales and processing of orders to ensure that expectations are met, implementation of systems to produce quality (excellence) in product and service, inclusion of after-sales service.

Grant describes the other component of capabilities (value-chain analysis) as including all the activities that can be considered as a 'sequential chain' that starts with 'R and D', includes development and continues through production to ensure that the customer receives what they expect (2002). It is divided into 'primary' and 'support' activities. Whilst Grant provides a 'hierarchy of capabilities', Lynch develops it into an extremely useful model of 'hierarchy of competences' (see below). This hierarchy, he explains, can be seen as a way of building on 'low-level individual skills' through group-task skills through to 'higher-level combined knowledge and skills' (2006, p.477). These are (in the order they appear in the notional hierarchy):

Single task which is based on the person's knowledge, skills and understanding of how to carry out their day-to-day tasks.
Group task which is based on the ability of groups/teams/networks to interact and cooperate successfully.
Middle management which is based on the ability of managers to solve complex problems using their experience.

As a consequence of these, Lynch suggests that by considering competences in this way the organisation can concentrate its efforts on 'integration' that will allow analysis of specialised competences in the following:

- function-related;
- cross-functional.

The former is based on a need to develop the knowledge and skills of all experienced people in every functional department of the organisation. The latter is based on the ability of middle and senior management to cooperate and coordinate the organisation towards its strategic objectives.

Additionally, Lynch provides a ten-point checklist for using resources in the consideration of options that might exist (2006, p.477); see below. Crucially, such a list should be thought of as merely the starting point for deeper analysis of what is unique about the organisation. The ten points are:

1. Technology – what is special?
2. Links between products and/or services and operations.
3. How is 'value-added' created and what is the difference from competitors?
4. Consideration of people skills. As Lynch explains, there are many aspects to be considered such as: how are their skills special, what do they contribute, and how unique are they (difficult to replace)?
5. The financial resources that allow the organisation to fulfil its vision through the ability to invest in future development.
6. Benefit to customers by the use of comparison of issues such as quality, performance and value for money.
7. Other skills that allow the organisation to differentiate products and services. These might be peculiar to the context in which the organisation operates and will be obvious if success is being enjoyed.

8. Innovative resources, skills and competences, especially those that will be developed in the medium to long term.
9. Consideration of how the environment may change and the impact that this will have on capabilities (resources and skills).
10. Comparison with competitors. What are they doing or appear likely to do which may allow them to enjoy a significant advantage?

10.8   Using the six criteria to judge strategy

According to many commentators on strategic management, in considering how evaluation of options might be carried out, there is much to recommend a systematic approach based on logic and facts. Such an approach will be entirely rational and assume that success that has occurred in the past can be replicated in the future. So, therefore, it is merely a matter of extrapolation of what is already known. In carrying out analysis, it is useful to develop a checklist of criteria that can provide the basis for carrying out the process systematically. As the consideration of other methods of analysing options should have demonstrated, there is no limit on the number of issues that might be considered; management in every organisation will be in the best position to judge what **they** consider to be the factors that relate most appropriately to its circumstances and expectations. Having reviewed the work of others (Day, 1987; Tiles, 1963; Rumelt, 1980; De Wit and Meyes, 1994 and Quinn, 1991), Lynch asserts that there are six basic criteria that may be used to evaluate options for strategy (2006, p.493):

1. consistency;
2. suitability;
3. validity;
4. feasibility;
5. risk;
6. attractiveness to stakeholders.

10.8.1   Consistency

This may seem obvious: do what the corporate mission says consistently. However, over a period of time there may be intended or unintended deviation. Therefore, in any consideration of options, it is necessary to realign either the mission (and any associated goals or objectives) or the actual practices used to produce output. Reconsideration of options will be important to ensure long-term 'fit', most especially with respect to the aspirations for potential growth or strategies for diversification. Those organisations that attempt to achieve too many things that are inconsistent will discover that there will be tension and difficulty in ensuring that management effort and resources are not wasted in trying to ensure that intended outcomes actually occur. Indeed, organisations with diverse portfolios of activities that experience periods of financial difficulty frequently see a return to consistency (by divesting) as their best strategy.

10.8.2   Suitability

Following on from consistency, suitability considers the appropriateness of each option to the prevailing circumstances (the environment). Using SWOT will undoubtedly assist in providing useful knowledge in terms of what options are most likely to take advantage of opportunities that may be identified whilst, of course, avoiding threats. There is wide scope for consideration of almost anything that is likely to be of interest so long as the organisation has the capabilities that exist (or can be developed) to respond. Like consistency, whilst diversity has been seen as a way of spreading risk, it brings with it the danger of being unsuitable to the skills and expertise of the organisation. And like consistency, such a strategy creates the risk that managers may be stretched to deal with potential conflict between different parts of the organisation trying to achieve radically different goals.

10.8.3   Validity

As all writers on strategy should readily acknowledge, the development of strategic options is about providing the best guesses of what is likely to happen. There are implicit assumptions in considering options. However, this does not mean that such assumptions are invalid, merely that they should be scrutinised by subjecting them to whatever tests possible to investigate the soundness of judgement. In economic terms, assumptions may normally follow a pattern of growth, stability and decline. As real estate developers have discovered at the time of writing (late 2008), the assumption that there would be a continued growth in property values (or at least a steady decline that can be 'managed') has been found to be severely mistaken. The corporate world is replete with stories of those who subjected their assumptions to what they thought were tests that demonstrated validity but were found to be entirely wrong when applied. Whilst the use of economic modelling is a useful tool for considering scenarios, they can be inconsistent.

10.8.4   Feasibility

Lynch makes the point that whatever options are considered must be capable of actually being applied and that there are three areas that may cause potential difficulties (2006, p.494):

- 'Culture, skills and resources *internal* [his italics] to the organisation';
- Consideration of the external aspects of the organisation's environment;
- Support not being forthcoming from people in the organisation (most especially where there will be a major change in routines, practices or processes used).

   In addition, he draws attention to the fact that some options may cause technical difficulties that require investment in technology and/or training of staff to deal with the challenges they will face, as happened in the 1980s during the widespread introduction of IT into all organisations. In order to deal with consideration of internal feasibility, Lynch provides a checklist that has been adapted (2006, pp.494–5):

1. Costs – where are the funds going to come from?
2. What are the likely returns to the organisation (projected profit or gain in terms of efficiency or improvement in terms of levels of service)?
3. Potential changes that will be required and the organisational impact.
4. Consideration of technological innovation and change.
5. Product and/or service development required.
6. Investment in marketing and sales techniques.
7. Consideration of alliances and partnerships to achieve change.
8. 'Communication of ideas to all those involved: how will this be done?' As he stresses, commitment from all those involved is an essential consideration.

Externally there will be those who are affected by the options being considered. Accordingly, it will be necessary for the reaction of customers, competitors and suppliers to be considered during this process.

10.8.5   Risk

Unless the option(s) being considered is a certainty (which is extremely unlikely), there will be risk. As we all know from personal experience, many things we do have levels of risk which we have different ways of assessing. Organisations must do likewise and consider the levels of risk associated with the various options. Therefore, a systematic evaluation of risk should be carried out which provides both quantitative and qualitative assessment. In doing this, it can judge the likely outcomes of particular strategic options against one another. In competitive situations the level of risk is usually seen as an indicator of the potential returns. However, with risk there is the corollary of failure if the intended outcomes do not occur. Therefore, it is sensible to judge the level of contingency that should be provided to deal with the consequences of failure (which would at least lessen the impact). There are various financial techniques that can be employed to carry out the evaluation of benefits and risks, all of which will provide particular **scenarios** (see below) based on the inputted assumptions.

10.8.6   Attractiveness to stakeholders

In dealing with options, there will be the likelihood of alteration to practices and processes that will impact on all of those involved: the stakeholders (shareholders, customers, people including all managers and operatives, suppliers). In implementing any particular options, the garnering of support from as many stakeholders as possible (ideally all) will be extremely valuable in terms of increasing the chances of success.

10.9   The ADL Matrix

This is a model that seeks to consider the particular state that the market is in and explore how the organisation can best cope with the circumstances that exist. Its name comes from the person who developed it, Arthur D Little, who believed that there are two important aspects to be considered by an organisation:

- Competitive Position – which can range from dominant (clear leader) to those who are weak;
- Industry Maturity – which ranges from very young (embryonic) to old (ageing).

By comparing these two aspects, a matrix is constructed that suggests a variety of alternatives that can be considered. So, for example, if an organisation is dominant and is considering a new product or service, it will probably seek to market it very vigorously. This position would be in contrast to an organisation that may be dominant but is in markets that are ageing and declining. In this situation the organisation would be better to seek newer markets that have greater potential for growth. Accordingly, the ADL Matrix provides the basis for considering the range of combinations that exist and the implications of implementing particular strategies.

Table 10.1 The ADL Matrix © for evaluation of the life cycle of product or portfolio (developed by Arthur D. Little, 1996)

Competitive position \ Maturity	Embryonic	Growing	Mature	Ageing
Clear leader	Hold position Attempt to improve market penetration *Invest slightly faster than market dictates*	Hold position Invest to sustain growth rate (and pre-empt potential competitors)	**Hold position** Grow with industry Reinvest as necessary	**Hold position** Reinvest as necessary
Strong	**Attempt to improve market penetration** Invest as fast as market dictates	**Attempt to improve market penetration** Invest to increase growth rate (and improve position)	**Hold position** Grow with industry Reinvest as necessary	**Hold position** Reinvest as necessary or reinvest minimum
Favourable	**Attempt to improve position selectively** Penetrate market generally or selectively Invest selectively	**Attempt to improve position selectively** Penetrate market selectively Invest selectively to improve position	**Maintain position** Find niche and attempt to protect it Make minimum and/or selective investment	**Harvest, withdraw in phases or abandon** Reinvest minimum necessary or disinvest
Defensible	**Attempt to improve position selectively** Invest (very) selectively	**Find niche and protect it** Invest selectively	**Find niche or withdraw in phases** Reinvest minimum necessary or disinvest	**Withdraw in phases or abandon** Disinvest or divest
Weak	**Improve position or withdraw** Invest or divest	**Turn around or abandon** Invest or disinvest	**Turn around or withdraw in phases** Invest selectively or disinvest	**Abandon position** Divest

10.10   Who makes the decision?

In their analysis of how strategy actually takes place, Wheelan and Hunger believe that there are five essential tasks (1992):

1. environmental scanning;
2. internal scanning;
3. strategy formulation;
4. strategy implementation;
5. evaluation and control.

In effect, the first considers the external aspects that create the likely opportunities and potential threats (OT). The second consider the strengths and weaknesses that the organisation possesses (SW). The third concerns what will be done in terms of creating the mission, considers the best strategies to be pursued and what the tactics are to carry out operating processes. Implementation and evaluation and control will be considered in greater detail below.

Lynch asserts that as well as putting mission and objectives as a precursor of environmental analysis, it is equally important that there is a clear indication of who takes the key decisions (2006, p.515). Whilst there is no set belief in what is best, the assumption is that those who have the best understanding of the options and their implications should be involved. In practice what this often means is that there is a hierarchy of strategic decision-making resonating with the level of apparent 'importance'. So, in terms of mission and objectives, senior managers with overall corporate responsibility will be assumed to have the greatest influence, albeit they will need to consult with those who have an intimate understanding of operations and, in particular, what can really be achieved. As the strategic decisions work their way down through the organisation, others who are less senior will probably be involved. Implicit in the consideration of options, most especially when one is chosen as the preferred way of achieving the desired objectives, is planning. This is considered in the chapter subsequent to this one.

10.11   Using scenarios

Using scenarios, according to Wilson (1978), provides alternative futures that may be considered. Langford and Male (2001) suggest that there can be three types of scenario planning. The need to develop scenarios is, according to Langford and Male, necessary to attempt to 'portray the future of the construction world' (2001, p.167). Their purpose is to use intelligence and data that allow managers to 'construct' potential views of how the external environment may alter. Strategic management is based on considering the range of scenarios that are feasible and, using experience and judgement, selecting the one that they believe is most likely. Developing scenarios, according to Langford and Male, is a vital part of forward planning, the objectives of which will be:

●   Creating forecasts of the economy that will impact on the demand patterns for construction;
●   Considering feasible 'alternative visions';
●   The identification of 'branching points' which will enable the organisation to make effective decisions about how to develop strategic capability through its resource base.

The use of PESTEL is valuable in that it considers many of the key influences that will impact on the organisation. However, if there are other factors that are worthy of consideration, they should also be included. Importantly, the key to success is to be able to assign values for the likelihood of factors altering in a particular way. By doing this, it is possible to assess the impact on the organisation. Clearly, the higher the probability for occurrence, the more likely the environment will alter in the way that the particular scenario assumes.

Using scenarios can generate a great deal of potential information which needs be 'manageable' (Langford and Male, 2001). In order to ensure that the process of using scenarios can indeed be managed in an effective way, Langford and Male offer the following advice:

> A simple way of filtering the information is to quantify the probability of an event occurring and multiply it by the importance of the event to the [organisation]

As they contend, 'weighting of trends' is vitally important in that if a particular scenario occurs, even though it may be extremely unlikely, its impact may be potentially devastating. For example, any speculative housebuilder will consider trends such as interest rates and how their alteration may affect the behaviour of potential purchasers. Accordingly, companies would be wise to develop strategies to cope, especially with downturn. However, the recent experiences of housebuilders has demonstrated that very few developers had envisaged the magnitude of the downturn in house-buying associated with what is commonly referred to as the 'credit crunch'.

10.12   The importance of context

In considering strategy, there is always the inherent difficulty of dealing with the uncertainty of the environment in which the organisation operates. This poses the dilemma for an organisation of taking effective decisions when the context may be far from certain. Indeed, as organisations frequently find, the assumption that the environment is likely to remain stable is, at best, optimistic and, at worst, hopelessly flawed. Therefore, the use of what is known as the classical model of strategy as a way of developing strategic options has been criticised as being too simplistic and inadequate in terms of ability to respond to the increasingly dynamic circumstances of the contemporary world of business.

Critics contend that strategic models should be able to be used in order to precisely react to the alterations in circumstances (context) that occur all too frequently. As Chaharbaghi and Lynch argue, there are two essential reasons why alternatives to classical strategy should be considered (1999c, p.47):

1. **The problems of uncertainty in the external context.** There are no guarantees that the world will remain as it does. Technology changes. Tastes alter. New laws are instituted. Sometimes the world order simply shifts in ways that no one could have anticipated (often at remarkable speed). For example, who would have predicted the rise of China, India and Russia as economic superpowers in the 1980s or early 1990s?[2]
2. **How to anticipate what will actually happen (as opposed to what is hoped for) in the internal context?** Organisational studies demonstrate the difficulties that occur in trying to predict the behaviour of people, most especially in how they work together in certain configurations (after restructuring) or in reacting to the implementation of new technology, regardless of the intention to improve processes or make the achievement of tasks easier.

---

[2]As the preface to this book describes, the financial world has, since the so-called 'credit crunch' of late 2007, been in turmoil. The ramifications that have stemmed from this have been felt in every context but most sharply in markets of construction that had become used to cheap and readily-available credit. Therefore, those organisations whose long-term strategies had been developed on the assumption that financial markets were stable and that growth would continue have found their plans are in disarray.

As Lynch states, context becomes an extremely important influence on the 'process of strategy development' (what he calls 'content') (2006, p.533). Therefore, he advises, the '*combination* [his italics] of context, process and content' provides the basis for developing five alternatives to the classical approach:

1. survival-based;
2. chaos-based;
3. network-based;
4. game-theory-based;
5. learning-based.

10.12.1   *Survival-based*

This approach to the development of strategy is one that is resonant with Darwin's theory that those species that are most likely to survive are usually the fittest in terms of their ability to procure and utilise resources most effectively. Additionally, those that are most nimble or flexible and can respond to changes in the environment will be best able to cope. In effect, such organisations are adaptive in nature. They are likely to look to the short term and are constantly scanning the context (both internally and externally) for changes.

Crucially, an organisation that is survival-based will consider many potential options but allow itself as much flexibility as possible in terms of actual selection. The critical test is knowing what is the option that best suits the 'current' context. Such an approach does mean that those involved in the organisation will have to cope with the consequences of being 'footloose' and able to contemplate and deal with constant change. People, in general, have a tendency to prefer stability and they find uncertainty stressful. Therefore, whilst a survival-based approach has advantage during times of turbulence in the context, it comes at a cost to people.

10.12.2   *Chaos-based*

This approach is based on the belief that as well as context being inherently unstable the outcomes of any strategy will be difficult to predict and possibly result in chaos (which is never the intention). Stacey (1993) believes that organisations are able to cope with unpredictability and chaos by their ability to transform and innovate. Thus, similar to a survival-based approach, a chaos-based way of deciding on strategic options will embrace uncertainty; it will develop techniques and a concomitant culture that learns to thrive on the need to constantly 'bend and shift' its objectives; it will carry out its processes to achieve them in such a way as to suit the prevailing circumstances.

In organisations that deal with chaos there is a need to develop methods of communication and feedback that allow evaluation of the impact of a chosen action. The aim is that monitoring carried out instantly will provide an immediate indication of success (or failure). There is a strong likelihood that the people employed will have a shared commitment to innovation and creativity. They will be unlikely to engage in the use of formal planning and long-term commitment to particular goals because they simply do not see the relevance of making decisions based on assumptions that may turn out to be unpredictable. Stacey provides eight suggestions of what he considers that an approach based on the acceptance of uncertainty should consist:

1. Loosen control and allow events to occur;
2. Make the organisation one that is more cooperative and less inclined towards internal competition that is divisive;

3. Create an environment that encourages groups to decide the priorities;
4. Engender a supportive culture;
5. Allow the challenges to emerge from people involved;
6. Make sure that the organisation is absolutely outward focused and willing to recognise changes that should be responded to as soon as possible;
7. Dedicate time, effort and money to ensure that new skills are inculcated in every person for the new approach;
8. Encourage and support an environment in which there is a willingness to experiment (success should be celebrated, and failure seen as unfortunate but as an opportunity to learn – certainly not to result in either punishment or blame).

10.12.3   Network-based

In many sectors the environment is one in which cooperation and collaboration become an option that provides attractiveness as the way to develop strategy. Therefore, the organisation structures itself in such a way as to achieve a network of relationships inside the organisation. This is known as 'network cooperation' and seeks to achieve improvement throughout all of its activities. In large organisations this will mean that any advantage achieved in one area will be transferred to all others through dissemination, training and best practice based on procedures that are implemented for all. Importantly, in most industries, an organisation's ability to produce value is based on the input of others who supply materials or components, labour or professional expertise and possibly all of these. Therefore, whilst it may not be strictly true to describe this as an internal network, it is logical to think of the totality of the supply chain as being thought of as a whole organisation. This lesson is one that has been learnt in manufacturing and, as many commentators argue, construction should emulate the example.

Important networks occur outside the organisation. These are usually based on links within the area of the market (and beyond) that are most likely to produce benefit. Such networks may be formal and imposed, through the use of regulation, or informal when there is agreement to implement changes that will provide value to those within it (the organisations) or without (customers). The important principle is that there is benefit to those who collaborate in the network and, usually, there is a willingness to share information and knowledge.

Standardisation of components is an example of how cooperation throughout any industry is advantageous to all; construction has many examples, most notably that of brick size. Once every organisation is agreed on what constitutes the standard size, it is less likely that there will be waste (from having to cut to size to fit) or confusion from uncertainty about what is actually available. By doing this, every organisation involved in the supply chain will be aware of what they should be able to do to cooperate and make collaborative decisions.

Lynch suggests that there are a number of ways in which an organisation can add value internally and externally which have been adapted as below (2006, p.548):

Internally
- The development of superior knowledge or skills;
- Investment in technology and innovation;
- Sharing of systems intended to achieve improvement through cooperation;
- Willingness to pool expertise.

Externally
- Logistics throughout the supply chain;
- Agreement between suppliers and subcontractors;
- Sharing of technical developments and expertise throughout the industry;

- Collaboration in terms of developing and implementing industry standards and codes (this would be especially useful in matters designed to increase health and safety);
- Willingness to invest in industry-wide representative bodies who are able to lobby at regional, national and international level.

Importantly, Lynch explains, networks are based on the willingness of people and organisations to collaborate and cooperate and, therefore, they can be inherently fluid and dynamic:

> In this sense, nothing is fixed and everything is open to negotiation. Therefore, objectives may need to be revised and selection may be compromised by the need to persuade groups to join or remain in the network. In a sense, the implementation process itself is now part of the selection process and part of the strategy. (2006, p.549)

The last part of this statement draws attention to the important link that needs to be made between the development of strategic options and how they are effectively implemented (which is covered in the next chapter).

10.12.4   Game-theory-based

This approach to strategy is based on the implementation of a process of negotiation between customers, suppliers and competitors in a way that is systematic and allows all options to be measured in an objective way. The assumption is that by using this approach all interested parties will seek to maximise their own outcomes. Whilst there may be cooperation, there may be what is known as the 'zero-sum game' in which one gains at the expense of others or 'negative-sum game' in which everyone loses (because of mutual destruction).

Game theory will, like any game based on judgement, involve an attempt to second-guess what your opponents are most likely to do. We do this in our everyday lives but not in the structured way or using mathematical models that game theory suggests are necessary. Game theory has few particular rules. However, some general principles do need to be accepted:

1. Consider all viewpoints – what do you think everyone wants?
2. The assumed rules at the beginning may be altered at any point to suit some or all of the players.
3. Constant re-evaluation of what is being achieved – if it is not giving advantage, withdraw!
4. After conclusion, consider what was actually achieved and whether the outcome could have been better.

The last can be seen as an acknowledgement of the importance of learning, a concept that we all experience in our progression through life. As Chapter 8 explained, the ability of an organisation to collectively learn is extremely useful and can be a particularly good way to ensure that strategic options are generated and evaluated. The next section considers what this involves.

10.12.5   Learning-based

Learning involves testing and evaluating the success or failure of a particular approach. We do it when we try out new products or services or use a new system. Sometimes it works out well, sometimes not. Whatever happens, we know better the next time. Therefore, an organisation can test strategic options to consider which is best. In order to do this, those involved should be willing to accept the state that it is in (and where it starts from) and how each option will affect it. The emphasis is on the sharing of all knowledge about all of the important factors and, crucially, continually adapting to the impact that each has (and developing an approach that is collective). People in the organisation should be encouraged to critically question and constantly evaluate everything that goes on. The key, of course, is to keep trying to develop solutions that make performance better.

Peter Senge's work on learning has become seminal in how to appreciate the way that principles can assist organisations to develop strategic options and implement the one most likely to produce optimal outcomes. He advocated what is known as 'double-loop' learning in which people are capable and certainly encouraged to reconsider methods of operation and to ask whether there are different ways of doing things. Learning, he stressed, should be a collective activity through teams or groups. Additionally, he asserts, it can be considered as being of two types: **adaptive** which happens when the organisation changes to suit circumstances in the external environment and **generative** which looks at ways to do things differently within the organisation.

In creating an environment that will enable learning to take place that is dedicated to the evaluation of strategic options, it is important that time and space are provided. It is important that people can meet as teams and that they are able to suggest how they (as a group) can change things for the better.

10.13   Conclusion

In making choices, managers should have carefully examined both the merits and disadvantages of each strategy. Choice is easy if there is one option that is clearly shown as being best all round. The reality of making a choice is that there is often more than one obvious choice and that benefits are often undermined by disadvantages. Sometimes there may be a possibility of combining more than one potential strategic choice so as to achieve a solution that is super-optimal. If this is not possible, the 'best' choice may be the one that has the least disadvantages associated with it. What will be important, therefore, is to successfully implement this strategy in order to ensure that successful attainment of objectives is achieved. As the next chapter describes, in implementing any strategy, there are frequently issues and dilemmas that should be contemplated and, if they arise, dealt with by judicious management.

As Chapter 5 explained, very successful organisations do not achieve what they do by accident; they do so by considerate management of people and careful utilisation of technology and other valuable resources. Most of all, the managers of such organisations constantly scan the environment in which they operate to search for opportunities and to avoid potential threats. Armed with the requisite intelligence, these managers will then consider what they must do in order to alter or adapt the organisation's resource base to maximise its potential. This process is continuous. It is not wise to be complacent. Success may be fleeting and today's exemplar may become tomorrow's white elephant that results in closure or severe curtailment of strategic intent. As Chapter 11 will describe, there may be a need to constantly adapt strategy in order to survive (usually short term) or radically change so that it best matches the conditions that are considered most likely to pertain.

Chapter 11
Implementing the strategy – issues, dilemmas and delivery of strategic outcomes

What's the use of running if you are not on the right road?
German Proverb

11.1 Objectives of this chapter

All management is academic until put into practice. That is where success (or failure) of choice truly can be judged. This is the case with strategic management and it is obvious that decisions will be based on what will be most likely to work, given the assumptions that will have been made with respect to the environment and resources that will be available (or likely to be procured or developed). It will require the organisation to engage in the planning for and implementation of the strategic option. It will involve sufficient support in terms of people and financial commitment. The process of implementation is an activity that might be considered as simply being the last logical stage in strategic management. Using a prescriptive approach, implementation is simply a matter of the imposition of whichever choice is deemed to be most likely to produce the optimal results.

However, the other approach, based on the belief that strategy is more appropriately achieved by being adaptive (emergent), is one that would eschew what might be criticised as a 'fire and forget' view of implementation. Instead, advocates of an emergent approach would argue, there is a need to continually monitor and refine tactics used in implementation and subsequent management. They would stress that because the environment is constantly altering, then so should the way that strategy is achieved.

Much depends on the circumstances in which the implementation takes place. Accordingly, if the objective is widespread change in a way that affects the whole organisation, such as when the market shifts significantly, it will be necessary for the implementation process to be carried out in a way that recognises this. However, it is entirely possible that during implementation the market may alter in such a way as to make the original plan redundant. It would be foolish to continue implementing something that will never work. Some rethinking (and planning) will be expedient. Using an adaptive approach might avoid the need to make significant shifts. Rather, the emphasis is on being able to constantly alter implementation and day-to-day management in order to be in alignment with whatever influences (internal and external) create the impetus for reconsideration of the organisation's strategy. As will be explained, such an approach is likely to require less turmoil and trauma amongst people than trying to achieve a 'big bang' in which there will be an expectation that almost everything will change.

11.2   Getting to the end – the difference between 'intended' and 'realised' strategy

Strategy can certainly be a process that involves a set of logical steps that are presumed to result in particular outcomes. This involves planning, which is described in greater detail below. Grant (2003) provides a model (see Figure 2.4) that shows there are nine of these steps to this process. Clearly, if regular monitoring of actions against the overall plan (box nine) results in what was intended, the plan worked. But, as personal experienced teaches us, what we expect is not what we get and events have a habit of intervening. This is something that affects organisations in the same way. Indeed, Johnson *et al.* (2005) explain that there is often a lack of congruence between what those managers responsible for strategy believe is happening and what is really occurring (p.565). As they contend, the disparity between what was **intended** which they define as 'an expression of desired strategic direction deliberately formulated or planned' and **realised**, which is what is actually being followed, is common. This, they believe, is because whilst what is intended tends to be written down so as to create a definite sense of purpose (see below), what is achieved (realised) is different (unrealised) because:

> The plans are unworkable, the environment changes after that plan has been drawn up and the managers decide the strategy, as planned, should not be put into effect, or people in the organisation or influential stakeholders do not go along with the plan. (2005, p.566)

As such, the actual strategy that is realised is one that is emergent and comes, they describe, 'through more everyday routines, activities and processes' (2005, p.566) which are based on what works. Implementation of strategy become a process of tinkering and a series of small adjustments to cope with the exigencies of the world as:

> ... a pattern of what has become known as incremental strategy develop-ment is apparent. Strategies do not typically change in major shifts or direction. They typically change by building on and amending what has gone before. ... An apparently coherent strategy of an organisation may develop on the basis of strategic moves each of which makes sense in terms of previous moves (2005, p.567).

Importantly, there is a need for strategic managers to acknowledge the fact that their plans have not worked out as intended. However, it offers the advantage that if the organisation is inclined to deal with uncertainty, its members – people – will be better able to adjust to the constantly altering circumstances.

11.3   The influence of purpose and dynamics on resources used

Purpose is something assumed to be embedded in the process of strategic development and implementation. However, purpose may change and, logically, so should the way that the strategy is achieved. The reasons that purpose changes, due either to internal or external factors, will create a particular combination that makes the purpose relevant or, potentially, irrelevant. It is important that those managers overseeing strategic implementation are aware of variations in any (or all) of these factors and ensure that purpose is aligned. The difficulty may potentially be that an organisation's purpose has become fixed and difficult to change. If this occurs, the organisation will suffer from a mismatch in what people believe the purpose is and what is really appropriate for the conditions that exist. At best, there will be levels of confusion that undermine confidence and credibility. At worst, the organisation is under serious threat because of its inability to respond to the contemporary conditions.

In considering alterations in purpose, it is important to appreciate the effects on resources being used in implementation. In terms of the combinations of resources that are used, it can be assumed that these will change in accordance with the particular purpose that is being pursued. According to Lynch, this will create three potential 'dimensions' that should be considered (2006, p.728):

1. time;
2. the advantage of being a second 'pioneer';
3. imitation.

11.3.1   Time

In implementing strategy, there are assumed to be changes that will occur as time passes. In effect, the external environment will be affected by events, some of which may be anticipated but many of which will not. However, as previous explanations have made clear, making plans on the basis of assumed trends or anticipated changes can be dangerous. At best, the resources which have been planned for may need action to remedy them so that they can cope with the new circumstances. At worst, though, they may become completely redundant which, of course, has serious implications for the organisation's coping or, even, survival.

Implicit in the consideration of resources is the way that they are used to create 'routines'. As Nelson and Winter (1982) argued, these routines can of course provide the basis of competitive advantage if they enable the organisation to achieve outputs and/or objectives more effectively and efficiently. However, they can cause difficulties if they have become so embedded or fixed that changing them becomes problematic. As they explained, routines frequently rely on specialised knowledge that may have been learned over a long period of time and which people do not want to 'unlearn'. Routines may be either 'sticky' (slow to change) or even 'blind' (where there is a belief that nothing better is possible!). Routines, it should also be pointed out, are frequently reliant on particular people whose presence is crucial. The challenge is to consider how these routines can be continued once these people leave either voluntarily or due to a sudden event (illness or worse).

Additionally, Joseph Schumpeter (1942) considered how resources can be dynamic over time. Most especially, they can undergo phases during which they experience rapid change or, alternatively, stagnation in terms of development and/or innovation. Recent events such as the 'credit crunch' are ample demonstration that what may seem good strategy during one period can prove to be a fallacy. The key lesson for those who make strategic decisions concerning resources is to carefully consider their continued effectiveness. Better still, they should think about how a sudden change can seriously undermine them.

11.3.2   The advantage of being a second 'pioneer'

Being the first to implement a radical strategy in a market can certainty be advantageous. This provides the opportunity to be a market leader which, of course, is very beneficial if the strategy works out as intended. However, there is the potential for things to alter and for markets to be different from expectations. Those who are first will learn these lessons (but at a cost). Being second into a new market or implementing something radical gives the organisation the potential advantage of being able to learn from mistakes or alterations that were not anticipated. It will probably mean that less investment will be needed on research and development than will have been spent by the pioneer. The easiest way to achieve this is to simply find out what the pioneer did, in order to imitate (copy). However, as the next section explains, those organisations that enjoy success will attempt to find ways to thwart imitation.

11.3.3   Imitation

In implementing strategy, the objective is success. Assuming that this has been achieved, a tenet of competitive advantage is attempting to deter others from gaining similar advantage by copying ideas, technology or processes. This can be done by considering certain tactics such as patenting, copyrighting, or use of routines (see above), based on internal knowledge, that are so complex or difficult that others will be deterred.

11.4   Planning for action

Planning is an activity that is intended to produce particular outcomes. We do it every day to achieve the mundane such as allowing sufficient time for routine activities. If we are ambitious we make longer-term plans to try to ensure that we will be able to realise our personal goals. Organisations that have made strategic decisions must plan how they are going to be able to realise the intended goals as effectively as they can. The problem with planning is that it is often seen as a 'fire and forget' activity that can be carried out in advance of action but that day-to-day tasks are left for others 'on the ground' to sort out. This is incorrect and there have been many examples where rational planning (all eventualities and possibilities can be considered in advance) created more difficulties than were solved (see Lenz and Lyles, 1985). Rather, as Lynch believes, it is an 'ongoing activity that responds simultaneously to pressure of events' (2006, p.633). On the basis of research into planning, he presents the main reasons that strategic planning carried out as a top-down and highly rational activity causes typical problems such as:

1. Lack of direction from senior management:
   - Too many ad hoc decisions;
   - In-depth consideration of issues is not being carried out;
   - Short-term focus is the priority;
   - Lack of discussion;
   - Resources are inadequate;
   - General inability of managers to cope.
2. Need for greater flexibility:
   - Budgeting becomes too rigid;
   - Bureaucracy is the main concern;
   - Resources are made to fit rather than being developed;
   - Staff are not being given the opportunity to improve themselves.
3. Problems within the organisation:
   - Infighting amongst senior managers;
   - Lack of clear direction;
   - Interest groups override the overall strategic interests;
   - Structure is rigid and needs reorganisation;
   - Communication systems are inadequate.
4. Organisational culture:
   - Is too rigid and not tolerant of change;
   - Short-term and risk-averse;
   - Blame culture exists;
   - Innovation is not encouraged;

- Risk taking is discouraged;
- People are not valued.

Planning should be seen as an integral part of implementation. Indeed, the mere act of planning will assume that there will be a template of activities that will need to be carried out and, as well as 'milestones' (predetermined points on the journey), there will be methods of comparing the rate of actual progress with the plan. As has been stressed in all previous chapters, strategic management is the act of planning ahead in the belief that what is likely to occur can reasonably be anticipated. There are a number of stages which will incorporate the key objectives which, it may be assumed, will come from corporate level, and will then be 'cascaded down' through the organisation and translated into long-term strategic plans (up to five years), thereafter into medium-term departmental/section plans (up to a year) and finally into day-to-day operational plans. All of these will be subject to constant feedback, review and, should the need arise (especially where the potential for failure [see below] is detected), alteration.

11.5   Making it happen – the influence of Kaplan and Norton

One way of planning that will allow future comparison of actual against intended is the use of what is known as the 'Balanced Scorecard', based on measures that may certainly be financial (which are important), but on others that are qualitative such as the views (and aspirations) of key stakeholders in the success of the strategy. This model was proposed by Kaplan and Norton (1996) who recognised that traditional approaches to strategic planning and its implementation tended to be over-concerned with financial aspects of the organisation (which are usually objective) but largely ignored measures that are more subjective (but which they felt were fundamentally important). For example, they believed that there was sufficient evidence that aspects of performance such as customer satisfaction and loyalty, employee development and motivation and organisational learning are crucial[1]. The Balanced Scorecard is resonant with much of what is contained in the literature that explains the principles of total quality management – an emphasis on making the measures effective in so far as they should be understood and used to monitor processes. Processes, they argue, are what an organisation needs in order to create action and, therefore, potential success. As such, processes occur at every level and with respect to every task carried out in pursuit of the strategic goals. Meeting financial goals, which they acknowledge as being important, is the consequence of success, not the starting point for making decisions. According to Lynch, there are four key principles (2006, pp.617–8):

1. Translating the vision through clarifying and gaining consensus;
2. Communicating and linking by setting goals and establishing rewards for success;
3. Business planning to align objectives, allocate resources and establish milestones;
4. Feedback and learning to review the subsequent performance against the plan.

The Scorecard is based on four key elements (perspectives) which incorporate the four principles:

- Financial – which would include ROCE, growth in terms of sales or profit, cost efficiency;
- Customer – which would be based on metrics designed to explore how well the organisation is doing in terms of satisfaction and loyalty;

---

[1]Examination of what are known as 'excellence models' (see EFQM) explicitly identify the importance of these.

- Internal – which should consider all activities that are integral to production (cost information, stock levels, quality measures and any relevant consideration of how people are used and their satisfaction levels);
- Future – which should explore what is required to cope with change.

The following diagram (11.1) demonstrates how the Balanced Scorecard is used to consider aspects of strategic performance that are dedicated to achieving outputs in a way that ensures that strategic intentions are appropriate:

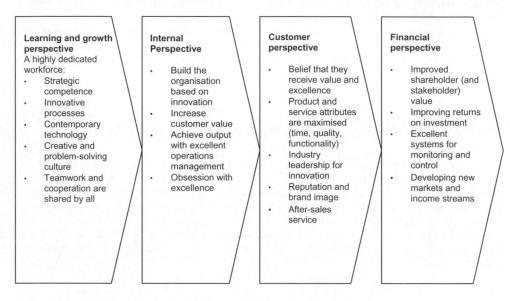

Figure 11.1   The link between the balanced scorecard and strategy development.

11.6   Communication and approaches

In implementing strategy, it is essential that all those affected are kept informed of what has been planned, why particular decisions have been made and what their roles and responsibilities are. In a small organisation it might be assumed that people will be informed throughout the process and that communication is continuous. Equally, in a very large organisation it could reasonably be believed that many people, and most especially at operational level, are remote and will not have been informed during the process.

The reality is that implementation may be carried out in ways that have no connection with the size of the organisation. Indeed, many very large organisations go to extraordinary lengths to engage with people, the benefit being that by so doing they will be likely to have created support and active cooperation (which is better than having to impose action on those who are sceptical or fearful of the consequences). Even more pertinently, as events change rapidly, people who can respond quickly and can alter action to suit are far more able to cope than those who are expected to comply (see Chapter 9 on change management). Communication, therefore, should be seen to be a continuous process that allows constant clarification and explanation. In effect, it should be a dialogue which implies a two-way process.

There are many approaches that can be adopted. Which is chosen will be dependent on aspects of the organisation such as the type of environment in which it operates, the technology, the culture and, to some extent, the relative size and structure. Based on research by Goold and Campbell (1987), there are particular considerations that should be included in the approach that is adopted:

- Bureaucratic – in which the emphasis is on rules (procedures) based on explicit goals and expectation of particular behaviour (compliance);
- Organised anarchy – where there is less certainty than a bureaucracy and the nature of the organisation is decentralisation, so-called 'ad-hocracy' and a disjointed style of management, decision-making and use of information;
- Political power – where those who have greatest control of resources (technology, information, people) can wield influence and, therefore, must be brought into the process at the earliest opportunity.

11.7   Strategic control

According to Capon, strategic control is concerned with the 'ongoing monitoring of staff, managers and activities or an organisation to evaluate **efficiency** and **effectiveness**' (2008, p.354, emphasis in the original). This, she states, is based on the organisation putting in place systems that can monitor the way that the strategy is being implemented and identify where problems or failures are occurring. In order to ensure organisational effectiveness (and competitive advantage), her advice is that strategic-control systems should be instituted which are 'the means by which managers evaluate whether [it] is achieving efficiency, quality, innovation and customer responsiveness' (2008, p.366). The most important thing that these systems should do is provide data that is up to date and gives an accurate assessment of exactly where the organisation is in its quest for success (potential failure is considered below). As such, Capon believes, there are four stages necessary to support an effective control system:

1. The setting up of performance targets;
2. Creating the control systems;
3. Comparison of actual performance with the targets that have been set;
4. Corrective action to deal with negative variance between targets and actual performance.

   According to Bungay and Goold (1991), there are a number of ways by which strategic control can be improved:

- Do not create too many measures as people will become disillusioned with what they see as over-obsession with performance targets;
- Distinguish between the different measures that are used at particular levels of the organisation;
- Relax measures as people become more proficient.

   Using these systems in a way that can detect problems as early as possible is the key to ensuring the strategy remains appropriate. As the next section describes, failure is something that may happen and, should it occur, it must be actively managed.

11.8   Dealing with failure

Failure is the worst possible outcome for managers who have taken decisions about strategy. However, the corporate world is one in which strategies do not always produce the results that were hoped for. A cursory glance at any daily paper or watching the evening news will usually reveal impending problems of a once-dominant company. The reasons for decline or failure may be complex but it can usually be assumed that those responsible will, if they can possibly antici-pate them, have attempted to avoid such problems. Sometimes, of course, events will change so fast that there is little to be done apart from reacting to them as quickly as possible to limit any potential damaging consequences. However, any methods that can be implemented to give advance warning of problems will be better than having nothing in place at all. This is the purpose of having control systems.

Those organisations that choose either not to use control systems (which is foolish) or disregard the signals will be likely to suffer the consequences of failure in such a way as to lead to potential terminal decline. Like a car driver who ignores the red, oil-warning light, the organisation may eventually decline so much as to make recovery impossible. However, taking appropriate and timely action to arrest failure may be sufficient to ensure that terminal decline is avoided, and will allow the strategy to be revisited and revised in such a way as to make eventual success more likely. As Capon argues, many of the symptoms of decline and failure (see list below) can be monitored against previous performance to enable managers to react sufficiently quickly (2008, p.376):

● Sales that do not match targets;
● Profitability that is less than expected;
● Dividends that are not sufficient;
● Increasing debt;
● High turnover of staff or reduced morale;
● Paralysis in terms of key decisions being made.

Once any early signs of problems have been detected, investigation should be carried out with urgency. This will elicit likely causes and, therefore, provide suggestions as to how the problem can be arrested and remedial action implemented. According to Capon, likely causes of failure are (2008, p.378):

● Poor management and leadership;
● Neglect of the core business;
● Inefficient operations;
● Poor financial management;
● Inadequate control over expenditure;
● Inability to compete with others or to react to changes in circumstances or the environment.

11.8.1   Poor management and leadership

Given that strategy is frequently decided by senior management, any failure in its implementation will be difficult to accept, especially if they are the cause. However, there are many examples of organisations that have experienced decline and failure due to the fact that the senior managers (and leaders) are no longer able to fully understand the market in the way they used to (and pos-sibly made it successful in the first place). As Capon asserts, managers who are no longer in touch will be very unlikely to develop strategies that are effective and, in turn, will create a sense (a culture) in which there is a lack of clarity (2008, p.378). However, whilst weak management and leaders may not be good, she also suggests that failure is often due to those who have most control over the organisation being autocratic. The inability of such managers to adapt, she argues, will create further decline which is to everyone's detriment:

> The unsuccessful autocrat refuses to recognise the need for change and loses the support of others, meaning business growth is jeopardised. Ultimately, this may lead to his or her downfall (Capon, 2008, p.379)

Another problem may be that the traditional origins of the organisation have created a legacy in which only one type of manager (or profession) is allowed to become a senior manager (such as an accountant or quantity surveyor). The consequence of this is that the mindset of these managers may be that they are unwilling to accept the need to alter their approach to markets or to change the way that the organisation is run. Most especially, there is a belief that British management has, in the past, been too male-oriented, excluding women and reluctant to embrace more flexible and people-centric practices. This has been a criticism levelled at the UK construction industry.

11.8.2   Neglect of the core business

Whilst innate conservatism in strategic management may be a bad thing, it is equally dangerous to pursue goals that are so ambitious and radical that the organisation loses its ability to carry out its core function (for which it became successful in the first place) any more. Peters and Waterman use the expression, 'sticking to the knitting' (1982), to mean that an organisation should concentrate on things in which it has expertise. As Capon believes, organisations that lose their focus on core business will be likely to suffer the consequence of serious decline because this is the part most likely to generate the cash that is vital to continued survival and investment in growth.

11.8.3   Inefficient operations

Chapter 6 described the need for effective operations management. Carrying out day-to-day tasks with maximum efficiency is something that requires continual monitoring and consideration so as to ensure that every possibility for improvement is considered. The use of performance targets and metrics that can be used to evaluate levels of efficiency, productivity, utilisation of equipment and people, and stock levels will assist in ensuring that effectiveness is as good as possible. But as Capon warns, care must be taken in the way that all of the measures combine to provide a full and accurate picture:

> Effectiveness is how well an organisation sets and achieves its goals. In considering capacity, utilisation, productivity, efficiency and effectiveness, thought should be given to how these measures combine. For example, high productivity is of no use if the quality of products [or services] produced is poor or if the products remain in a warehouse because there is no demand for them. (2008, p.381)

The use of lean production techniques will assist in identifying how production can be more closely linked to customer requirements and, in particular, alterations quickly made that will ensure what is actually being produced is exactly what is needed by those who purchase and/or consume the product. Like all internal aspects of the organisation, constant communication with those who carry out activities (operations) will be valuable in identifying improvement. As leading Japanese manufacturers of cars have shown, strategies based on the desire to engage with people at operational level to attain extremely high standards of quality (excellence) are very beneficial.

11.8.4   Poor financial management

Lack of good financial management is a common cause of problems in organisations. There is a need to institute accurate and appropriate financial controls that enable decisions to be made which will invest wisely in the parts of the organisation that require funds to continue activities effectively. It is recognised that there may be tensions between those who control funds and those who require them. The former will have a tendency to be conservative. The latter will naturally want as much as possible. Whilst compromise is the key objective, there will need to be useful mechanisms that allow financial management to work in a way that not only provide the right balance, but also enable constant monitoring to be carried out. It will also be necessary to attempt to evaluate the risk of becoming involved in major investments for expansion or diversification which, if they go wrong, will put the whole of the organisation at risk. According to Capon:

> Avoiding failure requires that companies [or any organisations] do not stretch financial or management resources with big projects as this can easily cause other healthier parts of the business to suffer. Therefore, it is essential to control expenditure and costs and to forecast potential revenues without being unrealistically optimistic. (2008, p.381)

11.8.5   Inadequate control over expenditure

Like the previous reason for failure, this occurs because management are unable to utilise the financial information being generated. Understanding the data being produced by cost monitoring, forecasts and budgetary control systems is a task that requires skill and a willingness to interpret so as to appreciate the true meaning and potential consequences. Having specialised managers who can do this will be very valuable (although their salaries may seem prohibitive in a small organisation). Such a cost should be balanced against knowing how particular parts of the organisation and product or services lines are performing (by comparing expenditure and revenue).

As Capon stresses, the inability to have possession of all the financial facts will very likely create the circumstances that lead to decline:

> Inadequate costing systems can result in [management] not being aware of the costs of their products and services, and hence if the mix and volume of products manufactured and sold changes, [the organisation] can move from profit to loss without realising it. (2008, p.382)

Cost structure has become increasingly important in the face of so-called global competition. Products from the Far East, where wages are lower, allow companies producing in such countries to have a much lower cost structure. As component suppliers look to save costs, it is increasingly likely that they will look to procure materials, components and larger prefabricated modules/sections from those whose costing makes the price unbeatable.

11.8.6   Inability to compete with others or to react to changes in circumstances or the environment

In strategic terms any organisation should be aware of the environment it operates within and be able to carry out activities in a way that allows it to offer potential products or services that match current expectations (whether in terms of quality or price). Being unable to anticipate how customers' tastes alter or shift by being offered something different is important. For example, Capon describes how UK carpet manufacturers were affected by the ability of laminate-floor companies to be able to supply a product that was seen to be a very fashionable (and often cheaper) alternative. As she asserts, complacency amongst managers and employees 'at all levels' together with poor marketing combine to contribute to failure (2008, p.383). Based on Capon's list, examples of poor marketing can include the following:

- Failure to develop the key features of what is on offer;
- Lack of understanding of current customer behaviour and taste;
- Misunderstanding of market 'segments';
- Advertising that is no longer relevant;
- Sales and/or advertising staff are poorly motivated;
- Customer services do not match what was promised.

11.9   How to recover

Sometimes, an organisation may have to recognise that its strategy is simply just wrong for the environment and will never be able to produce the anticipated outcomes. Therefore, the organisation must consider what to do. In particular, it should consider how it might change the strategy and whether this will allow it to 'recover'. As Thompson suggests, recovery may be possible where an organisation believes that the problems are not terminal and can be 'overcome', that the environment is still relatively 'attractive' and there is 'potential for creating or enhancing competitive strategy' (2001, p.640). Citing Slatter (1984), Thomson believes that there four options that are possible (albeit that it is only the last that will enable the potential decline to be sustainable in the long term):

1. **Non-recoverable situations**   in which it is believed that there is very little hope of recovery (either because of lack of the correct resources or the market is in serious decline);
2. **Temporary recovery**   in which there may be a period in which the strategy change may work sufficiently to allow time to reconsider options but ultimately the main emphasis should be on developing a completely new strategy;
3. **Sustained survival**   where the strategy alteration is sufficient to arrest any decline but will not result in further growth;
4. **Sustained recovery** (see below).

As the diagram might suggest, more than 'breaking even' (profit) may be considered as the basis for judging success[2]. Importantly, the strategy will require urgent attention when it gets into the critical zone which means that the organisation is making a loss. However, as was suggested in Section 11.7, discussing control, an organisation would be foolish to allow the strategy to continue to the point where it is actually in crisis. It would be presumed to have detected the early signs that there is a problem and to have taken evasive action which avoids getting to the point where action is urgently required and where options may have become much more limited This was identified by Weitzel and Johnson (1989), who suggested that there are three zones: 'success' (which is relative), 'action' (when something should be done) and 'crisis' when it is possibly too late to do anything.

---

[2]This could be adapted to provide a measure that suits the circumstances (for instance, if the organisation is non-profit-making), such as revenue or consumer satisfaction.

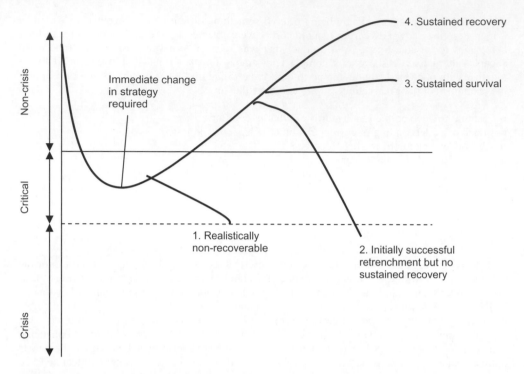

Figure 11.2   The feasibility of recovery.

11.9.1   Issues concerned with sustained recovery

Slattery (1984) believes that there are three distinguishing features of a recovery that is likely to be successful and sustained:

1. Money is made available to invest in creating and/or improving resources;
2. New ways of thinking appear which means that alternative leaders must be recruited (and the existing ones managed in terms of stepping down or leaving);
3. The systems used for management processes and finance are examined and improvements implemented.

In considering how to ensure that recovery is successful, there are, according to van de Vliet (1998), a number of questions that can usefully be posed:

● Which parts of the organisation are worth saving (and which are not)?
● What are the core activities in the organisation?
● What are the skills and abilities of people and are they appropriate to the strategy that will need to be implemented?
● What are the current limits to managerial discretion?
● Have all product(s) and/or allied service(s) been fully examined and potential improvements considered?
● If there are particular resources required to implement the new strategy, are they freely available and affordable?

Allied to sustained recovery will be the desire to use what are known as 'retrenchment' or 'turnaround' strategies.

11.10   Retrenchment strategies

Retrenchment strategies are essentially operational, according to Capon, and intended to make the organisation successful by concentrating efforts on activities that add value:

> The aim is for fairly swift improvement in the relationship between costs, prices and profits, with the objective of steering the [organisation] away from an immediate and more severe crisis. (2008, p.385)

As part of retrenchment the organisation may consider any measures that are likely to have sufficient short-term effects so as to produce cost reduction and allow immediate investment in relatively easy (and quick) improvements. As such, according to Thompson (2001, p.646), the following may be considered:

- Organisational changes;
- Financial changes;
- Cost reduction;
- Asset reduction;
- Revenue generation.

The first of these involves any alterations intended to produce changes in people's roles or responsibilities. The second will be intended to make processes more efficient and cost effective but could also focus on reduction in overhead costs which may have become a drain on productive capability. The third logically follows the second and could included widespread examination of all costs that contribute to the pricing structure. As well as this, the basis of contractual arrangements might be considered and potentially renegotiated (the alternative being that there will be no contract if the organisation ceases to exist!). The fourth contemplates raising cash by selling off assets that can be used for investment. This might consider sale and leaseback (if the asset is still needed), or complete loss if the asset is surplus or can be easily replaced by, for example, using cheaper accommodation. Finally, revenue generation will consider ways that funds can be obtained quickly by consideration of methods that will ensure a greater difference between costs and income. As Thompson suggests, revenue can be increased by 'management control systems' such as better stock management and reviews of production systems such as 'just-in-time' to reduce cost and improve quality and service (2001, p.647).

11.11   Turnaround strategies

Thompson (2001, p.648) believes that retrenchment tends to be about dealing with the short term and that if the recovery is to be sustained, effort needs to be dedicated to thinking about changes that will have longer-term effects. As he advises, it will be necessary to think about 'repositioning or refocusing' existing products and services or to think about alternatives that have not been attempted previously. Ultimately, he suggests, whilst the emphasis must still be on short-term avoidance of crisis, the long term must be the main focus. Whereas retrenchment was primarily about internal cost management, turnaround is fundamentally intended to increase customers and consumers through effective marketing and identification of needs. As such, Thompson suggests that the following methods of turnaround might be considered:

- Changing prices;
- Refocusing;
- New product development;
- Rationalising the product line;
- Greater concentration on selling and advertising;
- Rejuvenating mature business.

The first of these, changing prices, can involve either raising or lowering. A price rise will be difficult unless the customers can be convinced that they are getting more for their money, particularly difficult for the organisation to justify if it was in crisis due to intense competition. Lowering prices will, of course, reduce revenue which, potentially, may make the short-term situation even worse than it already was. However, this may attract new customers who are tempted to swap. It might also make the situation of competitors more precarious. The danger is that a 'price war' could be started in which only those who can sustain losses over a long period survive.

A better way to create turnaround is to engage in refocusing. This will involve concentrating on those customers who have the greatest willingness to consume and will be likely to remain loyal if, for example, the price was raised in the longer term to pay for improvement or increasing the features of the product and/or service. This approach is similar to developing new products. Care would be needed to ensure that, in managing this process, loyal customers continue to purchase and that the new product, as well as attracting new business, maintains the existing base.

The fourth on the list, 'rationalising the product line', would be carried out in a situation where there is a wide variety of customers and, it is believed, the complexity of attempting to serve all of them is counterproductive (and too costly). The fifth on the list is something that all organisations will do as a matter of course. The real difficulty is in knowing how much to spend in order to attract additional custom. Advertising can be extremely effective if it ensures that more products and /or services are sold or consumed. However, knowing how to make this connection is notoriously fraught and any expenditure might have been better targeted on a promotion to specific customers (marketing). Finally, in considering how to deal with a mature business (one that is believed to be at a stage where improvement and greater success are almost impossible), managers may need to be willing to consider radical steps. Baden-Fuller and Stopford (1992) argue that there are four stages in the process:

Stage one, in which it is recognised that change is urgently needed. It may involve new management (leaders), additional resources (including outside assistance), rethinking the strategy, and investigation of how the business can be stimulated by change in order to create rejuvenation.

Stage two, in which the organisation is able to concentrate on important things that will ensure added value, enhance reputation and consider how the way that customers perceive the organisation is improved.

Stage three, whereby the organisation develops new competences which will enable it to gradually develop competitive advantage in order to lead to the last step.

Stage four, in which the organisation has truly turned around (and rejuvenated).

As Thompson advises, the process of rejuvenation can be carried out in conjunction with other techniques intended to produce improvement:

> Total quality management initiatives and business process re-engineering programmes can make a major contribution [and ultimately should ensure that the] whole enterprise must become more customer focused, committed to efficiency and improvement and responsive to environmental demands. (2001, p.651)

In addition to the above, van de Vliet (1998) provides what are known as **turnaround themes** which include the following:

- Constantly looking for opportunity and niche(s) to develop;
- Ensuring that there is synergy;
- Committed management;
- Clear and focused structure that enables achievement of strategic objectives;
- Willingness to act (and certainly not paralysed by an inability to take crucial decisions).

11.12   Managing in recession and decline

Given the recent problems that have occurred in the global economy which have been caused by difficulties in the financial sector and by general uncertainty, it is useful to consider what any organisation (and construction is likely to be affected as badly as any other) should do to manage in a recession or when there is general decline. There is, of course, nothing new about this situation. Those who can recall the early 1990s will remember what a recession feels like, 'recession' being defined as two quarterly periods with negative economic growth. The particular problem for organisations is in attempting to accurately anticipate when a recession will occur (always difficult!) and, more especially, to adapt the strategy to cope with the fact that there is likely to be falling demand frequently coupled with inflation (due to rising commodity and wage prices) and increased interest rates for borrowing money[3].

The issue of finance, therefore, is the key to success in dealing with inflation. As Thompson believes, it is the aspect of an organisation that will largely determine its ability to survive (or certainly manage to continue to compete):

> The organizations that are best prepared to cope with a recession are those with relatively low borrowings. (2001, p.657)

Those organisations that have high levels of debt (normally referred to as being 'highly geared') will be likely to experience the greatest challenges. Their need to pay back loans will be a drain on revenue and reduce their ability to invest in measures that may be required to remain competitive and ensure survival in the recession. Clifford (1977) makes the point that management will need to concentrate their efforts on ensuring that the finances are robust and that margins are adequate so that investment can be made in the innovation and creativity that will provide distinctiveness. Thompson (2001) reinforces this point by arguing that 'effective cost control', 'innovative differentiation' and the desire to provide excellence in terms of service and product quality are what will enable an organisation to continue to build its reputation and ensure customer loyalty. Moreover, he stresses the need for investment in aspects of organisational development such as education and training to continue, regardless of any desire to consider them as part of a cost-reduction exercise.

Increasing emphasis on good financial management will assist in providing opportunities for avoiding unnecessary expenditure – or certainly in reducing it wherever possible – on things that might be regarded as non-essential. As earlier chapters have explained, using people as the basis for improvement is an entirely sensible approach. Indeed, as long as people are kept informed of why the organisation faces a potential crisis, they may be extremely willing to dedicate themselves to efforts intended to produce improvement and reduce cost. The important thing that an organisation should focus on is what its capability will be when the recession ends. Like a car

---

[3]These are the sort of conditions that are potentially facing organisations in the period during which this book is being written.

that runs out of fuel, an organisation will be effectively running on empty, and recovery will be much more difficult than if there was something left in reserve.

Thompson provides a list of things that any organisation might consider in the context of managing its strategy during a recession (2001, p.659) and which have been adapted:

- Be clear about what the priorities are;
- Management should be willing to make decisions that appear tough but can be justified on the basis of ensuring survival (especially where the livelihood of employees is concerned – such as redundancies);
- Keep abreast of developments that occur and which create the environmental conditions that will impact on the ability to compete;
- Monitor expenditure very carefully;
- Consider all options that will increase cash flow;
- Look at all overhead costs and consider alterations to reduce them;
- Reduce debt and borrowing wherever possible;
- Attempt to make the prices charged as competitive as possible;
- Look for opportunities to invest in staff (training, education and social activities) which will ensure cohesiveness, improve morale and maintain the innovative creativity of staff during and after the recession.

Thompson suggests that recession can provide opportunities for those able to maintain resources and expertise:

> Paradoxically, an economic recession is often an ideal time for [an organi-sation] to invest if it has the appropriate resources. If it can afford to hold onto its staff, the chances are that they will have time to deal with the implied changes. New plant, equipment and technology could then be in place in time for when the economy turns around – placing the organization in a strong position. (2001, p.663)

Some industries go into decline which, for those whose strategies are based on custom from them, is problematic. For example, the 1980s saw the decline of traditional industries which, of course, meant that their desire to purchase from construction also declined. The speed at which the decline occurs is probably the most important factor. If it is very rapid, any ability to alter resources and switch to other markets will be compromised. Managed decline will be preferable as it is possible either to switch resources and the targeted market or to gradually 'run down' resources. Harrigan (1980) suggests that if an organisation is facing decline it will need to adapt its strategy by taking into account the following factors:

- What is causing the decline and whether there may still be 'pockets' of the market that remain attractive;
- The potential to target any remaining segments that are still viable (and attractive);
- What are the costs that will be required to 'exit'? These would include things such as making resources redundant or switching (such as retraining of workers), and the potential damage to reputation caused by apparent 'desertion'[4].

The following diagram provides a summary of strategies that might be employed in declining industries:

---

[4] Some firms are required to ensure that there are adequate spare parts for customers in the future.

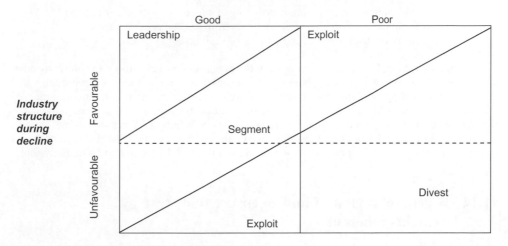

Figure 11.3 Strategies for declining industries.

11.13   What about turbulent markets?

If conditions are appropriate, there may be what is known as 'turbulence' where a change in strategy is uncertain and will need to happen very quickly to cope. This will provide both opportunities and difficulties. The former occur because, for those organisations that have the ability and expertise to respond sufficiently fast, there will be the potential to exploit any changes that present themselves. For those that are unable (or perhaps unwilling) to change, their traditional markets will either shrink or disappear. Such difficulties may eventually prove to be terminal.

Turbulence in markets, like recessions, even though frequently difficult to predict, should nonetheless be anticipated. Perhaps like an aircraft that has seatbelts to hold everyone in place, an organisation should consider what methods it can use that will ensure that employees can be supported through periods of difficulty and encouraged to consider how they contribute to efforts to innovate.

During periods of turbulence new ways of thinking about the existing resource base will be required. It is highly probable that innovation will cause accepted paradigms or models to be challenged. As a consequence, opportunities emerge for new 'players' to enter on the basis of being able to provide better value (by, for instance, cost reduction or by enhanced value through better supply-chain management). Normally, the innovative process will go through a period of relative stability; that is, until a new innovation is developed or introduced, after which the cycle begins again.

Lynch suggests that strategy dynamics should be considered in what he calls 'fast-moving' industries (2006, p.740):

- Bold initiatives should be considered as a way to develop market share;
- Use existing technology and constantly adapt to suit new customer tastes;
- Look for further opportunities to develop the customer base.

As Lynch further advises, an organisation should inculcate a spirit (a culture) of entrepreneur-ship which is defined as:

> ... a way of thinking, reasoning and acting that focuses on the identification
> and exploitation of business opportunities from a broad general perspective
> driven by the leadership of individuals or small groups. (2006, p.743)

The important thing, Lynch emphasises, is for the organisation to be willing to constantly adapt and to ensure that any plans that are being used are 'flexible' and that there is a culture in which learning from testing (and making mistakes) develops better practice and opportunities for advancement. Feedback, he stresses, is an essential part of this process. Most crucial, though, is the ability to give those who consume exactly what they believe is necessary in the 'current' market. By so doing, the organisation will survive the turbulence and, like the air passengers, emerge from the experience perhaps somewhat shaken but better able to cope with the future.

11.14   A general review of how to ensure that strategy remains coherent

Lynch, in his summation of corporate strategy, explains the use of the 'Seven S Framework' that was developed by seminal management commentators and writers, Richard Pascale and Tony Athos (who co-authored *The Art of Japanese Management* in 1981), and Tom Peters and Robert Waterman (who co-wrote the spectacularly successful *In Search of Excellence* in 1982). All of these were employed by the management consultants, McKinsey and Company, and wanted to produce a model which, according to Lynch, shows the 'interrelationship between different aspects of corporate strategy' (2006, p.792).

As Lynch believes, it is assumed that each S is equally valid and that, for the strategy to be successful, all have to work in harmony with each other:

> ... the framework provides a checklist of important variables for evaluation
> of proposed strategy developments [...] it provides a structure for the
> network of interrelationships that exist between the various elements, espe-
> cially when an organisation is ensuring that they are all *coherent* [his
> italics] during the strategy process. (Lynch, 2006)

Whilst much of the material stemmed from the work carried out by these writers (particularly Tom Peters), the importance of the Seven S model is that it emphasises the importance of the connection between the hard and soft elements of the organisation. The former are the strategy, structure and systems, and the latter are the style, skills, staff and super-ordinate goals (essentially the higher-order aspects that provide the values and concepts enshrined in the vision that leaders provide). As successful strategic management will demonstrate, having all of these elements 'just right' is the objective (albeit difficult to achieve).

11.15   Conclusion

This chapter has been concerned with what actually **doing** strategy involves. However, the really important test for strategy is whether or not it works in practice. All organisations have some form of strategy, no matter how rudimentary. They will pursue particular objectives that will be presumed to be in the best interests of all concerned, whether they are investors (stockholders) or those with vested interests such as employees or suppliers (stakeholders). Whether those taking key strategic decisions have been guided by theory is not the primary concern. What is important is to be able to understand and appreciate how the decisions were taken and to provide explanation as to what the particular circumstances tell us. This is the basis of developing more informative theory. The final chapter in this book presents a number of case-study contributions by managers in construction organisations who describe the basis of strategic decisions taken and subsequently operationalised. As their accounts make clear, there is no theoretical model that provides all of the answers for them, nor indeed that guarantees success. What they do stress is the need for hard work by all concerned and the importance of dedication to the achievement of satisfaction by those who either purchase or consume their products and/or services.

Chapter 12
Turning theory into practice – some empirical examples of strategy in construction organisations

> 'I claim not to have controlled events, but confess plainly that events have controlled me'
> Abraham Lincoln (1809–1865), 16th American President who led the country though the Civil War (1861–1865), preserved the Union and abolished slavery

12.1 Introduction to this chapter

This chapter provides ten accounts of how strategy is being carried out in construction organisations. One of the difficulties that has existed with respect to strategy in construction is the lack of agreement as to what theory is most applicable. The fact that organisations are dependent on external 'forces' adds another problem. This causes a difficulty for those trying to make 'neat' connections between strategic theory and its practice in construction. As Green *et al.* state:

> There is little agreement on how contracting firms, and actors within such firms, constitute and enact strategy. Understanding of the competitive strategy of construction firms has stagnated within recent years ... It is perhaps the overriding lack of homogeneity that has led to an increasing disconnection between the generic literature and that relating to the strategic management of construction firms. (2008, p.63)

Langford and Male, however, are more forthright in their belief that those construction firms which have 'good strategic management processes' are characterised by the following (2001, p.101):

- Formulation of strategy (at the 'apex') that is 'based on a combination of intuition and informed awareness';
- Allowing 'operating units' to develop their own plans which can be incorporated into an overall strategic plan;
- Using their planning function to 'provide contextual background information, undertake analysis' so as to ensure that strategic goals can be developed;
- Being able to alter their strategic plans as the circumstances alter, either externally because of, for example, the economy, or internally due to changes in personnel.

The central difficulty still remains. What do construction organisations really do to pursue strategic goals? The answer, it appears, is simple. They do what works but are able to cope with changing circumstances. Right back to the emergence of contracting firms in the great navigation boom when canals and railways were being built, the key to success was being able to adapt to the fast-changing market. Notwithstanding the legitimate criticism about the way in which workers were treated (so-called 'navvies'), the proto-contractors that evolved at this time demonstrated their ability to be dynamic and respond to rapidly changing circumstances. The sociohistorical analysis carried out in Chapter 3 described how a sophisticated system has developed which allows the main 'actors' involved to integrate their skills, expertise and resources to carry out construction. But as Groák contends, the difficulty of trying to analyse 'construction' is in defining what this means from a demand point of view, an essential prerequisite of formulating strategy ((1994). Indeed, his belief is that attempts to treat construction in a way that assumes homogeneity (like some sectors of manufacturing) are fallacious. Rather, he suggests, because of the diversity of participants, the fragmented nature of the processes carried out, and the increased use of 'unfamiliar' technology (which is comparable to technology fusion found in other industries), construction should be viewed as being 'organized [on the basis of an] agglomeration of projects' (p.287).

The case-study material presented in this chapter demonstrates the commitment that the contributors have to working with the particular participants that come together for the project (or series thereof). On the basis of the maxim that 'the greater the difficulty, the more glory in surmounting it' (Greek philosopher, Epicurus, 341–270BC), those involved in construction create success regardless of the obstacles. This complies with the belief of Green *et al.* who, on the basis of research they carried out, emphasise the importance of strategy being 'something that is enacted by a constantly evolving and loosely connected group of actors' (2008, p.76). Interestingly, they point out that their research indicated a lack of the use of formal planning techniques to develop strategy, something that Langford and Male believe is more likely to be standard practice. Crucially, they assert, what is most important in strategy is the ability to achieve an enhanced reputation among potential clients and to develop excellent relationships with those they serve. This is apparent in the material that follows.

12.2 An overview of the contributions of strategic practice

The first contribution describes the way that a traditional, family-based company, Thomas Vale, has developed its strategy around the belief in using excellence as a way to differentiate itself. In particular, the person concerned, Dr Richard O'Connor, explains the way that people are encouraged to learn how to use a range of quality-improvement techniques that will enable them to deliver value throughout all of the processes used in carrying out day-to-day tasks. Using experience gained in manufacturing, Richard's leadership has produced demonstrable benefit to the organisation in its quest to give customers both a better product and an enhanced level of service. By so doing, he believes, Thomas Vale has been able to enjoy the advantage of repeat business and long-term partnerships with clients.

The second contribution by Alex Housden concerns consutancy group Faithful and Gould and describes how it has dealt with the consequences of reduction in spending on construction by potential clients. As he asserts, whilst it is important to reduce cost, the key strategic objective is to continue to give 'best possible value'. Contribution three is construction contractor Adonis by Oliver Clements who examines the strategic decision-making process of how to ensure that it can identify segments in the market for which its resource base and skills are best suited. The fourth contribution is by Ed McDonald and demonstrates the continued strategic importance of effectively implementing and utilising quality management and approaches intended to produce organisational excellence and superior customer service.

Contribution five is by Daniel August of construction group Morgan Ashurst which carries out large-scale complex projects. He stresses how this organisation ensures that innovation and creativity are the 'keystones' of providing consistent 'exceptional customer service'. The next contribution is by Paul Williams of NHBC (National House Building Council) and explains the rational and purpose of being able to delivering excellent service to potential customers (house-builders and purchasers). Contribution number seven is an account by Ross Fittall of strategy and change by UK developer Argent. Once again the emphasis is on ensuring that Argent can achieve its strategy through excellent relationships and, crucially, delivering successful developments that potential investors and tenants see value in. Contribution eight by Daniel Weller explains the way that strategy has been developed and implemented by construction contractor Wates Group. The importance of customer-focus, he acknowledges is absolutely fundamental to Wates' continued success. Also crucial, he contends, is the excellent relationships that Wates enjoys through its supply-chain. In the next contribution, by Ian Davis of Birmingham Urban Design (a local authority organisation), the strategic importance of collaborative partnerships with organisations such as Wates is emphasised. Most especially, Ian stresses the ability to deliver value can only be effectively achieved by the willing involvement of such partners. Developing a 'partnering culture' based upon trust and the principles explained in *Rethinking Construction* is, Ian believes, essential.

The last two contributions concern organisations much smaller than those already described. The first of these, TRIQS is a small professional quantity surveying and project management organisation. In this contribution its managing director Mark Monaghan explains how the company has coped with rapid expansion in workload and, more latterly, reduction in demand which is a consequence of the uncertainty caused by the global credit crisis. Finally, Geoff Badham presents highly personalised insights into the challenges that he has faced as the sole person responsible for strategy in the specialised steam boiler market. Even though Geoff must take key strategic decisions on his own, their impact and consequences are intended to deliver quality and excellent service to all potential clients. This maxim equally applies to all of the other organisations described in the contributions that follow.

12.3   Continual improvement in Thomas Vale as a way of life

Richard O'Connor, Director of Business Improvement

12.3.1   Overview

The strategic importance of customers can never be overestimated. In Thomas Vale this commitment is central to every process carried out by every person from the top right the way down to those carrying out work on site. Constancy of purpose to deliver the vision of excellence is achieved through a business-improvement team that consists of four directors representing training, HR, operations and the commercial part of the organisation. This group provides a strategic focus which is intended to demonstrate the unwavering dedication that senior managers have for the desire to become the best.

Throughout the organisation there are coordinators and 'champions' whose task is to assist others to learn how they can consider the day-to-day processes. This is done through the use of 'mapping' which allows people to carefully consider what they do, how they do it and to look for ways to improve. A vital part of improvement is the setting of targets by which it is possible to measure the impact of change and to learn what works best. Like any contractor, Thomas Vale acknowledges the need to constantly improve the effective delivery of projects to customers that ensure we give them exactly what they want, on time and within budget. The key lesson that we wish to learn from elsewhere is that a vital part of this objective is dependent on learning how to reduce waste both in terms of materials used and effort by people. Our stated aim is to ensure that we add value through making the processes used more efficient. For example, we aim to reduce the occurrence of not getting things right first time by 15% year-on-year. On the basis of lessons learnt in manufacturing we know that this is a proven way to improve client satisfaction and reduce cost.

12.3.2   *Altering culture*

Undoubtedly the need to get people to alter the way that they consider how they work is difficult. People do not automatically think about using measurement as something to be applied to everything they do. In order to encourage this we provide training in the use of a wide variety of tools and techniques which they can apply to all of the processes they carry out. The effectiveness of the use of these tools and techniques is constantly monitored and reported on by the champions to a steering group which can get an overview of progress. This will help the steering group to dedicate resources to areas that champions consider to be a priority and to see where benefits are being achieved. It also allows them to appreciate what the barriers to success are. An important development we use is known as the 'dashboard' which is used as a method of measuring a variety of KPIs and the trends that they are following and which is reported to the quarterly construction board.

Important to us is that everyone regards the data as having integrity. The culture is about encouraging honesty and allowing people to suggest what they think is important. We also have to recognise that different parts of the business have different priorities. There is no point in forcing people to do things if they have other pressures. As we happily accept, what we are doing is a journey which takes time and patience. It cannot be imposed and certainly will not be achieved instantly. My own experience outside construction shows that benefits come slowly at first but that you build success upon success. In automotive industries the development of lean processes and continuous improvement must become accepted as being the norm, not something to be done after, when the proper job is complete.

12.3.3   *A personal perspective*

Being involved in business improvement is a job that requires skill to deal with the following tasks:

- Encouraging people to believe in the benefits of improving customer service to both the organisation and them as individuals;
- Encountering and managing resistance to change;
- Streamlining activities and associated processes;
- Being able to identify champions and supporting them in facilitating others to alter their practices;
- Implementing objective measures;
- Using others outside of Thomas Vale, such as subcontractors and suppliers, to become part of an integrated supply chain that delivers genuine value to our customers.

Business improvement is certainly one that requires dedication to the overall aim and, at times, the ability to cope with scepticism from those who are unconvinced of the need to change.

12.3.4   *A practical example of using tools and techniques to achieve improvement in a social-housing project*

This has been a project to refurbish homes which requires Thomas Vale to carry out full-scale improvements with maximum efficiency, least time and in a way that limits the disruption to tenants as evidenced by customer feedback. Importantly, a number of 'toolbox' techniques have been used to explore ways of achieving these objectives. These are delivered by a group of 'stream' managers who, in conjunction with representatives from all of the key trades, consider all of the processes used in carrying out the work. Importantly, our training department, in collaboration with a local, further-education college, has provided NVQ training for all staff in using improvement tools and techniques.

Typically, we have a group of 15 people working on each 'stream'. This team will consist of our employees and staff from the subcontractors who actually carry out the work. They meet regularly in order to both reflect on experiences and lessons learnt and to proactively consider changes in processes for the future. The tools used will aim to achieve three objectives:

1. Diagnose;
2. Improve;
3. Sustain.

Given the repetitive nature of the work being carried out, it allows us to use techniques such as the 5 Ss, 5 Cs and 5 Zs. These are as follows:

The 5 Ss

SORT   (SEIRI) Eliminate what is not important but keep what is needed.
STRAIGHTEN   (SEITON) Put whatever will be needed where it can be easily accessed when required.
SHINE   (SEISO) Ensure that the workplace is clean and tidy; mess is symptomatic of confusion and contributes to inefficiency.
STANDARDIZE   (SEIKETSU) Maintain order and encourage perpetual neatness.
SUSTAIN   (SHITSUKE) Which reinforces the message that personal commitment and pride in the job are the key to success and maintenance of excellence.

The 5 Cs

These reinforce the absolute importance of workplace organisation and are based on the following principles:

CLEAR OUT   Separate the essential from the non-essential.
CONFIGURE   Have a place for everything and make sure that everything is where it should be.
CLEAN AND CHECK   Constantly monitor and check the workplace for deviation.
CONFORMITY   Make sure that all standards are consistently maintained by every person.
CUSTOM AND PRACTICE   By making the obsession for excellence the standard way that things are done, everyone is part of the effort to achieve strategic success.

The 5 Zs

ZU UKETORA   Do not accept defects.
ZU TSUKURA   Do not make defects.
ZU NAGASA   Do not supply defects.
ZU BARATSUKASA   Do not create variation in processes.
ZU KURIKAESA   Do not repeat mistakes.

On this project we have been able to reduce the time of 'turnaround' for each dwelling from an average of 35–40 days when we started in 2002 to 22 days in 2008[1]. Our satisfaction levels

---

[1] Continuous improvement effort means that at the time of writing average turnaround is now 15 days.

are on average 94% which, whilst being good, suggests that there is room for improvement. This is particularly so, as a competitor working on the same scheme has been able to achieve 96%. As I am happy to stress, we may have come a long way on the 'journey' to excellence. However, as car manufacturers have consistently demonstrated, perfection is a never-ending quest. This objective, we believe at Thomas Vale, is what will give us competitive advantage over our competitors. In the current climate the ability to deliver consistent levels of quality and value is key to our continued success.

12.4   Past, present and future – survival strategies in the face of a global downturn in construction work

Alex Housden, Faithful and Gould

12.4.1   Introduction

Strategy is explained by Clegg *et al.* (2005) as requiring managers to be able to respond to turbulence. In particular, they state, '[It] is supposed to lead an organisation through changes and shifts to ensure growth and sustainable success.' Accordingly, strategy is the means by which the management of any organisation makes decisions as to future direction, alteration of its mission and development of values consistent with the changing operational environment. Strategic management, therefore, is crucial to the success (or otherwise) of any organisation.

Against a backdrop of unprecedented growth and optimism in the construction industry over the past decade, consultancy firms, such as Faithful and Gould, have expanded, diversified and stretched their horizons in an attempt to satisfy seemingly limitless demand. However, 2008 saw the economic downturn gather pace across the world and such consulting firms found themselves at a 'crossroads'. As existing projects came to an end, and new projects were being curtailed or postponed, there was a need to avoid being over-resourced and to develop strategies that would enable them to survive the recession. This section describes the way that Faithful and Gould has reconfigured its strategy. Most especially, it explains the difficulties and dilemmas of doing this when the present climate and the future are so uncertain.

Faithful and Gould's history will be described. This will provide the background to the organisation's current position. The options being considered to deal with the future will be examined and the questions of whether the company should expand, contract, reposition, relocate, specialise or diversify will be looked at. Ultimately, the success of the chosen strategy will determine whether the correct decisions have been taken.

12.4.2   A brief overview of Faithful and Gould

Faithful and Gould (F+G) is an international, multi-disciplinary construction consultancy, employing some 2,000 staff predominantly in the UK (27 locations) and USA (23 locations). It also has offices in Dubai, Hong Kong, Shanghai and Singapore. The structure of F+G is organised along two axes, by services offered and by market sectors. The 'headline' services offered are commercial services, project management and consulting. The main market sectors served are general industry, property and transport.

F+G was founded in 1947 and was acquired by WS Atkins in 1996. It achieved rapid growth and also saw major expansion through a series of acquisitions. These included Silk & Frazier in 1998, Yeoman & Edwards in 1999 and then Hanscomb in 2002. During that time turnover of the

UK business increased from £38.5 m in 1998 to £94 m in 2007. Over the same time period the number of UK-based employees has increased by 16% from 1001 to 1163.

F+G's clients come from both the private and public sectors. Private-sector clients are based in a huge variety of sectors and locations, including finance and banking, property development and airport operators. Recently, F+G has increased its public-sector workload considerably with its appointment onto the 'Building Schools for the Future' framework as well as involvement in high-profile projects such as the redevelopment of Birmingham New Street railway station and consulting roles within PFI hospital projects. This growth of profile in the public sector has gained vigour and importance with the deteriorating economic circumstances in the UK which is seeing a contraction in overall private-sector construction activity.

12.4.3   Faithful and Gould's strategy

The core strategy of F+G during recent years has been: 'To use our size, breadth and scale to create sustainable competitive advantage'. This strategy has been presented to stakeholders as *Vision 2010*, a 5-year plan which was launched at a series of events at conference centres across the business in 2005. The headline features of the 5-year plan are titled 'Excellence', 'Identity' and 'Growth'. Under these banners F+G intends to achieve the following by 2010:

Excellence
- Through the effective leadership of services and sectors we will deliver best practice underpinned by in-depth market understanding;
- Having the best people supported by best-in-class training and development;
- Ensure we have market-leading knowledge management, IT systems and business processes.

Identity
- Reinforcing our national identity through a consistent brand, 'global identity, common values and goals, modern image';
- Organisational clarity ensuring greater integration and cooperation through reducing our geographical structure;
- All staff having a clear message of who we are and what we do and where we want to get to.

Growth

- A substantial market in the USA delivering significant growth;
- A mature market in the UK providing modest growth;
- Protectionist Chinese market allowing moderate growth.

Logically, strategy should be based on a combination of retrospection (looking back at past performance), which allows managers to know what worked, and the use of skill and judgement to identify potential market opportunities. The latter, of course, is difficult and predictions are fraught with danger. Until very recently, private-sector clients accounted for some two-thirds of revenue, the remainder of work coming from the public sector. This division represents the huge growth in commercial development which has happened in the developed world in the last decade or so. However, the onset of the 'credit crunch' in 2007 has caused the market for commercial property development to decline very rapidly. Indeed, as *Building Magazine* in May 2008 remarked:

> The Bank of England's credit conditions survey in the first quarter confirmed that corporate credit availability had reduced over the past three

months; a further reduction in credit availability was predicted over the next three. Specifically, credit availability to the commercial property sector was reported to have fallen. At the same time, demand from this sector fell. (Fordham and Bauldauf, 2008)

Recent indications suggest that the deterioration in commercial work will continue unabated for the foreseeable future. Therefore, any predictions for the short to medium term cannot anticipate increases in 'traditional' markets. This has caused a difficulty for strategic thinkers. Has the very success enjoyed made the organisation unable to think beyond the traditional markets where work was plentiful? Newer and more dynamic companies are seen to be more able to innovate and identify new opportunities. So, in order to break out of the established mindset, F+G has acknowledged the need to make crucial strategic decisions to avoid being left with insufficient work to keep the organisation intact. This has created the need for a survival strategy based on four key principles:

- **Specialism** – which requires F+G to differentiate itself from its competitors by offering innovative services, capitalising on niche skills such as dispute resolution or sustainability.
- **Speculation** – which requires F+G to become a dominant market player by expanding through acquisitions of smaller firms and maintaining a diverse presence across the spectrum of services.
- **Migration** – in which F+G looks to capitalise on the strength of international construction economies which are faring better than the UK, relocating staff and resources.
- **'Weight loss'** – by which F+G explores opportunities to cut costs (especially where there is no 'value-adding') from every area of the organisation.

Specialism

Strategy is often referred to in the context of four operating arenas, which are product quality, cost and price, service, and strength of sales channels. Where a competitive marketplace exists, firms look to product differentiation. F+G looks to provide clients with service and expertise that others cannot match. In the context of the *Vision 2010* (an internal plan) and under the banner of 'excellence', F+G has pursued strategic differentiation through depth of knowledge, high-class training and development, and innovative IT systems.

F+G has invested heavily in training and development over the past decade. One very public commitment is its desire to support new members of staff on courses that will assist their ability to think creatively. This is particularly demonstrated by the benefit of employing non-cognate graduates from other disciplines including law and modern languages. The fresh perspectives and skill sets that these people bring has been very useful in giving F+G a competitive edge by maximising innovation and creative thinking.

F+G has invested heavily in IT infrastructure, including the provision of CATO software throughout the commercial services team. This expert system provides cost planning, procurement and drawing-measurement assistance to the quantity surveyors, streamlines processes and provides greater cost and measurement certainty.

Examples of how F+G is attempting to specialise are:

- A new carbon-management function within the building surveying department offering consultancy advice in relation to sustainability and, in particular, the provision of Energy Performance Certificates, which became a legal requirement of all business premises in England and Wales from October 2008;
- Reinvigoration of the consulting department to maximise the exposure of its specialists in construction-related insolvency advice, public-sector procurement, and contracts and dispute resolution;

- The opening of a new office in Canary Wharf in 2006 which, it was intended, would provide consulting services to the financial sector, particularly in the field of sustainability.

Speculation

In the context of acquiring smaller firms, F+G could be described as a major speculator in the past decade. Its growth in size and visibility can be judged by the acquisitions it has made: Silk & Frasier (in 1998), Yeoman & Edwards (in 1999) and Hanscomb (in 2002). Each time this has occurred there has been a step-change in size and, in the case of Hanscomb, entry into the USA.

There are no known plans for F+G to expand further at the present. Even though others may view this as a beneficial strategy (Franklin & Andrews, together with its parent company Mott Macdonald, plan to quadruple the size of their Reading office), F+G believes that it should concentrate its activities on what it does best already.

Migration

The downturn in construction activity in the UK has caused many firms, including F+G, to consider entering emerging markets such as those of developing economies. In the short to medium term, the Middle East appears to provide the best opportunities for revenue with both Dubai and Abu Dhabi seeing continuing strong growth in construction output. In a presentation, Professor David Boyd claimed that UK consultants have been vying for a slice of the lucrative Middle East market for half a century but, as the economic recession causes the volume of work to reduce domestically, the region has become a sponge for surplus staff.

Within F+G there is evidence of staff relocating to the Dubai office, either as a positive outcome of redundancy consultation or in recognition of the increased scale of projects in that region. The potential to use UK-based staff to provide input on such projects is being actively explored. This has the benefits of minimising non-productive time for UK resources and minimising the costs associated with relocating staff across the world. Furthermore, this approach can be seen to be consistent with F+G's 'identity' stream of the *Vision 2010* programme.

The objective is to remove the effect of the geographic dispersal of the firm by adopting consistent procedures and integrating offices through technology and commonality of approach. It is believed that there should be no practical reason why projects based in other parts of the world cannot be resourced from any other part of the world. This should avoid the expense associated with relocation of staff. The success or otherwise of this integration strategy will be a crucial factor in the ability of F+G to survive the downturn.

'Weight loss'

In any period of difficulty in which the objective is survival, cost reduction is likely to play a part. Unfortunately, this is what is currently happening at F+G. As senior managers realise, the ever-present danger is in cutting out so much cost on essential elements of the business as to put the whole organisation at serious risk. This is analogous to a slimmer becoming anorexic. Thus, a distinction must be made between 'keeping fit' and 'losing weight'. The downturn within the UK business has required F+G to adopt a number of proactive cost-cutting measures which would never have been contemplated during periods of stability and growth. During October 2008, F+G announced the desire to reduce staff by means of a redundancy programme. As a consequence, over 40 members of staff have identified themselves as being willing to leave. F+G is not alone and many other major consulting firms are using a similar approach.

F+G is also reducing its need for office space in an attempt to cut costs. Its Birmingham office is due to reduce its net area by one-third by the end of 2008. All staff are briefed with regard to saving space, reducing paper use and being prepared to work in closer confines to one another.

All of these can be achieved very easily and do not require any investment by the company. It can certainly be observed that the active reduction in the physical size and asset base of the UK business was not contemplated when *Vision 2010* was formulated. The growth potential identified in both the USA and the UK markets has proved to be somewhat optimistic in the short to medium term. But nor could the potential severity of economic change in late 2008 be foreseen. Therefore, the question should not be, 'Was F+G's strategy wrong?' but instead, 'Can F+G's strategy be flexible to accommodate the new uncertainty of the future?'

The notion of emergence, based on March and Simon's research (1958), is that decision-making is always undertaken under conditions of uncertainty and imperfect knowledge. The concept of 'bounded rationality' suggests that firms, when making decisions, 'satisfice'. This means that managers, when they reach conclusions based on the evidence available, acknowledge that a degree of uncertainty as to the outcome must always exist. To this end, strategists must work with whatever assumptions exist. The present assumption in F+G is that the UK (and world) is entering a period of economic contraction, the duration of which is uncertain. The need for proactive cost reduction in order to survive until recovery commences is therefore important.

12.4.4 The future

Having examined the present-day circumstances in which F+G operates and considered how its strategy has been adapted to cope with the deterioration of the market, the question turns to the future. How can F+G effectively form a strategy for the future given the inherent uncertainties? The technique of scenario-planning enables managers to assess a range of external influences to determine the likely best and worst case situations and then to strategically plan what factors and decisions may be critical to steer the business through any permutation of events. As Clegg *et al.* (2005) state:

> Armed with a range of scenarios, and a range of strategic responses to deal
> with them, organisations could be more fleet-footed and nimble in respond-
> ing to their environments than if they just assumed that the future was an
> extrapolation from the past.

Scenario-planning is framed by four factors for consideration: social dynamics, economic issues, technological issues and political issues. The technique often involves devising a comprehensive set of questions relating to the four factors, covering aspects such as:

1. The vital issues – what variables does the firm need to factor into its decisions?
2. A favourable outcome – what would be the best possible set of circumstances delivered for the firm in terms of performance?
3. An unfavourable outcome – if things took a turn for the worse, what factors would the firm act on to mitigate and prevent loss?
4. Internal systems – how can the firm add to the favourable outcome?
5. Lessons from past: successes and failures – how has the current situation come about?
6. Decisions have to be faced – what should be prioritised at this point in time?

Chandler (1962) suggested that the strategic plan has influence over its organisational structure. This implies that the structure should be fluid so that it can easily adapt to changing conditions. Past experience of strategies used (particularly when they have proved successful) is instructive. For example, the construction of railways in Britain allowed many small businesses to be transformed by their ability to serve markets over a much wider geographical area than hitherto (and at greater speed which was crucial for produce such as fresh food, e.g. fish). This still remains a good place to commence strategic considerations. So, in order to remain competitive, strategies for growth, vertical integration of supply chains and distant sourcing of raw materials are viable strategic options.

Accordingly, globalisation has transformed the dynamics of the UK-based construction organisations. Domestic consultants such as F+G can no longer rely on the strength of the domestic market but should widen their services to emerging markets. Currently the Middle East is seen as the best short-to-medium-term option to maintain revenue and protect against the downturn in traditional markets. Even this strategy is dangerous. There is evidence to suggest that the Middle East is witnessing a peak in construction activity and may well experience a property downturn similar to the UK and the USA. Even the newly-emergent economic behemoths of China and India are starting to experience decline. There is, it seems, no escaping what is clearly a world recession. Nonetheless, F+G must continue to operate as an organisation that offers its customers the best possible value. To do otherwise will not allow it to emerge with the skills and expertise that will be essential to provide the basis for clients to rebuild and improve their real-estate assets.

12.4.5   'Added value' – a strategic imperative

Michael Porter (1980) is considered to be a leading influence in strategic thinking. One of his key contributions was to introduce the concept of the 'value chain'. This principle states that a firm differentiates itself strategically by the added value of its services from the cost of production to the revenue realised.

In the context of a construction consultancy, added value may be realised through offering multiple services to a client or providing innovative services alongside traditional project and cost management. Understanding how added value adds to the margin between cost and profit is key to strategic management. For example, F+G publishes a quarterly magazine for clients entitled *Solutions*. This features expert opinion and advice on a range of topical issues.

In an ever-more-competitive market, adding value will be one of the key criteria by which clients will choose F+G over its competitors. Strategic managers must make the right decisions about how to add value, both in terms of what clients value and where costs can be reduced through activities such as outsourcing.

12.4.6   Conclusion – emergence

Much has been said within this piece about the fundamental considerations which F+G must make whilst positioning itself to remain competitive and deliver the core aims of the *Vision 2010* programme without being derailed by the economic downturn. But a crucial feature of an effective strategy is that it is emergent – that is to say that something additional can be gained from it, over and above the simple input of resources and output of profits. Weick (1979) suggests that strategy can provide meaning and direction for organisations and give employees confidence. Moreover, Clegg *et al.* (2005) argue that strategy provides a 'picture' of where the organisation is and assists in providing a sense of what is going on and why. As a consequence, it helps in 'orientation'. Importantly, they assert, it should be remembered that strategy is 'a social construction of reality, constructing a terrain it ostensibly mirrors.'

Weick insisted that strategy is distinct from planning. Whereas planning is a means of addressing the future through creating visions, assessing opportunities and threats and so on, strategy is emergent because it motivates and animates an organisation. It will be interesting to observe whether, after five years, the business plan and the positive messages contained within *Vision 2010* will prove emergent. This was considered to be something that would be difficult to measure. Experience shows that this was not the case. The emergent qualities of *Vision 2010* may be that a workforce of enhanced skills, equipped with innovative technology and positioned to operate efficiently in the world's growing economies, enables the firm to emerge from the downturn stronger than its rivals.

12.5   'Specialisation vs generalisation in the construction industry – a strategic overview': Adonis construction

Oliver Clements

12.5.1   Who are Adonis?

Adonis is a family-owned construction company based in the west midlands. Founded in 1978 by the current chairman, Neville Clements, the company has been active in a variety of construction markets. However, the last twenty years have seen Adonis foster a strong association with the automotive retail sector (construction of car showrooms and workshops). The advantages and disadvantages of pursuing this strategy are discussed below. Using inspiration drawn from elsewhere, the company has engaged in organisational change intended to strengthen its strategic position. This has resulted in the relaunch of the Adonis brand.

12.5.2   Theory in practice: choice of strategic process

Senior management at Adonis have traditionally followed a strategic process based on an **emergent** approach (rather than that which is **prescriptive**). It was thought at the time that Adonis' operating environment was the reason behind taking this direction. However, recent thinking has exposed the problems with that assumption. To explain this point, it is worth looking at the positioning of the building contractor in the value chain. Traditionally, the contractor is located 'high up', adjacent to the final customer who takes delivery of newly constructed (or refurbished) properties. Further down the chain there are many levels which consist of subcontractors and/or suppliers. Additionally there are manufacturers, wholesalers and, potentially, the producers of the raw materials used. All of these are part of the totality of the processes that allow Adonis to serve its customers. This offers Adonis many opportunities to explore **vertical integration strategies**. Equally, there can be a downside if too much investment is made which restricts manoeuvrability and capacity to adapt to market forces.

  Like any other contractor, Adonis will only be able to operate successfully if it achieves a constant flow of work from clients. However, the ability to attain this objective is often dependent on market trends which, ultimately, rely on the vagaries of consumer preference. These underlying factors often change unexpectedly resulting in potential future work being lost. This could leave a gap in the forward-order book worth many millions of pounds. It is worth bearing in mind that, even though a £50m-turnover company may seem like 'big business', a construction company of this size will perhaps start only 20 contracts in any one financial year. In this sector, the ratio of 'average sales transaction size' to 'total turnover' is much higher than in many other sectors. This highlights the importance of converting potential orders into real projects that will occupy resources.

  The uncertainty of the marketplace, coupled with the importance of securing large one-off projects, is the main reason why a prescriptive approach to strategy had thus far been ignored.

Adonis has adapted its working practices to suit the reality of its environment. Most importantly, the board of directors set the objective of producing high levels of customer satisfaction (and customer awareness). A balance must be struck between the two approaches of specialisation and generalisation. This can be done (under the emergent approach) with reference to **customer analysis** (see Levitt, 1960) and making the resulting adaptations to the business strategy to suit. In recent times, Adonis has refreshed its approach to this dilemma, which is discussed below.

12.5.3   Theory in practice: Porter's Five Forces

The construction industry has traditionally been parochial. The reason is that in the past most organisations have tried to keep their work within easy reach of head office, usually no more than 50 miles away. Spreading resources across too wide an area was thought to be incorrect. If enough work can be secured, this would be a sensible strategy. However, as contractors have become more specialised, there has been a need to travel further to secure sufficient projects to survive. Potentially, this has a detrimental effect on the controllability of its key resources and it might be argued that specialisation and resource control are inversely proportional to each other.

Whilst this is certainly another factor in the 'specialisation vs. generalisation' debate, it also leads naturally to a discussion of the **forces** which influence a construction company. Porter's Five Forces Model, and similar approaches, can assist in developing a strategy to gain competitive advantage, and a decision of where to strike the balance between becoming a specialist contractor and remaining as a general contractor.

Firstly, there is the relative bargaining power of suppliers. Porter tells us that suppliers are more powerful if there are only a few of them, or if there are no substitutes for the supplies they offer. In a specialist-market sector, both of these factors are present, as detailed, and repetitive specifications will often restrict the range of potential suppliers. This factor is more acute than in the 'general' scenario, where a greater pool of suppliers is present, due to the sheer variety of building types in myriad different markets.

Porter argues that clients (or 'buyers' in his terminology) have more power when the selling price from the organisation is unimportant to the total costs of the client's company. This will often be a much smaller consideration in construction, because of the large, one-off nature of projects. However, if the client were a larger organisation with very deep pockets, the cost of one building project could be immaterial compared to the amount of trade it undertakes in a financial year.

A weakness of specialisation is revealed in another of Porter's 'client strength' conditions when the product (or service) that is being sold is undifferentiated. This is quite a subtle point. It may seem that the specialist contractor could use this factor to its advantage. However, in some specialist markets, such as that in which Adonis operates, the actual nature of the product (the type of building, the technologies used) is only one of a number of contributing factors in how the specialisation has been established. Another, perhaps dominant, factor will often be the retention of ongoing relationships with certain clients or professional practices. This makes it easier for the client to 'spread his net wide', as he can construct a tender list comprising companies perfectly capable of undertaking the works, even if they are not all specialists! This factor enhances client power and forces prices down.

Although most of the points highlighted thus far have seemed to favour a generalised approach, the area where specialisation scores heavily is in 'barriers to entry'. Porter argued that there were seven major sources of barriers to entry, and most of these can be utilised (to a greater or lesser extent) to forge a specialisation. Exploiting economies of scale and product differentiation are the very essence of specialisation.

Another important factor is described by Porter as 'cost disadvantages independent of scale'. Porter states that when an established company knows the market well and has the confidence of major buyers, it can become daunting for new entrants to gain a foothold in the market. A parallel is often drawn to the attempted entry of Korean and Malaysian manufacturers into the

European car market. This factor is certainly in evidence in specialist construction markets, and is one of the key benefits of specialisation, as general construction projects (and, indeed, whole sections of the market) can be 'bought' or, in the case of markets, dominated, by the very large construction companies if they so choose. They can afford to undertake a loss-leading project much more easily than a smaller specialist outfit. This factor may also be the reason why many smaller construction firms use specialisation as a 'lever' for growth, ultimately broadening their operations once they reach a critical mass.

'Capital requirements' is often a less significant factor in construction. This is because the nature of the typical payment cycle lends itself to low capital requirements on behalf of the contractor. The downside of this is the increased volatility in the industry – starting in business as a contractor and accelerating turnover growth is arguably made too easy.

12.5.4   *Theory in practice: the Seven Ss (implementing strategy)*

At some point in the strategic planning process, the focus will shift to the implementation stage. To ensure a successful implementation, the strategic plan needs to be compared to the organisation to determine whether it 'fits'. To help in this process, Peters and Waterman (1982) advocated the Seven S model which suggests that strategy should be 'interwoven within the fabric of the organisation'. The Seven Ss consist of the following:

- Structure;
- Systems;
- Skills;
- Style;
- Staff;
- Superordinate Goals/Shared Values;
- Strategy.

It is important to note that there is no implied hierarchy of these elements. In an 'excellent' organisation, Peters and Waterman argue, each S complements all of the others to help it achieve its objectives. While a full examination of each S in the context of Adonis is beyond the scope of this section, careful thought of how the company should approach its Ss has been carried out. Additionally, the process of 're-engineering' has been carried out within Adonis. The driving force behind this change was an unexceptional period of trading in 2006 when the company returned what was regarded as a very small profit. The board of directors needed to find a solution to inject fresh enthusiasm into a thirty-year-old company and decided that there was a need to embark on a new strategic direction.

After extensive internal and external consideration during which the plan was endorsed by the company's accountants and management consultants, BDO Stoy Hayward, the directors decided to split operations into two separate divisions. In the new arrangement one division works nationally and undertakes specialist work (automotive retail); the other division operates locally and undertakes all types of work (general works). Each division would be allowed to develop two completely different cultures and ideals.

Application of SWOT analysis has been used in considering how to manage each of the divisions. For example, by forcing the general works division to consider its 'opportunities', strategies that utilise the company's understanding of local markets are emphasised. The danger of attempting to try to focus on particular markets has been avoided. Importantly, the new configuration allows each division to forge its own future and not be held back by trying to be 'all things to all people'. As an example of this principle in action, the general-works division no longer needs to be constrained to suppliers in the specialist markets, and is free to develop wider links with suppliers, which could expose more value-engineering opportunities. This demonstrates the way that the company believes it will best serve all of its stakeholders.

12.5.5   Conclusion

In an age when innovative technology is used as standard in many other sectors, people's fascination with **categorisation** can be properly fulfilled. Putting 'things' (people, companies, projects and so on) into neat 'pockets' or 'categories' is facilitated by a multitude of database products. This is true both at work (think of customer databases) and at home (Apple's iTunes is a good example). In the past, Adonis has often suffered from being typecast as a result of its specialisation. 'They are the car showroom people' used to be a common reason why the company failed to make a tender list, perhaps for a health centre, school or industrial development.

Most of the strategies discussed in this case study refer to the levels of functional strategy and business strategy. With its two new divisions, Adonis' corporate strategy is based on a belief that it can continue to build its brand 'the old way' (through high levels of customer service and delivery), but that this same brand can represent different things to different people. Through this concept, Adonis will always be put in the right 'box'; they will always seem like a good choice for a particular construction project regardless of its nature. This enables the company to cast away the ties that are associated with being a specialist contractor, whilst retaining the benefits.

Adonis has made a commitment to never surrender the entrepreneurial spirit (and the associated emergent approach to strategy) that lies at the heart of the company. In such a volatile industry, the road ahead is never certain (although inevitably very bumpy!). However, with turnover growth currently standing at a healthy 30% year-on-year (as of 2008), the first signs that the new strategic approach is working are beginning to appear.

12.6   Interserve

Edward McDonald

12.6.1   Overview

The construction industry today is not as it was 25 years ago, with a massive emphasis on H&S, quality and customer service. Over this period professionals in construction have seen leaders in other sectors implement quality-based philosophies to transform their enterprises into market leaders. The desire to become 'world class' is now common in sectors such as manufacturing and electronics. Construction, on the other hand, appears to have struggled to appreciate that it needs to implement quality techniques so as to provide a product and service that can be regarded as being consistently excellent.

Improvement of quality requires every member of an organisation to embrace the principles of continuous improvement. Quality management is a customer-oriented approach that stresses the effective use of people and emphasises the application of process improvement to continually improve everything it does. In addition, the importance of cultural change should be stressed by leaders who must demonstrate a sincere commitment to supporting the efforts of every person. The payoff, it is believed, will be the ability of every person to derive benefit from carrying out work that is 'right first time' and to produce higher levels of customer quality.

12.6.2   What have we done?

Interserve has various protocols and policies in place to support its improvement efforts. One very important feature is a feedback questionnaire used for both clients and employees. The client questionnaire was structured to provide information from clients which feeds into three of the Construction Best Practice Key Performance Indicators (KPIs). The success of the client questionnaires is strongly correlated to the results from 'employee perception survey', based on the well-established EFQM Excellence Model (see 12.6.3, below).

It is important to recognise the difficulty of getting accurate measurements of client satisfaction. Given the diversity of clients and types of work carried out, no one proforma will be entirely appropriate. However, whatever information and data that can be collected which gives an indication of performance are better than nothing, which used to be the norm. The key objective is to help the business to better understand its strengths and weaknesses. The outcome will provide us with indicators of what we can do to enhance our expertise and work on our weaknesses. This, together with the employee survey, will assist us in understanding if employees are being given the desired level of respect and recognition, and enjoyable, challenging work.

The employee perception survey acts as a tool to measure how satisfied employees feel. This should determine and influence performance which can be used to inform our understanding of how we can produce consistent client satisfaction. Interserve has undergone tremendous change in recent years, especially in terms of its partnerships with key clients. This has been coupled with new staff entering the company who must be integrated and assisted. The EFQM model helps to understand the culture and type of leadership that currently exists at Interserve. Consideration of factors contained in this model has helped shape culture, generate a motivated workforce and achieve improved customer satisfaction.

12.6.3 Using EFQM

Adopting the EFQM Excellence Model is a commitment by Interserve to continue to achieve improvement in every aspect of the business. Excellence and improvement are vital parts of the quest to use TQM to drive the business forward. In particular, it assists in annually assessing how employees perceive the company's performance in key business areas. Surveys, a vital part of developing intelligence, pose questions based on the model that include the five enablers: leadership, people, policy and strategy, processes, and partnerships. Such surveys enable Interserve to benchmark and compare current performance with historical data and identify areas where there has been demonstrable improvement and, of course, the need to improve in the future.

The importance of customers is fundamental to using EFQM. Accordingly, client feedback is essential because it demonstrates how acceptable the physical product is and where we should use our quality systems. It also provides a clear indication of how effective our processes have been from the client's perspective. Any negative feedback indicates where amendments to processes should be actively considered. The objective, of course, is that Interserve improves its ability to provide a superior service.

12.6.4 The importance of people

Interserve views its people as being its most important asset. Therefore, effort has been dedicated to improving all aspects of the way that the HR department carries out its function. At Interserve, line manager are considered to be leaders. Organisational transformation cannot take place without their active cooperation. Culture is based on a 'can do' attitude and dedication is recognised and rewarded. Traditionally, Interserve was a company run by engineers, not managers. Whilst many managers were able to make the transition successfully, some did not. In recent years many of the older board members have left which has provided the opportunity for change at that level which, in turn, has cascaded downwards in terms of beliefs (culture). Most importantly, Interserve has moved from committing to short-term objectives to committing to long-term objectives. In order to achieve this transformation in culture, Interserve trains the workforce accordingly so that they can rise to the challenge of becoming known for excellence in a service-based environment.

The HR function at Interserve is evolving. This is demonstrated by Interserve's purchase of a new HR IT self-service management system which came into action in January 2009. This system will work throughout the company highlighting the required training for staff. This is yet another step towards developing quality through culture. In support of this, the HR department has also engaged recruitment specialists to ensure the best people available (or who can be persuaded to be available) are employed in the roles needed.

Previously, it had been usual to simply train people so that courses were full. Courses are now targeted to the needs of people to improve their ability to carry out their roles. In addition, employee comments are acted on to ensure that training is continually enhanced. The policies initiated by the new generation of directors have developed the 'internal organs' of the company. This allows Interserve to continue to create a motivated, people-oriented culture through effective leadership.

12.6.5   *A personal reflection on the influence of quality management*

On commencing at Interserve, my first two roles allowed two weeks' preparation before the project commenced. For a variety of reasons, the project started off behind programme. Because the project was trying to catch up and meet targets, health and safety, quality and customer service were jeopardised. This meant the site team had to put extra effort into making sure that these essential elements were carried out. Typically, plans and drawings arrived on site at the last minute and there was insufficient time for preparation or control over the design process. Using quality management and teamwork enabled the site team to overcome these difficulties. To everyone's satisfaction, the project was a success.

As a site manager, my experience over the past five projects has been that more time is now given to preparation. More time is allocated to setting up the site, reviewing drawings and involvement with subcontractors and cost plans. There is more overall control throughout a project because site staff are usually put under less pressure; there is less 'target date manipulation' or ad hoc decision-making. The organisation is undergoing transformation and listening to its workforce. A major effort is being made to alter the values that drive it and encouragement given to support for the use of TQM. A survey was carried out which summarised the values that Interserve recognises as being fundamental to service excellence:

- Provide opportunities for staff to function as human beings rather than as resources in a process;
- Provide opportunities for each organisational member, as well as for the organisation itself, to develop their full potential;
- Attempt to create an environment in which it is possible to find exciting and challenging work;
- Provide opportunities for people in the organisation to influence the way they relate to work, the organisation and the environment.

12.6.6   *Conclusion*

Interserve is an organisation that has experienced much change in the past decade. This has come in the form of brand and image change, and the shift towards framework arrangements (often called 'partnerships'), as well as the reorganisation of all functions. Whilst employees do not have a responsibility to manage change, they must be included in considering how best to achieve the desired objective. Planning, implementing and managing change in a rapidly transforming environment such as the construction industry are increasingly challenging. Dynamic environments such as these require dynamic processes, people, systems and culture.

12.7   Innovation and creativity in Morgan Ashurst

Daniel August

Morgan Ashurst is a leading full-service construction business with more than 2,000 employees in the UK. Its activities range from small works and maintenance services to large-scale complex projects across the commercial, government, leisure and retail sectors. In 1994, the constituent parts of the regional construction businesses had a turnover of £30 m. Following considerable growth, acquisitions, and a large-scale restructure (which took place in 2001 to merge six regional divisions under the brand of Bluestone), a £400 m business emerged. The restructure was designed to bring a consistent approach to the regional businesses. Since then further acquisitions have strengthened the business in key areas. In 2007, Bluestone acquired AMEC Design Project Services and the Major Projects division, which has enhanced the company's project portfolio and target range of projects from anything between £50,000 and £300 m. Turnover of the business now stands at approximately £850 m.

The belief that business strategy is frequently influenced by its history is true in the case of Morgan Ashurst. Graham Shennan, managing director, a Scotsman, strongly believed that the Scottish building market would not accommodate the entrance of what had been the Bluestone business, especially as it had no previous presence. The acquisition of the former AMEC offices in Glasgow provided a direct strategic link into the Scottish market.

12.7.1   Leadership in Morgan Ashurst

The corporate mission statement reads:

> We will be recognised as the best in industry through consistently providing an exceptional customer service. Perfect delivery will be our clear differentiator and conduit for achieving market leadership.

This form of strategy tends to adopt both an emergent and prescriptive approach respectively, by continually adapting and improving the customer experience, but also by presenting clear service-level statements to the customer.

The business's definition of future state is an emergent process, with systems, processes and procedures refined through feedback and self-developed by the business units rather than imposed.

12.7.2   People in Morgan Ashurst

Morgan Ashurst has identified four guiding principles which underpin the company's strategy and objectives: firstly, through recognition of 'people' and their rights to be safe and to be listened to; secondly, through 'ownership' by taking responsibility for our actions; thirdly, by 'enterprise' which means challenging how we do things through investment in people and innovation; finally, through the level of 'quality' in everything that is done and how we do it.

Perfect delivery culture is not yet uniformly widespread. However, the merging of project teams and focus-group activities is, and this is likely to help create the concept of 'synergy' which Ansoff (1965b) termed as the '2 + 2 = 5 effect'. A recent 'Being Morgan Ashurst' presentation and activity have been rolled out to each of the six regional Morgan Ashurst teams led by an ITV news presenter to help engage and shape the mindset of staff to think and work as 'one'.

General systems of management for ensuring operational tactics are purposely designed to match the espoused mission. Supply-chain management emerged as a key strategic objective in

2001 and high-value contributions have been derived from this strategic asset which continues to go from strength to strength. Ultimately, the business is looking to create a team of 'strategic partners' built up of specialist subcontractors and suppliers which share in each other's success through continual development. The recent acquisition has served to strengthen its immediate customer base and its wider customer franchise, hence enhancing a strategic objective which is to carry out at least 60% of works derived from partnering or framework agreements by year-end 2008. The business now has a wider portfolio which can accommodate frameworks that include a diversification of works and construction-project value.

12.7.3 Culture in Morgan Ashurst

Morgan Ashurst has set measurable objectives based on the four 'cornerstones' of 'perfect delivery', each included in a document referred to as the organisation's health check. The four cornerstones include: handing over a project 'on time'; 'snag free'; complete with 'O&M manuals'; and to a 'delighted client'. A future 'headline' objective is to achieve 'perfect delivery', which is awarded by the client based on complete satisfaction of the four cornerstones, on 60% of projects completed by year-end 2008. Between 2001 and 2005, projects achieved just 34%, however, the core value was relatively new and seen as an emerging strategy. Future strategic objectives will be set at a higher level and measured accordingly. This will continue whilst the culture is being fully embedded into the organisation. The hope is that strict measurement will become something that is the norm, not something that has to be enforced. The company's bonus scheme incorporates resource-based theory, which highlights appropriateness by rewarding business regions and projects which meet cornerstone targets.

12.7.4 Ways of delivering improvement in Morgan Ashurst

The business has issued a CSR (Corporate Social Responsibility) policy statement to external clients which acknowledges the long-term consequences of decision-making, their continual improvement toward sustainability issues, whilst also seeking to develop an approach to CSR which is emergent rather than process-led. The business regularly utilises the established tools at a higher (board) level, such as the BCG matrix to monitor the performance of the company and competitors, and environmental analysis tools to direct marketing efforts essentially toward relationship-based projects.

At individual project level, attitudes to strategy and objectives tend to vary, with some site management staff preferring to manage as they always have done, potentially neglecting to value what actually makes the client 'tick'. At departmental level, for example, the procurement team have engaged in the emergent nature of perfect delivery, and have published a regular supply-chain newsletter to engage other staff with the key strategic importance of the internal supply chain.

12.7.5 Change in Morgan Ashurst

The previous Bluestone strategy, structured in 2001, maintained a portfolio of small and medium-scale works (circa £50,000–£10m) across the regions to spread the risk if markets or regions suffered financially. In 2003 the organisation centralised certain key functions such as IT and procurement to effectively maintain consistent standards and build on the creation of the supply chain as a major long-term strategic asset. Current objectives seek to undertake sustainable relationship-based work using the portfolio of skills, for example, maintenance, design and mid-to large-scale project management.

The quality and volume of relationship-based projects which the business is involved in and the strength of core competences, including the supply chain and recent resource acquisition, have placed Morgan Ashurst in a favourable position to undertake sustainable workloads. Morgan Ashurst has reacted proactively to market opportunities. For example, the 'Building schools for the future' initiative has provided a high proportion of recent revenue whilst also seeking to expand the brand and reputation in new areas such as Scotland.

In addition, Morgan Ashurst has recently been awarded a public–private partnership scheme in Scotland and is working with joint-venture partner, AMEC Building and Facilities, together with MUSE, an acquired property-development firm, to develop a number of schools and mixed-use commercial and residential development on the surrounding land owned by the client.

12.7.6   *Organisational development for ensuring a match between social and technical systems in Morgan Ashurst*

Morgan Ashurst's 'perfect delivery strategy' (and core value) seeks to identify as early as possible the client's key requirements. This is passed on to the site team and assessed during semi-formal review sessions between the marketing and client teams. In addition to assessing the other three cornerstones, the acquisition of design-project services is now used to provide the customer with access to a specialised internal design and build team. This maintains effective communication and facilitates a close appreciation of each other's needs.

12.7.7   *Innovation and creativity in Morgan Ashurst*

The vertical integration process to acquire AMEC is a resource-based approach, which has effectively provided the company with an opportunity to enhance its strategic asset value. Bluestone's strategy has always sought to be 'light on its feet' and, consequently, took the decision to minimise direct overheads by outsourcing where appropriate. This included its own dedicated plant desk, manned predominantly by external 'Speedy Hire' staff. As part of the acquisition arrangement between Bluestone and AMEC, Morgan Ashurst acquired a team of approximately 250 full-time staff who had acted as AMEC's single point of supply for the provision of hire assets including tools, lifting equipment, general plant, temporary accommodation units and survey equipment. A decision was made to sell existing stock and expand the ex-Bluestone hire desk, with existing internal staff transferring across to Speedy Hire Ltd, forming a preferred supply agreement for the business.

Morgan Ashurst currently utilises the expertise of supply-chain partners to highlight new innovations and advise on regulatory compliance, which demonstrates to the client a forward-thinking approach. For example, the supply-chain manager has coordinated a recent visit for internal design, environmental and procurement staff to the Wolseley Sustainability Business Centre, to highlight the latest product innovations, key issues and relevant legislation.

12.8   As safe as houses – the importance of the NHBC

Paul Williams

The National House-Building Council (NHBC) is an independent and non-profit distributing organisation. As the UK's leading warranty and insurance provider for new homes, NHBC's Buildmark Warranty covers around 80% of new homes built in the UK and currently protects approximately 1.6 million homes. NHBC protects consumers (homeowners) by providing the ten-year Buildmark warranty insurance and by setting construction standards for new homes. NHBC's commitment to working with the housebuilding industry to raise standards is demonstrated in every part of its business, from maintaining the register of almost 20,000 builders through to its **Key Stage Inspection** regime, Building Control work and continually evolving risk-management strategies.

In 2007–8 the organisation carried out nearly a million inspections on over 29,000 sites. NHBC provided 15,000 staff days of training both on and off site. Additionally, it has undertaken nearly 18,000 health and safety inspections, answered approximately 275,000 telephone calls in its customer-service centre. NHBC has a research 'arm' in partnership with the BRE Trust called the NHBC Foundation. The Foundation's research has been well received and is highly regarded in the construction industry. The research conducted throughout the period 2007–8 has added value by bringing independent thinking and clarity to the issues of today.

12.8.1   NHBC's corporate history

NHBC started life as the National House-Builders Registration Council in 1936. It was conceived by Sir Johan Walker-Smith who was at the time director of the National Federation of Building Trade Employers (NFBTE) and proposed that an overarching association would be a good idea. Actual implementation was by a group of far-sighted builders who believed in the objective of giving the public a fair deal at a time when 'jerry-builders' were common and a matter of national concern. In 1936 the government gave the association its support on the condition that the members were nominated to represent the interests of various official bodies concerned with housebuilding and that builders were not in the majority. The name of House Improvement Association eventually became the National House Builders Registration Council and in 1973 was changed to NHBC.

12.8.2   The key objective of the Association

> To encourage and promote a high standard of design, workmanship and materials in housebuilding in the United Kingdom and to take all steps and measures requisite to that end.

Measures used included the establishment of a register of qualified housebuilders, adopting (and revising where necessary) a model specification for homes (now called the NHBC Standards) and creating the administration to oversee the inspection and certification of all new homes where a guarantee is sought.

The company was founded as a company limited by guarantee, one of the two principal methods of constituting and owning a company. A company limited by guarantee is a common vehicle for business activities that are not intended to generate a profit or where surpluses are not distributed. The guarantors of the company, by agreeing to be bound by the memorandum and articles of association, undertake to guarantee the debts of the company if it is wound up. In NHBC's case each guarantor's liability is limited to £1. Any profits generated are reinvested and used in improving and developing the products and services NHBC provides to the industry to further improve housebuilding standards. Importantly, and despite some misconception, NHBC is not part of the government nor is it a charitable organisation.

Today a Council governs NHBC with representatives from organisations or groups who have an interest in raising the standards of UK housebuilding. They include mortgage lenders, consumer groups, architects, surveyors, the Law Society and housebuilders. The chairman of the Council is also the chairman of the board. NHBC's board of directors is responsible for developing NHBC's strategy and managing its financial resources, particularly the business planning and budgeting processes. The board has 15 members, who are ultimately accountable to the Council. The board has eight committees reporting to it: these are consumer, standards, finance, audit, remuneration, nominations, Scottish, and Northern Ireland committees. The purpose of these committees is to monitor, review, and make recommendations to the board on their subject areas.

The company also has two subsidiary companies, each with its own board. These are:

1. NHBC Building Control Services Limited, established to manage the Approved Inspector Licence for the purpose of undertaking building control.
2. NHBC Services Limited, established to manage NHBC's non-insurance services – primarily training, air pressure testing, Energy Performance Certificates, Home Information Packs, Code for Sustainable Homes, Health and Safety and environmental services.

12.8.3   Delivery of NHBC's services

NHBC works with the housebuilders and the wider industry to raise the standard of new homes and to provide customer protection for new homeowners. Fulfilling the organisation's purpose is achieved through applying NHBC's Virtuous Circle business model (see Figure 12.1). This is a model the organisation has developed to help assist with driving quality and improving industry standards through the following:

1. Profitably providing the most comprehensive home warranty and insurance range in the UK housing market, together with a comprehensive portfolio of complementary services to the UK housebuilding industry;
2. Being a national and an international authority on standards and risk management in housebuilding;

Figure 12.1   NHBC's Virtuous Circle.

3. Continually improving standards of new homes;
4. Ensuring sufficient funds are available.

These four key strategic objectives of the organisation are summarised in the figure.

12.8.4   The current market conditions and the challenge for NHBC

Significant change is currently taking place within the UK housebuilding market. This creates challenges for NHBC in achieving its strategic objectives. The changes involve:

● Increased competitive activity in the home warranty market and in the Building Control market with a significant increase in the number of approved inspectors;
● Challenging customer profile through an increased level of activity and change in the supply-side structure of the UK housebuilding market;
● Changing product–build mix by introducing more high-risk build projects – for example, multi-storey residential towers;
● Heightened awareness of sustainability;
● Increasing homeowner expectations (particularly in the industry downturn).

The economic downturn, with its catastrophic effect on the housing market and construction industry, has increased the difficulties faced by NHBC whose revenue is dependent on the level of housing 'starts'. Out of the five areas shown above, the organisation has identified that its main focus must be on competitive advantage which can be achieved by improving customer relationships, ensuring the products and service offerings meet customer needs, making it easier to buy from NHBC and delivering better customer service.

NHBC has, until relatively recently, enjoyed the benefit of being the only approved inspector to have a licence to carry out the Building Control function on new-build dwellings. The government subsequently changed this situation in that any licensed, approved inspector in the UK can now carry out the Building Control function on new-build residential dwellings. This meant that NHBC had to carefully prepare a marketing strategy and closely monitor its competitor activity for new Building Control. This involved a significant change in NHBC's approach to its residential Building Control.

The speed and severity of the economic downturn starting in 2008 has meant the company has had to quickly reshape the organisation in order to operate more efficiently. Whilst it is important to ensure the organisation's focus is maintained on the six areas, this focus has been affected by the unprecedented downturn in new-build registrations. This has resulted in the need for swift decision-making on how to reduce the size of the organisation (number of people) in order to reflect the projected reductions in business for the foreseeable future.

One of NHBC's most successful years for new-build warranty and Building-Control volumes was 2007. Current forecasts suggest that the coming years will not be so buoyant. Therefore, an organisation which has been planning to grow its business must now look to retrench and concentrate on giving itself 'edge' (though value and quality) over the newly emerged competition. In order to do this, training of staff is seen as crucial. This is carried out through the 'Training Academy' which has been established in recent years. NHBC has invested heavily in structured training programmes for trainees as a result of the national shortage of skilled construction workers, particularly qualified Building Control surveying staff. NHBC identified, through its strategic business planning, that in order for the organisation to service its customers and support growth, it had to select individuals with suitable academic (preferably construction-related) qualifications and good interpersonal skills, which could be developed further via a structured training programme. This training programme involved the organisation sponsoring its trainees to attend courses at colleges and universities. By developing its own staff, there is a belief that NHBC will have an excellent skills base that can deal with all types and levels of work once the market for housing starts to improve.

12.8.5 Current market

Unfortunately (as discussed earlier), the organisation has seen a massive reduction in new-build warranty and Building-Control registrations based on figures taken from the same period of the previous year. The downturn in some areas of the business has been significant which has meant that the organisation has had to reduce staff accordingly. In some areas this has resulted in training and development programmes for assistants and trainees to be suspended as part of an economy drive.

The current downturn in the economy has also meant that the organisation has had to make significant changes in overall financial operations. Whilst it has been necessary to reduce the size of the organisation in the short term, management requires the resources to be in place when recovery starts. With this in mind, the company is prepared to carry a notional 15% surplus in staffing numbers so there is some immediate 'flex' within the workforce to deal with any unexpected increase in business. However, there have been many other sensible money-saving exercises implemented by management in order to protect the business and staff.

12.8.6 Overall strategy

NHBC's strategic objectives are reviewed regularly to take account of the ever-changing market conditions and, where change is necessary, to ensure the continued achievement of the objectives. Key business strategies are implemented to steer the organisation accordingly. There are five key business strategies that the company had set in the previous year's business plan but, owing to increased competitor activity and the rapidly changing marketplace, these strategies have had to be revised. The current key business objectives are:

- Build competitive advantage – ensuring that our customers view NHBC as the first-choice supplier;
- Improve organisational efficiency and effectiveness – by optimising our business activity, operations and processes;
- Take a proactive approach in the sustainability agenda – by working with government, maintaining consumer protection and providing assistance to the industry;
- Improve industry support – through the provision of research and information to enable builders to build better homes;
- Adapt risk-management processes – in response to the changing risk profile of new developments.

At present the main focus of the senior management team has been on how best to plan on steering the organisation through the current recession. Modelling carried out in 2007 predicted a reduction in business volumes and models of 10%, 20%, and 40% were analysed to establish what effects this would potentially have on the business. At the time it was not envisaged that the organisation would have to consider using the 40% model – something that is now reality. The economic downturn and change that the organisation is experiencing in its market have required the management team and staff to focus on winning business, delivering excellent service and retaining and developing customers. It is extremely important that this is carried out in the most cost-effective way.

12.8.7 Culture

NHBC is a medium-size but complex organisation with many business units and approximately 120 managers looking after approximately 1,100 staff. These employees work in all regions of the UK carrying out a variety of jobs. Many of these employees work from home. There are also office-based employees in Amersham, Milton Keynes, Droitwich, York and Edinburgh. Many of these will talk to other staff members regularly in other regions of the country but may not actually meet them face to face. This dispersion of staff around the UK is just one of the challenges that management face when tasked with building a culture that drives the purpose and function of the organisation.

NHBC's Virtuous Circle strap-line, 'Raising Standards to Protect Homeowners', creates the foundation for the organisational culture within. The organisation's main purpose is to continually promote improvement, quality and standards in new housebuilding. NHBC is looked upon as the construction industry's standard-setting body. Therefore, to enable the organisation to meet and maintain its reputation, the culture within the company has to reflect this in order for it to achieve its purpose and goals.

Leadership plays an important role within any organisation and the development of its culture. The chief executive and senior management team of NHBC establish the purpose and goals of the organisation and communicate this message through the management structure to staff. The management team plays an important part in the creation of the culture of the organisation as these individuals are tasked with shaping and focusing staff in line with the message filtering down from the chief executive. Managers are also tasked with recruiting new staff and the recruitment process is a crucial tool used by the organisation to select individuals that fit or can be shaped and developed within its culture.

NHBC is very focused on training and qualification programmes that support the improvement of housebuilding standards, risk management and consumer protection. It is also important for the organisation (as industry leaders) to ensure that staff are continually developing and up-to-date with the latest technology, products and issues that affect the industry. This is achieved by encouraging attendance of internal and external training events and ensuring staff become professionally qualified within their field.

Staff are encouraged to feed back any ideas or thoughts they may have on ways to improve business and customer relations. The culture within the company is focused on discussing issues and problems rather than shying away from them. The management team is encouraged to maintain an open-door policy when it comes to their relationship with staff. It is widely accepted throughout the company that if someone makes a mistake they are expected to bring it to the attention of their manager or colleagues. This early identification of issues allows the problem to be dealt with in a timely manner as a team and, by doing this, allows improvements within the business to be made if required. This promotes and establishes a culture of openness and honesty which is essential within any culture striving to continually improve and develop its business. The continual feeding back of information and review of issues in any part of the business allow corrections to be made; the process is then re-implemented, then monitored and then implemented if re-monitoring runs successfully.

NHBC has its own staff association to ensure that consultation and feedback is gained on any major changes that the organisation plans to make. The association is also involved with negotiation of working hours, annual pay reviews, representation of staff on disciplinary matters, and terms and conditions of employment. This demonstrates the importance the company places on its workforce and it identifies that consultation with the association plays an important role in forming its culture. NHBC is an organisation that listens to and learns from what its staff say.

Owing to the nature of NHBC's business, i.e. insurance-based, this has created a risk-averse culture within its business planning and decision-making process. NHBC has a complex and large management structure. Careful consideration is given to making key decisions, and the implementation of these decisions does take time. This approach suited the organisation when it had the luxury of enjoying the benefits of limited competition and a dominant market position and when the speed of its decision-making didn't affect competitive advantage. However, it has been identified in the company's strategic business plan that one of its main focuses is the increase in competitor activity. In order to stay one step ahead of its competitors, NHBC needs to be able to ensure it can react quickly to business opportunities within the market.

NHBC has a number of employees who have stayed with the organisation for many years. The quotation below has been extracted from an e-mail that was sent by a retiring member of the senior management team. It is possible to create a culture within an organisation that people are proud to be a part of:

> I joined NHBC in 1984 because I believed in its philosophy and its purpose. I have never regretted it. We make a very real difference in thousands of people's lives and our unique company structure has its benefits.
>
> At one time I wondered if NHBC, as it expanded, would lose the family spirit which I so admired when I joined. But it's clear to me that it lives on with our chief executive, the senior management teams and, of course, with you all. The family survives.

12.8.8 The future?

NHBC faces challenges which are no different from those of any major organisation affected by the current economic climate. However, the years of risk-averse, strategic business management by NHBC have ensured that it can remain solid through even the toughest fiscal periods. The significant insurance reserves which NHBC built up and the retention rather than distribution of historical profits have meant that the organisation has the financial strength to survive the current savage retraction in the housebuilding industry. This management philosophy is underpinned by ensuring the culture within the organisation is one that is stable and encourages staff development, improvement and business excellence.

12.9 Strategic management and change in construction: 'The Argent Perspective'

Ross Fittall

12.9.1 An overview of Argent

Argent is a well-known developer which specialises in city-centre regeneration projects. It currently has a number of high-profile developments around the country. Argent was formed in 1981 and has successfully completed over two million square feet of city-centre development, notably including Brindleyplace in Birmingham, and Piccadilly Place in Manchester. Argent is involved in London's largest single development site and one of Europe's largest regeneration projects: the 67-acre, 8-million-square-foot mixed-use redevelopment of a brownfield site at King's Cross.

The current economic climate has made property development a much more difficult prospect in the last year. As any cursory reading of the press makes clear, a number of companies have experienced a rapid downturn in work and have had to react with cuts in staff and expenditure. This presents a testing time for any company involved in the property and its associated construction industries. The main emphasis for Argent is to ensure that its strategy enables it to survive and, when workloads improve, be in a position to react to increased demand.

12.9.2   Leadership, culture and people in Argent

Argent's ultimate objective is to be the most widely respected and successful developer in the UK. In so doing, it is anticipated that the company can deliver its shareholders with an excellent return on their investment. This objective originates from the original two founders of Argent, Michael and Peter Freeman, whose desire was to create a significant business which provided excellent service (and 'product') and had a reputation for fair dealing.

Argent's strategies remain heavily influenced by the two current figureheads of the company, the joint chief executives, Roger Madelin and David Partridge, both of whom lead by example. They continue to espouse the culture created by the Freeman brothers. Developing teamwork and fostering excellent working relationships are crucial to the running and success of any Argent development. Their example of 'work hard, play hard' is found throughout Argent and their principles of trust, respect, reward, mutual support and mutual criticism override formality and sustain enthusiasm throughout the company (Argent, 2007).

The leaders of the company, which applies to the managing directors of the newly formed Argent (UK Developments) Ltd and other directors of the company, do not follow what might be seen as standardised textbook strategies and processes. Their knowledge, combined with their experience working for Argent, enables them to formulate the company's strategies, taking approaches from different strategic management models. Whilst there is evidence of what might be called 'command and control', they encourage other senior managers to play a significant role in forming the decisions of the company. Additionally, the symbolic mode, where the company's vision and overall mission established by the company's founders are shared by the staff, still pertains today.

Importantly, formal documents to communicate the mission to new or prospective employees are not used. Instead, simple principles are established and communicated within the organisation by teamworking and face-to-face cooperation. Fundamental to this is that Argent want to work with people with similar attitudes and with whom they can get on. One document that does exist is *Argent Attitude*. Written in the early days of the company, it set down principles, and what it meant to work for Argent, or be part of an Argent project. This document remains available to employees and is still regarded as the Argent way of doing business, still relevant, years after it was written.

In addition to *Argent Attitude*, Argent's culture is also demonstrated through a number of statements that have been communicated verbally to staff on a regular basis and are recognised by the majority of staff. Following these will set the company on its way to achieving the main aim identified above. Some of these include:

- We will focus on a limited number of significant and meaningful projects:
- in market areas with strength and diversity, reducing the locational risk;
- in a great place or a place big enough to create a new place and create values in unproven locations;
- in accordance with or with the thrust of government policy;
- We will work on projects that are stimulating, fun, rewarding and a source of pride, reward and involve people we like working with;

- We will add value by retaining an equity interest and asset management in our multi-phase projects;
- We will put sustainability principles into practice;
- We will try to be open, honest, fair and good team players.

The culture established through the *Argent Attitude* document is visible throughout the company and its working practices. It is worth noting, however, that certain aspects involved in running the company are carried out in a more formal and structured way, though not necessarily explicit to staff. This includes how the board conduct their meetings, sign off papers and make important decisions. This is a necessity in the modern business environment due to the size and importance of these decisions and being accountable to the shareholders.

12.9.3   General systems of management

To a casual observer, it might be considered that 'established' management theories are not significant in the implementation of Argent's strategies. For example, a mission statement does not exist in a formal, written format. The definition of strategic management illustrates that it is not a discipline that exists within an organisation but a process that managers go through to determine a company's strategy. Goldsmith (1995, p.1) defines strategic management as:

> ... the process by which managers set an organisation's long-term course,
> develop plans in the light of internal and external circumstances and under-
> take appropriate action to deliver these goals.

The statements listed in an earlier section are clearly not random and provide focus for the type of projects Argent want to be involved in. These help steer the company in one direction: to achieve the stated aim of becoming the most successful and respected developer in the UK. There is certainly a wide range of established theories that academics and business individuals have developed over the years, including Drucker's Management By Objectives, Porter's Five Forces Analysis, McKinsey's Seven S Model, (Thompson and Martin, 2005, p.139). A common set of tools that encompasses elements of these theories and binds them together, plus a number of others, can be presented in the following process:

1. **Identify the mission and goals of the organisation** – giving a sense of direction to people in the organisation;
2. **Provide an analysis of the external environment and internal resources** – through SWOT analysis, PESTLE analysis, Five Forces, Portfolio analysis and the Ansoff Grid, for example (Thompson and Martin, 2005, pp.138–41);
3. **Formulate a strategy to meet the objectives** – blending the development of missions and goals with environmental scanning and internal auditing;
4. **Implement the selected strategy in an economical and timely manner** – formally within the organisational structure and informally within the organisation culture;
5. **Monitor and evaluate performance and make adjustments in light of experience and changing conditions** – taking measures to reposition the organisation if gaps are found between goals and performance.

Figure 12.2 illustrates how following the points above is an ongoing process and that there is interaction between all five of the categories.

Argent's senior management may not directly consider such established theories when making decisions on the future of the company but it is envisaged that they may run through a similar

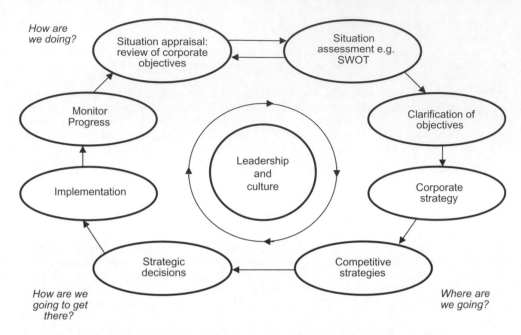

Figure 12.2   A diagrammatic representation of analysing practice in Argent.

process to that outlined above. The objectives and the statements set a clear vision for the staff, encouraged through the teamwork principle, giving a common effort and joint goals.

12.9.4   Creating improvement and sustaining benefits

Argent's philosophy of conducting business with contractors and consultants is implemented around trust and mutual respect. A conscious and strategic approach to engaging with members of a project team is the early involvement of contractors and subcontractors. The right people are therefore involved at a stage where they can appropriately influence projects, and projects can be properly planned and coordinated, maximising the time spent on the project and avoiding wasteful activity. Another practical example of a strategic decision by the company is the negotiation of the price with contractors, with the intention to minimise the final account, as opposed to the belief that minimising the contract sum is the overriding objective. The intention of this is to have a less confrontational environment and a more risk-averse project than has traditionally been seen in construction projects. Such an approach avoids confrontation with the construction team, which can equal less risk and more ownership of the project on their part. This, in turn, equates to less managed time from within the Argent team and therefore also has financial and time benefits.

Building relationships is key to achieving results and repeated success. Argent use a small number of select contractors that they have worked with on numerous projects and who understand not only the Argent way of working but also the standards expected. Since appointing a contractor for the first building at Brindleyplace, Argent has not tendered the contract for another building for this development. This, combined with the use of a standard Argent specification (amended for each individual project), has resulted in high quality and excellent buildings that deliver value and demonstrable benefit to clients and occupiers alike.

A useful practice employed by Argent after each project is conducting a full review of the project and a 'lessons learnt' exercise. These are then incorporated into future projects creating benefits in efficiencies of time spent on formulating solutions and on duplication of work, and minimising risk on future projects. This, in turn, has financial benefits for the company.

12.9.5   Coping with change

The property sector has been buoyant in the last decade or so. However, 2008 has been a year when the global financial crisis has hit many sectors of the economy, particularly property and construction. Argent has survived earlier recessions and was able to change and respond quickly to opportunities and threats. In doing this, it has been prepared to take calculated risks. The current climate of the property market demonstrates just how quickly the climate can alter. Commercial development pre-lets are a great way to reduce the development risk but these can be hard to achieve. Argent has therefore regularly undertaken risky large-scale speculative development.

Responding quickly to changes in the market has helped Argent succeed. A number of company characteristics have facilitated this. No long chain of command exists. Major strategic decisions can be made due to the flat management structure and, significantly, by the way Argent raises its finance to undertake developments. Backed by BTPS, the UK's largest pension scheme, there is a significant pool of capital available which can be accessed quickly rather than having to go through a lengthy approval process to obtain funds from a bank or other financial institution. Development sites can subsequently be acquired quickly even at times when competitors may be finding it more difficult to obtain funding (precisely the problem at the moment for many developers).

When the economy is buoyant occupiers are more likely to consider additional expenditure relocating to new premises when their leases expire. Undoubtedly occupiers, many of whom will be experiencing reduced revenue, will be seeking to negotiate their leases downwards. Sadly, some may go out of business which leaves a surplus of space. Argent has attempted to differentiate its 'product' so that potential clients enjoy higher value. For example, in our development in Birmingham, Eleven Brindleyplace, a different sector of the occupier market has been targeted. The building is being developed with small floor 'plates' of 8,500 square feet, rather than large floor plates of 20–25,000 square feet that exist elsewhere on the Brindleyplace Estate and continue to be developed elsewhere in Birmingham city centre. This could be classed as a change in the strategic plan for the company and has been made after careful analysis of both the market and the letting strategy for Brindleyplace. Half of the office lettings in 2007 were under 10,000 square feet and this trend looks set to continue with 43% of the lettings completed in the third quarter of 2008 also under 10,000 square feet.

Not only is brand-new, grade-A office space being provided for a very active area of the occupier market but Eleven Brindleyplace also provides office space in-between the small space offered at Regus' serviced offices in Three Brindleyplace and the larger floor plates in the other buildings on the estate. This enables tenants to grow, or perhaps even reduce in size within the estate, without having to break the terms of their lease, offering a bold and innovative landlord–tenant arrangement. This strategy was devised before the building was even designed but actually fits well with the current economic climate. Flexibility of lease terms and cost savings generated by maximising office space provide an efficient solution when moving offices. This is partly achieved by reducing the need to provide boardrooms and seminar rooms on each floor with Argent providing serviced meeting rooms on the ground floor of the building.

The recession and subsequent property crash in the early 1990s resulted in the only significant change in the objectives and strategy of Argent. At the time, the organisation was managing a large and diverse commercial property portfolio. Since then, Argent's strategy has changed and gone back to the basics of focusing solely on development and disposing of its investment properties. There is little other evidence to suggest that the organisation has altered its objectives significantly. Argent has approached developments on a long-term basis, creating and sustaining value, acknowledging the lengthy development process from assembling sites, obtaining planning permission and financing development through to building, letting and selling in order to maximise values. By taking a lead role in forming partnerships with local service providers and landowners, it helps deliver lasting benefits for residents, businesses and visitors.

One significant change and challenge that faces Argent is how it copes with rapid expansion whilst maintaining its core values and working practices. Argent was always a small company delivering large projects. Undertaking the King's Cross project has seen Argent increase from 30 employees to over 70 in eighteen months which presents challenges. Until recently the directors were all able to have close working relationships with all of the staff but this becomes increasingly difficult when numbers grow beyond a certain point. However, the joint chief executives are determined to keep the company feeling like a small company and not hierarchical, and the next few years will be an important transitional period.

As a property developer, if Argent becomes too inclined to avoiding risk then it may become too conservative; this may undermine its ability to be dynamic and, as a consequence, potentially less competitive (Millington, 2000, p.228). Argent has stated that they expect to continue to focus on just a few, exceptional, schemes at any one time (Argent, 2007). Operating in such a risky and volatile sector may require a change in strategy at certain points in the coming years. One recent objective, driven by the King's Cross scheme and its mixed-use characteristics, was to develop as much residential as commercial space by the end of the decade (Madelin and Partridge, 2007). Considering the economic climate and the current state of the housing market, particularly city-centre living, this may well be put on hold for a few years. Similarly, if the economy continues to deteriorate, Argent may look to reduce the number of speculative developments carried out at any one time in order to reduce their exposure.

Ultimately, Argent's strategy and decision-making will be opportunity-led. There has been a desire to take forward residential-led projects which have been included in various appraisals of potential schemes. The decision has so far been made not to take the scheme forward in that way for a number of reasons, including the state of the market and doubts as to whether the scheme would be financially viable. In support of this, Millington states that:

> no matter how competent an organisation is and no matter how exhaustive and thorough initial research may be, no one can control the future and eliminate uncertainty and even well thought-out plans can be thwarted when circumstances move against them that they cannot control. (2000, p.246)

This is unlikely to result in a change to the overall strategy for Argent, however. What it may do is contribute to the next stage of its emergent strategy, changing in response to the varying market conditions so that it can emerge successfully and maintain the competitive advantage that it currently enjoys. One example of this may be buying more sites, as the land values drop and competitors struggle to raise the necessary finance.

12.9.6   Innovation and creativity

Argent does not undertake simple projects. We have seen that the focus is on long-term developments, creating destinations and environments that businesses want to be in and the public want to visit. However, this means taking some risks, such as obtaining sites that are perhaps not seen as usual destinations. The inspiring individuals who lead the company, and its employees engaged in managing projects in-house, combine a wide range of skill sets to solve problems and ultimately deliver successful projects. Attention to detail, taking ownership and striving for perfection ensure that the company is getting the product it set out to achieve and continues the high standards established on previous developments.

Argent has been innovative in the way it has gone about raising finance from external sources. A 'revolving loan' from a number of different parties enabled the speculative development of different projects at a time when finance was difficult to secure in the mid-1990s. Finance has also been raised from BTPS, third parties and through joint-venture projects. This illustrates that Argent recognises that some companies have expertise in areas that Argent may not and can ultimately add value to a project, such as the Miller–Argent joint-venture land-reclamation scheme in Fyos-y-Fran.

The approach to partnering with Argent's construction partners aims to create ownership of the project throughout the supply chain with everyone involved working to achieve the same objectives: delivering buildings on time and on budget. Partnering remains not the norm within the construction industry but Argent has tried and tested results with this philosophy and continues to work with companies from its earliest developments through to those currently on site in Birmingham, Manchester and London with measurable financial and quality benefits.

12.9.7   In summary

Argent has operated a long-term strategy for a number of years and has the ultimate aim to be the most successful and respected developer in the UK. There is a wide range of external factors and influences that will affect Argent's success and its strategy for the future. Argent operates in a particularly informal manner and as such does not have a conventional suite of corporate documents identifying its strategy. The company's joint chief executives drive the organisation forward, continuing the values established by the founders. Argent's structure is such that it has the ability to adapt and make important decisions quickly, yet its approach to development has been consistent for a number of years. Market conditions in the coming months and years will no doubt affect Argent and their decision-making. Established management theory will underlie how the company is taken forward and certain strategic decisions may take Argent in a new direction; but the overall values and culture that have created the Argent way of approaching development are unlikely to ever change drastically, albeit they may need to be adapted to cope with the changes to the internal structure.

12.10   Wates Construction

Daniel Weller

12.10.1   Wates Group: an overview

Wates is a family-owned firm that dates back to 1897, a time when Queen Victoria was still the reigning monarch. It is a firm that has been able to demonstrate longevity because of its commitment to carrying out construction work that gives the customer what they want. However, over the period of time since its founding, it has experienced and learned to deal with rapid change and challenges. Today, its turnover is almost £1billion per annum and, should the intended strategy be achieved, this will double by 2012. Currently, the Wates Group consists of five core businesses. This provides a structure that affords the company a specialist approach to meeting the needs of its clients in those targeted sub-markets of the industry in which Wates operates. These five businesses are:

1. Construction;
2. Living space;
3. Retail;
4. Interiors;
5. Developments.

Construction

Wates Construction carries out building works for both public-sector and private clients within the UK. Focusing on targeted markets such as government frameworks, education, custodial, commercial offices and industrial developments, Wates is able to offer services which reflect the market procurement trends, including construction, design and feasibility.

Living space

Wates Living Space carries out new build, refurbishment and regeneration projects throughout England and Scotland, providing affordable housing in partnership with Registered Social Land-lords (RSLs), Local Authorities, and Arms Length Management Organisations (ALMOs).

Retail

Wates Retail works with a wide range of retail customers in the UK, Ireland and the Channel Islands. The Retail business provides a full range of feasibility, design, construction, fit-out and specialist services.

Interiors

Wates Interiors provides specialist fit-out and refurbishment services for office environments, targeting customers in the government, legal, IT, finance and corporate sectors.

Developments

Wates Developments is a land-management business that acquires land or promotes sites on behalf of landowners, and optimises the value through gaining residential and mixed-use planning consent on sites throughout the south of England.

These core businesses provide the backbone of the Wates Group. Importantly, they should not be seen as the sum of its business activity. Wates continuously reviews opportunities within the construction sector, particularly within new and growing markets. A good example of this would be 'Needspace?', a relatively new business, which acquires, renovates and lets underutilised office or industrial buildings in Greater London and the south east.

12.10.2 Leadership within Wates Group

Values

A feature of Wates Group as an organisation is the separation of the ownership and executive functions. The Wates family is engaged in the professional ownership of the firm through Wates

Family Holdings. As a collective, the family is not directly involved in the organisational strategy of the company. Instead, they regard themselves as the custodians of the Wates values: integrity, intelligence, performance, teamwork and respect for people. This structure provides the company with an essence of longevity and a permanence of values which underpin business activities and transcend organisational strategy and business objectives. These are core-instrumental values that govern the manner in which Wates as a company will carry out business affairs in the long term.

Business leadership

> Wates will be the first UK construction company to deliver for customers
> on time and on budget. Every Time. (Wates corporate vision)

The Wates corporate vision can be seen as a targeted level of performance, which will distinguish the company from its competitors. It is long term, but also short term, and is the starting point from where all aspects of company strategy are devised. Central to overall strategy are the organisation's business goals. These goals primarily constitute financial targets for the medium term but also reflect the company values by way of reference to 'supporting the communities and the environment in which we work'. The importance of this is the recognition that simply achieving financial targets at the expense of Wates' values will not constitute success.

This inference is easily justified since financial targets – particularly the levels of profit that the company aims to achieve – are a reflection of the operational strategy of Wates. The company actively embraces close working relationships with its clients. Within such working relationships, trust between the parties is paramount. It is vital that clients of Wates Group understand its core values, are comfortable with them, and have faith that these will be upheld in business operations. In relation to overall strategy, the business goals must be seen as being within a separate strategic sphere to the corporate vision in so far as a failure in, say, achieving turnover targets, would not by default constitute a failure in delivering for the client. Business goals must instead be seen as a reflection of market conditions and an appraisal by Wates' business leaders of what the company can achieve given its strategy as a whole.

In addition to these long-term operational objectives and medium-term business goals, Wates has provided key priorities for the short term which are seen as vital to fulfilling the overall strategy. Five critical business activities have been identified:

- Marketing;
- Operational excellence;
- Corporate responsibility;
- People;
- Supply-chain management.

What will become clear in the following description of these functions is their interdependent nature, their strategic importance in achieving improved business performance, and their criticality in terms of delivering long-term strategic objectives for Wates.

12.10.3   Marketing

Marketing is about establishing how Wates positions itself within the marketplace and, in particular, how it is able to distinguish itself from competitors. Wates' marketing strategy is defined through the corporate vision of the company and is linked very strongly to the preferred business model of partnership and collaboration with its clients. Through forging a close working relationship, Wates strives to understand the needs of its clients and through this understanding to be able to offer a range of services which enhances the client's own needs. This marketing strategy is vital in achieving the stated objective that 90% of turnover should come through repeat business.

12.10.4   Operational excellence

Operational excellence is the de facto core of Wates' vision for the business. Delivering projects on time and to budget is fundamental to operational excellence. This objective also expresses the essential business requirement of delivering not only a quality finished product, but also a quality service to the client. Achieving operational excellence requires joined-up strategy. Without a clearly defined marketing strategy, Wates would be less able to understand what constitutes outstanding service for a particular client. Furthermore, if Wates is unable to achieve the ratio of work envisaged from repeat customers they will not benefit from the valuable intelligence that is derived from customer feedback, especially in how improved performance could be achieved. A good illustration of this point is the implementation of Wates' mandatory policy on handover procedures. This was developed on the basis of client feedback on operational performance.

12.10.5   Corporate responsibility

The positioning of corporate responsibility within Wates' overall strategy is interesting in that it appears to constitute a business goal, when one might expect it to be aligned more with the corporate vision or indeed the core values of the Wates Group. The reason for this could be the increased importance of corporate responsibility in recent times, particularly with regard to the 'emerging' environmental agenda. The way in which Wates is strategically positioned as a partnering contractor clearly adds to the importance placed on corporate responsibility. Crucially, Wates has focused its corporate-responsibility agenda in three areas:

- minimising waste;
- reducing carbon;
- enhancing and developing skills and employment in the community.

The response to the environmental agenda has been to set ambitious, industry-leading targets on reducing waste, with the ultimate aim of eliminating waste sent to landfill. Further, as part of the initiative to improve skills and employment in the community, Wates offers apprenticeships either through themselves directly or through the supply chain. For Wates Group, corporate responsibility is high on the agenda and this can be justified as a clear business need. As a contractor that works with clients to provide buildings which are vital to communities, it is essential to be a responsible partner, who improves the environment and community in which they work.

12.10.6   People

For any organisation a key criterion for success is having people of the right quality to deliver business goals. The human-resource strategy is made up of a number of measures based on recruitment, retention, and identifying and nurturing skills within the company. Establishing a working environment which allows employees to develop their skills and fulfil their potential within the organisation is critical to fostering employee engagement. Furthermore, Wates regularly surveys all employees as a means to better understand the issues that are deemed important.

As a consequence, it can assess where and how these can be acted upon and, potentially, changes implemented. Wates also has an annual 'roadshow' where the business leaders and owners brief employees across the country on the performance of the company as a whole over the past year. This focuses on the performance of the group, the challenges ahead, and how people can contribute to promoting the Wates' values internally and externally.

It is important to Wates that employees can relate to the values of the group. These people will be practising and advocating these values every day through their business activities, through the business community and through communities as a whole. Wates, therefore, encourages employees to incorporate these values through, for example, establishing behavioural competences that reflect the manner in which Wates would like to do business. Importantly, it encourages people to participate in the annual 'work in the community day'. This is a valuable way for people outside the organisation to connect and see Wates as a virtuous organisation. Apart from the excellent publicity this generates, we hope that Wates is seen as a company that is strongly connected to its local community base.

12.10.7   Supply-chain management

The primary function of Wates as a business is to procure construction works from potential clients. As such, a robust strategy for supply-chain management is crucial to the business. The 'tag-line' for Wates' supply-chain strategy is 'closer to fewer', suggesting a close working relationship with a group of preferred suppliers who understand the approach that Wates endeavours to adopt in delivering for its clients.

The model Wates has adopted for its supply chain consists of a first tier of preferred subcontractors, and a second tier of suppliers of key products with whom Wates seeks to establish national trading agreements. A recent initiative, allowing Wates to drive value through the supply chain and to give a clear indication of the working relationship they wish to foster within it, has been the extension of these trading agreements to benefit the first-tier subcontractors.

Working closely with the supply chain is vital to achieving business goals and again links very closely with the other key business areas identified. Collaborative working and early involvement of suppliers provide Wates, and ultimately its clients, with the benefit of the expertise inherent within the supply chain. Another key benefit is that it fosters innovation and efficiency in providing creative solutions for customers.

12.10.8   Overall

What should be apparent from Wates' strategy is that it has been designed around an ethos of a company operating in a way that is both customer-focused and wishes to develop partnering arrangements with its suppliers (and subcontractors). This has occurred in the context of market procurement trends that have seen public bodies adopting partnering as a preferred way of working. The result is that all aspects of the strategy are geared towards collaboration with the clients on the projects and achieving the best result for all concerned.

The benchmark for an organisation's strategy is the way in which it allows reaction to changes in market conditions. Wates has taken the approach of developing close links with customers in order to establish themselves as that customer's contractor of choice. Key to this strategy is ensuring a sufficient diversity and range of customers to enable a constant stream of work as the demand for construction work from various customers fluctuates. Further, where Wates works extensively with public bodies, it is critical that they are able to offer the expertise to deliver for a client whose spend profile can be heavily influenced by the prevalent political mood and tide of economic change.

12.11 Strategic collaborative framework partnerships in Birmingham Urban Design

Ian Davis, Divisional Director

12.11.1 Background to the organisation

Birmingham City Council is the largest local authority in the UK, administering to a population of over one million people. The city council has a significant spend on construction and impact on the construction industry and regional economy. In 2008–9 the council's revenue budget was £3.2billion and capital expenditure was £448m. The council has an extensive public-building portfolio including over 64,000 council houses, over 400 schools, offices, and buildings for leisure facilities and social care.

In 2005 the Office of Government Commerce (OGC) and the Local Government Task Force (LGTF) chose the Birmingham Construction Partnership (BCP) as an exemplary best-practice case study in local government construction procurement. BCP was the first construction collaboration of its kind in the UK and came about from Birmingham City Council's commitment to modernise its procurement processes in order to achieve best-value outcomes in the delivery of services. Fundamentally the objective of BCP was a move away from traditional lowest-cost tendering and, instead, to concentrate on applying a more value-driven and integrated strategy to achieve efficiency, value for money and sustainability in its construction procurement.

12.11.2 Birmingham Urban Design (BUD)

Birmingham City Council's Urban Design (hereafter referred to as BUD) is the internal multidisciplinary consultancy service responsible for construction procurement. BUD was created in 2000, bringing together construction, design and maintenance functions (excluding housing maintenance at that time) under one management team. BUD is a trading organisation with financial targets and provides the city council with the expertise to procure, monitor and operate construction works. To support BUD's 200 or more professional staff, a consultancy framework has also been created to provide additional resources, capacity and expertise.

12.11.3 Internal and external environments

External

The construction industry has traditionally been structured on a perceived status of various professional roles and trades. Consultants maintained an independence whilst contractors competed for work. Specialist supply chains have struggled to provide a skills and knowledge base in a market driven by lower cost and faster delivery. Clients utilised robust forms of contract to safeguard their interests and competitive tenders to drive down costs.

Generally, project teams assembled for the first time have relied on professional and trade practice to coordinate work. This approach has often led to poor performance and adverse relationships between client, consultant and contractor – often with the development of a 'blame culture', associated with expensive disputes over financial claims, delays and quality issues. In the public sector this was frequently associated with time and cost overruns, adversarial relationships between the parties involved and a loss of any continuity of the project teams.

Two reports – *Constructing the Team* by Sir Michael Latham (1994) and *Rethinking Construction* by Sir John Egan (Construction Task Force, 1998) – have fundamentally influenced the approach to construction procurement. To improve the efficiencies of working within the construction industry, they advocated collaboration between clients, contractors, consultants and supporting specialist supply chains. Improvement, it was argued, should be implemented around enhanced performance and delivery in terms of time, cost, quality, client satisfaction, health and safety, and reduced defects. This has been the philosophy developed by BUD.

Internal

After an extensive consultation process with elected members, clients, users and contractors, it was deemed that a Framework Partnership would deliver the city council's needs. This encompassed new-build construction and major refurbishment works inclusive of a substantial Council Housing Decent Homes modernisation programme.

12.11.4   Adding value/benefits

All construction work is procured via 'term contracts' up to a value of £100,000 and via a Framework Partnership above that value. The city council does not operate an approved list of contractors. Term contracts fundamentally provide for emergency day-to-day and small, planned works for building, electrical and mechanical, legionella treatment, domestic lifts, electrical and mechanical testing, and fire alarms. In 2006–7 over 30,000 orders were raised to a value of £20million. The Framework Partnership delivers the city council's new-build and major refurbishment projects inclusive of Council Housing Decent Homes and structural-repairs programme, new schools, new offices, historic buildings, and new leisure and social care facilities.

The development of a Framework Partnership can also add value and benefits by increased local employment, development of supply chains for materials and specialist contractors, sustainability (economic, social and environmental) and generating whole-life cost options for clients to consider. There is also the reduction in procurement costs, the opportunity to maintain project teams' learning and knowledge, and striving for continuity to allow for economies of scale and reduction in costs. The Birmingham Construction Partnership (BCP) comprises Urban Design and three contractor partners: a national contractor, Wates Group, and two regional contractors, Thomas Vale Construction plc and GF Tomlinson Birmingham Limited.

Contractors were appointed in December 2003 and the partnership became operational in April 2004. BCP consists of a five-year partnership, extendable by a further two years. Crucial administrative changes were necessary to enable BUD to trade with external clients when the opportunity arose (and was considered to be in its interest). Initially, the forecast value of work over the life of the partnership was £350,000. It is now forecast to be in the region of £600–700million. Within the contract documentation, there were clauses on the need for the partnership to develop a supply chain to deliver the city council's capital-works programme.

12.11.5   Procurement of the Framework

After extensive consultation with clients and the market, a pre-qualification document requested specific business information, technical expertise, capacity and examples of collaboration. Short-listing was undertaken by BUD managers, a procurement specialist and an auditor. The NEC Option C with partnering principles was adopted as the contract basis. The Framework award was based on an 80 : 20 cost : quality assessment. This involved a robust process of evaluations including detailed written submissions to meet specific criteria, interviews and visits to contractors' offices to substantiate contractors' work bases, claims and resources. The cost assessment was based on cost-plan pricing of specific model projects. In addition, there was a specific whole-life cost and sustainability exercise to assess contractors' knowledge and expertise.

12.11.6   Tendered overheads, profit, and retention

There were originally differences between the tendered overheads and profits submitted by each of the contractors. During the early stages in the development of the Framework, it was agreed by the three contractors to pursue common levels. This allows project-allocation decisions to be based on specific criteria and key performance indicators which are not influenced by a contractor's overhead and profit percentage. The original contract included a retention figure for each specific project. This was withdrawn at an early stage of the Framework as trust and collaborative relationships developed between the parties.

12.11.7   Establishment of an operational manual

The early development of the Framework included the creation of working groups (see list below) with specific tasks set by the Framework Strategic Board to be achieved within a timeframe. These working groups have been critical to the development of working processes, procedures and relationships. Each working group includes experts and representatives from each of the partners and reports quarterly to a Partnership Framework Feedback Panel. The Feedback Panel coordinates tasks, monitors outputs and reports back to the Framework Strategic Board. The outputs and ongoing developments have been included in a Framework Operating Manual. This has been a key element in the success of the Framework in cascading information to all levels of staff within each partner's own organisation.

Original working groups

1. Marketing/PR;
2. Process mapping;
3. Project allocation;
4. Programming;
5. IT systems;
6. Dispute resolution;
7. Open book/target cost and 'pain/gain';
8. Overheads/preliminaries;
9. Technical development costs;
10. Forms of contract;
11. Supply chain;
12. Health and safety;
13. Training;
14. Whole-life cost;
15. Design quality/value management;
16. Key performance indicators;
17. Risk;
18. Customer care;

19. Value for money (VFM);
20. Innovations;
21. Continuous improvement.

All of these report to a Framework Feedback Panel which reports to a Framework Strategic Board. In April 2008, the Framework Working Groups were rationalised to reflect the areas for further development and to improve the efficiencies of the collective Framework resources.

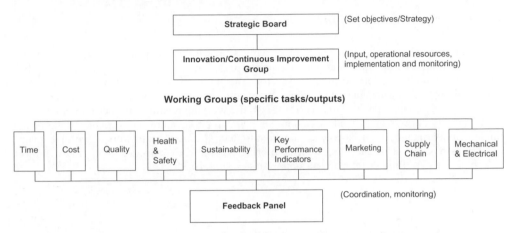

Figure 12.3   Structure of BUD.

12.11.8 *Evolution of a supporting supply chain*

Operating within the constraints of procurement legislation, Birmingham provides the principle of creating job opportunities for local citizens. This was indicated within the Framework documentation along with the need to promote local businesses through appropriate supply chains. The terms of the Framework Agreement determined the need to develop an integrated supply chain of supporting contractors. Initially, each of the contractors' own supply chain was utilised.

The Framework Strategic Board established a specific working group to develop a supporting integrated supply chain. This would be unique in as much as all three contractors would use the supply-chain pool to deliver all of the Framework projects with the remit of utilising other full-range local businesses. At an early stage, specific work categories were identified that would be needed to deliver the city council's major programme of construction works. These work categories were in mechanical, electrical, roofing, landscaping, and aluminium windows and doors; all other categories of work would be delivered via each contractor's own supply chain.

The development of an integrated supply chain began with open, public consultation in November 2004 to establish principles, realise opportunities and strive to understand how added value could be obtained. A thorough and robust three-stage procurement process led to the appointment of 61 companies in May 2006. The Framework supply chain developed now captures:

a)   Some of the contractors from the city council's original approved list;
b)   Some of the main contractors' own supply-chain contractors;
c)   New contractors successful in completing the procurement process.

Generally, all the companies had turnover in excess of £300,000 and identified specific overhead and profit levels with various project-value bands. The supply chain is incentivised to be innovative and provide value engineering to schemes. This is not seen as a cost-cutting exercise during the development of a scheme but a real opportunity to bring forward best-practice solutions to projects which may result in financial savings for the client. The total of supply chain companies has now been extended to 64 to incorporate a waste-management category. This has been necessary to satisfy some of the requirements of the Waste Management Regulations and the city council's sustainability agenda.

12.11.9 Project allocation

Project allocation to supply-chain companies involves a representative panel of all the Framework partners and an auditor. Operating within a strict protocol, packages of work are allocated following the appointment of the main contractor. The initial allocations were to provide work opportunities to all of the appointed supply-chain companies, followed by allocations based on value whilst KPI data is developed. At all stages, close working relationships between the Framework partners allow regular feedback about supply-chain companies' performance. One of the fundamental principles of the supply chain is to introduce specialist contractors into project teams as soon as possible. This allows specialists to bring specific knowledge and skills to the project before designs are fully developed and opportunities are missed.

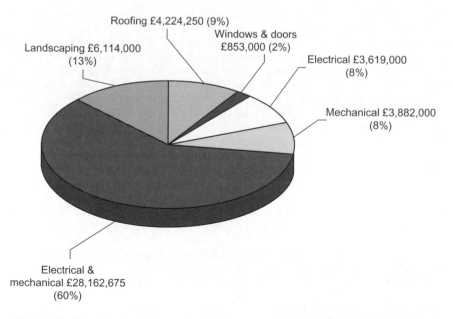

SUPPLY CHAIN PROCUREMENT PROCESS

Supply Chain Allocation by value £46,854,925
(September 2008)

Roofing £4,224,250 (9%)

Windows & doors £853,000 (2%)

Landscaping £6,114,000 (13%)

Electrical £3,619,000 (8%)

Mechanical £3,882,000 (8%)

Electrical & mechanical £28,162,675 (60%)

Figure 12.4   Indication of the breakdown of work to contractors involved in the supply chain.

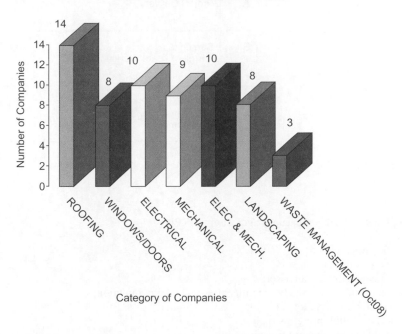

CATEGORY BREAKDOWN

Figure 12.5   Breakdown of categories of contractors engaged in supply-chain work.

12.11.10   *Further development of the supply chain*

It is envisaged that the current supply chain will be regularly reviewed in order to:

1. Develop common levels of overhead and profit across the various project-value bands for each contractor;
2. Rationalise numbers of existing supply-chain contractors;
3. Determine supply-chain contract commitment to the Framework Partnership;
4. Develop other work categories subject to future work programmes.

Future Framework Partnerships may restrict the numbers of supply-chain contractors in each category. This may have the effect of improving work flow, continuity, economies of scale and input in project teams. The main contractors have generally developed good working relationships with the supply chain. However, restricted input into project teams has reduced efficiencies.

12.11.11   *Performance measurement*

The efficiency of the Framework is reviewed utilising headline national Key Performance Indicators (KPIs). Performance is measured on a quarterly basis based on the preceding twelve months' data (not financial year) and compared to data from the same period twelve months previously. Each contractor's performance rating is assessed as a single weighted figure, using performance data from three sources:

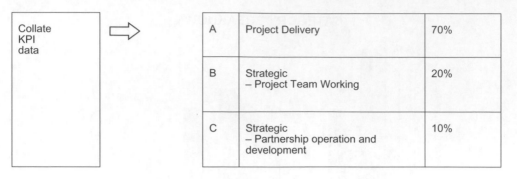

Figure 12.6 KPIs in BUD.

Project delivery KPIs

Each completed project is measured against eight KPIs:

- Cost efficiency;
- Cost predictability;
- Defects assessment at handover;
- Rectification of defects twelve months after practical completion;
- Client satisfaction at practical completion;
- Time predictability (pre-construction);
- Time predictability (post-construction);
- Safety (reportable accidents on site).

Strategic KPIs

Data for strategic performance indicators are obtained from two sources:

(I)   Project-team working KPIs
      These are derived from nine KPIs incorporated in performance questionnaires involving:
      a) Urban Design project lead assessment of client and contractor contribution to project-team effectiveness;
      b) The contractor's assessment of the client and Urban Design's contribution to project-team effectiveness.
(II) Partnership operation and development KPIs

  These are derived from Urban Design's senior management assessment of each contractor partner's effectiveness at delivering key strategic partnership objectives. Five KPIs are incorporated within performance questionnaires and completed at six-monthly intervals.

Work allocation – application of weighted KPIs

Weighted KPI scores are used to determine work allocation to contractors. This involves using the contractor's performance rating derived from the combined weighted scores for project and strategic KPIs, the contractor's current workload (weighted to a percentage of certainty) and payments made to contractors. This is all factored against an indicative work-allocation limit (WAL) calculated for the subsequent twelve-month period. New work is allocated with reference to each contractor's current workload as a percentage of the individual WAL.

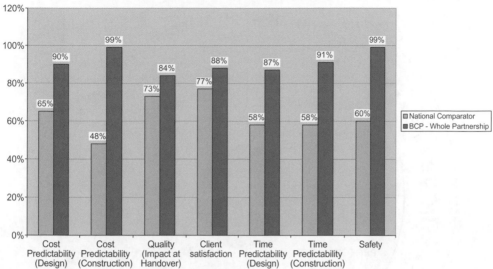

BCP KEY PERFORMANCE INDICATORS - 2007/8

Legend: National Comparator; BCP - Whole Partnership

Category	National Comparator	BCP - Whole Partnership
Cost Predictability (Design)	65%	90%
Cost Predictability (Construction)	48%	99%
Quality (Impact at Handover)	73%	84%
Client satisfaction	77%	88%
Time Predictability (Design)	58%	87%
Time Predictability (Construction)	58%	91%
Safety	60%	99%

Figure 12.7 Performance data in BUD.

12.11.12 Partnering culture

Prior to the start of the Framework, a period of three months was spent in collaborative working groups defining how the partnership would operate; these workshops facilitated the process of relationship building between partners.

A partnering culture can generate significant improvements in performance and the delivery of high-quality construction projects within agreed programme and budget (see Figure 12.8 that shows 'Partnering Culture'). One essential factor is collaborative team working at both strategic and project level.

At strategic level, it means representation from all partners at the highest level and agreeing strategic objectives, monitoring, controlling and striving for continuous improvement against KPIs in the interest of all parties and demonstrating value for clients. At project level, it means carefully establishing a competent project team that will work together, inclusive of client, project manager, contractor partner, designer, consultants and supply-chain specialist. The project team needs to be assembled from the outset and allowed to work together efficiently and effectively to develop the client's brief and successfully deliver the client's outcomes. Apart from an improved end product, this can also include more environmental sustainability features and lower life-cycle costs.

To change from a traditional approach to a partnering culture takes time. Each partner may have to reorganise internally to effectively support individuals and teams involved with partnering. There needs to be a committed leadership to ensure managers and teams 'buy into' the features of partnership and the operational requirements to achieve collaborative working. Creating teams of people to operate strategic- or project-level partnering ideals may exclude some individuals who are unable to adjust to change. Trust, open communication and a willingness to share and develop collectively are key features to achieve collaborative working.

Initially, at the early stages of a partnership, extensive training may be required to embed the culture and features within partner organisations. The partnering relationship and operational

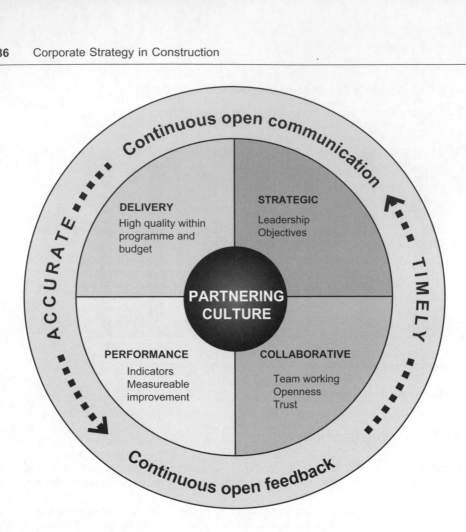

Partnering Culture:
Essential Features at Strategic and Project Team Level

Figure 12.8 The culture of continuous Improvement in BUD.

processes can be further developed by means of workshops or specific working groups to develop initiatives with all partners represented. Ongoing, joint training and innovation from specific working groups are a feature of collaborative working throughout the life of the partnership to improve performance. At all levels, communication and regular feedback will need to be accurate and timely to allow for the appropriate response to achieve successful outcomes.

A feature of BCP are the weekly 'performance board' meetings. BUD's senior managers meet with all senior contractor-partnership managers to discuss projects, operational processes/procedures, partnership initiatives and development, best practice, future workload and other general feedback. In addition, co-location at Urban Design's offices of contractors' management and technical staff means daily interaction is possible to exchange information and respond to project-team matters.

12.11.13   Conclusion – delivering better value

Against the backdrop of the government's Rethinking Construction agenda, BCP adopted the OGC's Achieving Excellence in Construction (AEC) guidelines as a comprehensive and strategic management framework for the implementation and measurement of a best-value approach to construction procurement. The BCP issued a Best-Value Report in May 2008. The purpose of the report was to provide evidence of the added benefits being delivered by the Framework Partners for the city council's capital-works programme and citizens of Birmingham. The report assesses the value and measures the benefits of moving from traditional lowest-cost tendering to partnership working.

The report details the cashable and non-cashable benefits of the partnership in its first three years of operation to be in excess of £100million. The AEC guidelines were used as a means of measuring the value of the partnership, because they:

1. are a standard set of principles for achieving excellence;
2. reflect the performance measures being recorded by the partnership;
3. are recognised throughout the construction industry.

The report identifies a value (cashable and non-cashable benefits) in the order of 30% of the capital spend which reflects favourably with theoretical improvements identified by Sir John Egan in *Accelerating Change* (2002). Eleven AEC guidelines have been used. Performance figures published by the OGC compared to Birmingham Construction Partnership are as follows:

Table 12.1 Achievement of strategic improvement targets by Birmingham Construction Partnership

Strategic Target	National Average	Birmingham Construction Partnership
Rethinking Construction principles	20%	78%
Client satisfaction	79%	81%
Time predictability	60%	69%
Cost predictability	29%	62%
Zero defects	70%	91%

These figures compare very favourably with the principles that Egan and Latham advocated. The report demonstrates that an essential key to success is developing a collaborative working culture within specific performance measurement and to drive for continuous improvement. This gives a basis for other benefits such as local employment, business development, training of young people and sustainability (economic, social and environmental). The report has been recognised by OGC and Constructing Excellence to be at the forefront of the industry in demonstrating the improvement and delivery of construction projects.

12.12   Is it possible for a small quantity-surveying consultancy to survive and thrive in an economic crisis?

Mark Monaghan, Managing Director of Triqs Quantity Surveyors

The purpose of this contribution is to consider the survival of Triqs Quantity Surveyors in the current economic turmoil and downturn in construction development. Triqs Quantity Surveyors is a multidisciplinary quantity-surveying consultancy currently operating within the UK. The author is a director and founder of the business. Triqs Quantity Surveyors began in 1998 as a sole trader with a view to expansion. The business steadily developed to 17 quantity surveyors by 2007. However, this number has recently fallen due to the effects of the recession. Nonetheless, future growth remains a priority. Phil Seymour, a Triqs director for four years, remains positive about the future and deems diligent client selection and meticulous research into prosperous markets necessary analysis for business buoyancy. He further insists that business can continue to excel in torrid times provided that certain markets in the public sector continue to receive government support and funding.

Current business pressures and outside economic issues are inhibiting growth. Cash flow is one aspect which is affecting stability and confidence. The Engineering Contract Group states there is likely to be a sharp rise in contractors reporting that public- and private-sector clients are now delaying payments.

Good management plays a key role in the survival of business in general, though in a recession it is an essential requirement. Simon Brittain, a partner at business psychologists, Kiddy & Partners, identifies aspects of management. He says that managing any business in recession puts focus on:

- Leading and listening;
- Optimisation and innovation;
- Pragmatism and ambition;
- Doing and thinking.

He further believes that managers who succeed in a recession are stable, mature and can balance conflicting pressures. Brittain advises managers to:

- Decide how much capability one has;
- Make the most of what you have and develop what you do not;
- Hang on to managers you need.

Planning and preparation are paramount to survival. Setting objectives and influencing the people involved will aid in changing a business environment. The process of change is an intricate blend of setting objectives, planning the change and managing the people involved, with all three areas involving gathering information and making decisions about what is important. Change will occur long after the actual programme for change has been successfully implemented.

12.13   Strategic management in a micro-organisation

Geoff Badham

When I chose to retire from an administrative management job in a university almost ten years ago, my personal and business strategy was twofold. Firstly, there was a need to enhance income. Secondly, I wanted to do it in such a way as to provide a working environment totally dissimilar to that from which I had just escaped. The nine years within the university had been exciting, productive and enjoyable, but for me, the future now needed to be more challenging and importantly, independent. Other than these objectives, the strategy had no detail. Assisted by a varied career in water treatment and pump technology before I worked in the university sector, this seemed to provide opportunity. Additionally, flexibility of approach appeared very useful and after a number of discussions and explorations the future began to take shape.

As is often the case, the right opportunity presented itself. Very soon after retirement, I was invited to become involved in two unrelated short-term facilities-management-consultancy assignments. Additionally, and importantly, I was invited to plan and launch a new specialist engineering company in the water treatment engineering and service sector. None of this was planned and was an example of how luck plays its part in personal or professional life. So far, these unplanned outcomes of retirement had been most satisfactory from the perspective of income and, of course, the change in working environment. However, short-term assignments do what they imply: they terminate. The important thing was to generate longer-term opportunities. This involved developing a small consultancy business.

Once the business was 'up and running' and developing well, I could devote energy to it. I had formulated the strategy and business plan for this new business in conjunction with two partners. The instigator of the new business and the provider of initial capital was a steam-boiler design, supply and maintenance company. For some time, they had recognised the opportunities to widen their product range into water-treatment equipment and chemistry for steam boilers. My availability, relevant technical and managerial background, together with being a 'known quantity', made me a good fit to create this new business. Another partner was established as a small, water-chemistry company. They expressed willingness to provide chemical supplies and support analytical services when necessary. My own part in this was, as managing director of the new business, to develop sales. I was also responsible for administration and delivery of services in accordance with the agreed business plan. That plan provided for business growth to break even within a two-year timeframe.

12.13.1   *The strategy*

The strategy of the business was essentially to provide water-treatment equipment, chemicals and chemical-analysis services primarily to the steam-boiler market, but with a particular emphasis on using the skills we possessed. A useful agency agreement had been reached with an Italian manufacturer of water-treatment equipment which provided us with good pre- and after-sales technical support on a wide range of competitively priced relevant equipment manufactured to ISO 9000 certified quality standards.

Progress was satisfactory in the first months and sales were generally ahead of budgets. The market response had been positive and sales growth primarily resulted from resolving problems which our competitors had failed to do. The company's blend of skill and experience together with an intense problem-solving focus were showing results and considerable promise. However, all was not well. The chemical-supply partner exhibited capacity problems in terms of chemical-analysis support and product supply. This was adequately managed for most circumstances but to some degree the additional demands drew the chemical-supply partner's attention to their shortcomings and indeed the implied inherent business strategy dictated by their organisational shape and size. Their culture, very admirable in the past, was not able to cope with rapid change and development. After some internal and family reflection, they concluded that they had no wish to expand the company in the way implied by the growth aims of the partnership company.

With the benefit of hindsight, it is certain that our business objectives were not common from the outset, and with this fact now being clear to our partner, it contributed to a rapid demise of the relatively newly-formed company.

At the same time, the steam-boiler company, or at least some influences within it, began to question the business plan in terms of the break-even strategy, agreed by written business plan at inception to be two years. As a background to this, there was an intimated preference for withdrawal of dividends from the main parent company rather than continued investment in the new company.

The stark choice either to break even in the first five months of operation or to close the company was essentially a 'no-brainer' and the company was quickly terminated, with the agreement of all parties concerned, including myself. There is, of course, no point in pursuing objectives which are not common, understood, agreed and supported by organisational, cultural or financial resources. This was a case of a marriage made in haste, or at least without sufficient fundamental thought, and about to be repented at leisure, which happened to be mine, and unpaid at that.

From my point of view, this was not a happy outcome. The first 12 months of retirement had started well, been a roller coaster of totally surprising success, considerable personal effort, followed by huge disappointment. The question now was – what should be my new personal and business strategy? Happily, my relationship with the steam-boiler partners – with one or two individual exceptions – was very sound. No hard feelings, but my view was that a good opportunity for the future had been sacrificed for short-term advantage; regrettable, irritating and inconvenient, but certainly not the first example in human history or indeed my own. But now that the company had folded, in what with hindsight was a beckoning 'eureka moment', why should I not continue with the business on my own individual account?

There was no objection to this from my erstwhile partners, and, in fact, a strong wish for us to cooperate where possible. Their feelings of guilt undoubtedly played a helpful part in our negotiations. Accordingly, I commenced operation as a specialist water-treatment engineer with the specific exclusion of chemicals in the product range, this to both simplify the offering and to be more appropriate to a sole-trader operation. In any case, the chemical-supply partner was clearly relieved at a reduction in demands made of them. Any attempt to continue with this relationship would have been entirely counterproductive. A series of discussions with the Italian equipment manufacturer arranged that the UK agency agreement passed to me, with full agreement of other previously involved parties.

Luck played another part. Gloomily, surveying relationships with potential customers, the prospects were promising, but all outstanding work had been allocated to and completed by the now defunct company. There were no orders to fulfil. The preparations had been made, but the opportunities to supply had not yet arisen. The old wisdom, 'luck is where preparation meets opportunity', was at the forefront of my mind at this time. And from nowhere came a telephone call seeking my urgent help. Some weeks before, I had quoted a very large and prestigious organisation for two large water-softeners. These were very urgently required for confidential safety reasons. Could these items be supplied quickly? Well, yes, of course, provided that there was no objection to a change of supplier's name, although the product would be exactly as specified. The order was duly placed with my new company and passed immediately to the Italian supplier. The required upfront funding was in place (and arose from retained consultancy fees of several months ago) and the equipment was supplied, installed, commissioned and, importantly, invoiced.

The profit arising from that transaction gave me the breathing space to consider strategy anew and to reflect on a business plan. However, that breathing space soon disappeared. My erstwhile steam-boiler partners had passed some of my earlier technical assessments and proposals to one of their local food-factory clients. Essentially, the water treatment provided to the boilers on these premises was totally ineffective. The then current chemical suppliers had not deduced from the symptoms, a dangerously scaling boiler, that the cause was an inadequately sized water-softener for a 24-hour-operation site. My company, myself that is, dealt with this by rebuilding the water-softener to a larger and improved specification to be appropriate for purpose.

This task was technically demanding, physically challenging, practical and as dirty as only a boilerhouse can be. But from my new perspective it was totally enjoyable and to a large degree a new experience. If a change in working environment was what I sought, I had surely found it. There is a considerable difference between a university management role with its attendant political, social and financial elements and the lonely, physical and spanner-wielding task which now engaged me. And a quite specific business strategy began to emerge from that time:

- The business should be service maintenance and problem solving in orientation and flexible in delivery.
- Technically and financially appropriate opportunities should be taken as arising, not necessarily limited by a 'comfort zone' mentality.
- Creative challenges to be both sought and considered as future opportunities, in particular where existing or novel technologies have new application.

- In all things, the business is about excellence, fair dealing and sound competence. We will engage in nothing which knowingly compromises those standards. This translates as paying all suppliers generally before payment is due, and certainly no later. To be meticulous in business relationships – what is promised **must** be delivered. To do the apparently simple things well, we will not expand the business to the point where managerial, financial or regulatory issues disrupt the owner's personal involvement in technical and operational issues on behalf of clients.
- We deal with commercial/industrial companies of good standing only.
- Profit is necessary to fund the future and provide modest reward. We will not engage in any particular project in its absence, and in the event that the business generally falls to a non-profitable level, it will be wound up rather than funded from outside sources.

12.13.2   Conclusion

That general outline, subject always to continuous review, has served the business well. Our end client list, generated from work as a specialist subcontractor to major engineering or building services maintenance companies, is surprisingly heavyweight. New product sales arise as a result of replacement of water-treatment units that are uneconomical to repair. Our products are also specified by an increasing list of consulting, building and mechanical-services engineers for new build and refurbishment of commercial premises. In this regard we supply and erect our systems during the course of construction of the building. This is invariably an entertaining experience and we have learned to be cautious with regard to the potential for destruction of our equipment whilst on site.

Our technical remit, based on proven success in small but interesting projects, is now quite wide. We can offer high-purity water production, ultraviolet disinfection and other technologies within integrated transfer and delivery systems of our own design and supply. We enjoy what we do, mostly. The pension is enhanced by profits. The work environment is considerably changed from my previous careers, but is variable almost daily. Nevertheless, I am often too busy for comfort, and perhaps it is time to consider retirement. Again, it is now almost ten years since retiring from the respectable 'proper job' in a university.

I therefore seek an alternative strategy to meet the original and unchanged objectives, but so far without success.

References

Aaker, D. (1992), *Strategic Market Management*, Wiley, New York.

Andrews, K. (1971), *The Concept of Corporate Strategy*, Irwin, Homewood, Illinois.

Ansoff, H. I. (1965a), *Corporate Strategy*, Penguin, Harmondsworth, London.

Ansoff, H. I. (1965b), *Corporate Strategy: An Analytical Approach to Business Policy for Growth and Expansion*, McGraw-Hill, New York.

Ansoff, H. I. (1968), *Corporate Strategy: An Analytical Approach to Business Policy for Growth and Expansion*, Penguin, Harmondsworth, London.

Ansoff, H. I. (1987), *Corporate Strategy* (revised edition), Penguin, Harmondsworth, London

Ansoff, H. I. and MacDonnell, E. (1990), *Implementing Strategic Management*, Prentice Hall, Englewood Cliffs, NJ.

Argyris, C. (1993), *On Organizational Learning*, Blackwell Business, Cambridge, Massachusetts.

Baden-Fuller, C. and Stopford, J. (1992), *Rejuvenating the Mature Business: The Competitive Challenge*, Routledge, London.

Baker, M. (1992), *Marketing Strategy and Management* (second edition), MacMillan, London.

Ball, M. (1988), *Rebuilding Construction, Economic Change in the British Construction Industry*, Routledge, London.

Balogun, J. and Hope Bailey, V. (2004), *Exploring Strategic Change* (second edition), Prentice Hall, Harlow, Essex.

Barney, J. (1991), 'Firm resources and sustained competitive advantage', *Journal of Management*, vol.17, no.1, pp.99–120.

Beer, M., Spector, B., Lawrence, P. R., Quinn Mills, D. and Walton, R. E. (1984), *Managing Human Assets*, The Free Press, New York.

Bennis, W. and Nanus, B. (1997), *Leaders: Strategies for Taking Charge*, Harper Collins, New York.

Bertelsen, S. (2003), 'Construction as a complex system', *Proceedings of the 11th Annual Meeting of the International Group for Lean Construction*, Blacksburg, Virginia.

Bevan, J. (2006), *Trolley Wars: The Battle of the Supermarkets*, Profile Books, London.

Blythman, J. (2005), *Shopped: The Shocking Power of British Supermarkets*, Harper Perennial, London.

Boyd, D. and Chinyio, E. (2006), *Understanding the Construction Client*, Blackwell Publishing, Oxford.

Bracker, J. (1980), 'The historical development of the strategic management concept', *Academy of Management Review*, 5 (2), pp.219–24.

Briggs, S. and Keogh, W. (1999), 'Integrating human resource strategy and strategic planning to achieve business excellence', *Total Quality Management*, July.

Brown, A. (1995), *Organisational Culture*, Pitman Publishing, London.

Brown, T. (1993), *Understanding BS 5750 and other Quality Systems*, Gower Books, Aldershot.

Brown, S. L. and Eisenhardt, K. M. (1997), 'The art of continuous change: linking complexity theory and time-paced evolution in relentlessly shifting organizations', *Administrative Science Quarterly*, 42, March, pp.1–34.

BS EN ISO 8402BS EN ISO 8402 (1995), *Quality Management and Quality Assurance – vocabulary*, (formerly BS 4778: Part 1, 1987/ISO 8402, 1986), BSI, London.

Bungay, S. and Goold, M. (1991), 'Creating a strategic control system', *Long Range Planning*, June.

Burnes, B. (2004), *Managing Change* (fourth edition), Prentice Hall Education, Harlow, Essex.

Cain, C. T. (2004), *Profitable Partnering for Lean Construction*, Blackwell Publishing, Oxford.

Capon, C. (2008), *Understanding Strategic Management*, Pearson Education, Harlow, Essex.

Carnell, C. A. (2003), *Managing Change in Organizations* (fourth edition), FT/Prentice Hall, Harlow.

Chaharbaghi, K. I. and Lynch, R. (1999a), 'Is the resource-based view a useful perspective for strategic management research?' *Academy of Management Review*, 26, 1, pp.22–40.

Chaharbaghi, K. I. and Lynch, R. (1999b), 'Dynamic strategy development as an endless journey in a fast changing environment', in *Proceedings of the 1999 Association of Global Business Conference*, Las Vegas, US, November.

Chaharbaghi, K. I. and Lynch, R. (1999c), 'Sustainable competitive advantage: towards a dynamic resource-based strategy', *Management Decision*, 37(1), pp.45–50.

Chandler, A. D. (1962), *Strategy and Structure: Chapters in the History of the American Industrial Enterprise*, MIT Press, Cambridge, MA.

Child, J. and Faulkner, D. (1998), *Strategies of Co-operation: Managing Alliances and Networks, and Joint Ventures*, Oxford University Press, Oxford.

Christopher, M., Payne, A. and Ballantyne, D. (1993), *Relationship Marketing – bringing quality, customer service and marketing together*, Butterworth Heinemann, Jordan Hill, Oxford.

Clarke, l. (1981), 'The transition from a feudal to a capitalist mode of building production in Britain', in *The Production of the Built Environment*, Third Bartlett Summer School, London.

Clarke, L. (1994), *The Essence of Change*, Prentice Hall Education, Harlow, Essex.

Clegg, S., Kornberger, M. and Pitsis, T. (2005) *Managing and Organizations: An Introduction to Theory and Practice*, Sage Publications, London.

Clifford, D. K. (1977), 'Thriving in a recession', *Harvard Business Review*, July–August.

Coleman, T. (1965), *The Railway Navvies, A History of the Men Who Made the Railways*, Hutchinson, London.

Collis, D. and Montgomery, C. (1995), 'Competing on resources: strategy in the 1990s', *Harvard Business Review*, July–Aug, pp.119–28.

Connor, K. (1991), 'A historical comparison of resource-based theory and five schools of thought within industrial organisation economics: Do we have a new theory of the firm?' *Journal of Management*, 17(1), pp.121–54.

Construction Task Force (1998), *Rethinking Construction*, DETR (Department of the Environment, Transport and the Regions), London.

Cooney, E. (1955), 'The origins of the Victorian master builders', *Economic History Review*, VIII, pp.167–76.

Cowley, U. (2001), *The Men Who Built Britain, A History of the Irish Navvy*, Wolfhound Press, Dublin.

Crainer, S. (1996), *Key Management Ideas: thinking that changed the management world*, Pitman Publishing, London.

Cyert, R. M. and March, J. G. (1963), *A Behavioural Theory of the Firm*, Prentice Hall, Englewood Cliffs, NJ.

Dale, B. G., Lascelles, D. M. and Boaden, R. J. (1994), 'Levels of total quality management adoption', in B.G. Dale (ed.), *Managing Quality*, Prentice Hall, Hemel Hempstead, pp.117–27.

Davenport, T. H. and Prusack, L. (1998), *Working Knowledge: How Organizations Manage What They Know*, Harvard Business School Press, Boston, MA.

Day, G. S. (1987), *Strategic Market Planning*, West Publishing, St Paul, MN.

Deal, T. and Kennedy, A. (1982), *Corporate Cultures: The Rights and Rituals of Corporate Life*, Addison-Wesley, Reading, MA.

de la Bédoyère, G. (2001), *The Buildings of Roman Britain*, History Press, London.

De Wit, B. and Meyes, R. (1994), *Strategy: Process, Content and Context*, West Publishing, St Paul, MN.

Dixit, A. K. and Nalebuff, B. J. (1991), *Thinking Strategically*, W. W. Norton, New York.

Doyle, P. (1997), *Marketing Management and Strategy* (second edition), Prentice Hall, Hemel Hempstead.

Drucker, P. (1964), *Managing for Results*, William Heinemann, London.

Dunphy, D. D. and Stace, D. A. (1993), 'The strategic management of corporate change', *Human Relations*, 46(8), pp.905–18.

Edvinsson, L. (1997), 'Developing intellectual capital at Skandia', *Long Range Planning*, 30(3), pp.366–73.

Eisenhardt, K. M. (2002), 'Has strategy changed?' *MIT Sloan Management Review*, vol.43, no.2, pp.88–91.

Farkas, C. M. and Wetlaufer, S. (1996), 'The ways chief executives lead', *Harvard Business Review*, May–June.

Floyd, S. W. and Wooldridge, W. (1996), *The Strategic Middle Manager: How to Create and Sustain Competitive Advantage*, Jossey Bass, San Francisco.

Fombrun, C., Tichy, N. M. and Devanna, M. A. (1984), *Strategic Human Resource Management*, John Wiley and Sons, New York.

Fordham, P. and Bauldauf, M. (2008), *Market forecast*, Building Magazine, Issue 17.

French, W. L. and Bell, C. H. (1984), *Organizational Development* (fourth edition), Prentice Hall, Englewood Cliffs, NJ.

Galbraith, J. R. and Kazanjian, R. K. (1986), *Strategy Implementation* (second edition), West Publishing, St Paul, MN.

Garvin, D. (1993), 'Building a learning organization', *Harvard Business Review*, July–Aug.

Ghemawat, P. (1991), *Commitment*, The Free Press, New York.

Gleick, J. (1988), *Chaos*, Penguin, London.

Goldsmith, A. A. (1995) *Making Managers More Effective: Applications of Strategic Management*, University of Massachusetts, Boston, MA.

Goold, M. and Campbell, A. (1987), *Strategies and Styles*, Blackwell, Oxford

Grant, R. M. (2002), *Contemporary Strategy Analysis, Concepts, Techniques and Applications*, Blackwell, Oxford.

Grant, R. M. (2003), 'Strategic planning in a turbulent environment', *Strategic Management Journal*, vol.24, pp.491–517.

Green, S. D. (1999)a Partnering: the propaganda of corporatism? *Journal of Construction Procurement*, 5(2), pp.177–186.

Green, S. D. (1999)b The missing arguments of lean construction, *Construction Management and Economics*, 17(2), pp.133–137.

Green, S. D. (2002) The human resource management implications of lean construction: critical perspectives and conceptual chasms, *Journal of Construction Research*, 3(1), pp.147–166.

Green, S. D., Larsen, G. D. and Kao, C. C. (2008) 'Competitive strategy revisited: contested concepts and dynamic capabilities', in *Construction Management and Economics*, January, 26, pp.63–78.

Groák, S. (1994), 'Is construction an industry?' in *Construction Management and Economics*, 12, pp.287–93.

Guest, D. E. (1987), 'Human resource management and industrial relations', *Journal of Management Studies*, vol.24, no.5.

Guest, D. E. (1989), 'Personnel and HRM: Can you tell the difference?' *Personnel Management*, January.

Guest, D. E. (2000), 'Human resource management, employee well-being and organizational performance', in *Proceedings of CIPD Professional Standards Conference*, University of Warwick.

Haberberg, A. (2000), 'Swatting SWOT', *Strategy*, (Strategic Planning Society), September.

Hamel, G. and Prahalad, C. K. (1994), *Competing for the Future*, Harvard Business School Press, Boston, MA.

Handy, C. (1984), *The Future of Work*, Blackwell, Oxford.

Handy, C. (1993), *Understanding Organizations* (fourth edition), Penguin, Harmondsworth.

Harrigan, K. R. (1980), 'Strategy Formulation in Declining Industries,' *Academy of Management Review*, vol.5, no.4, October, pp.509–604.

Harris, L. C. and Ogbonna, E. (2002), 'The unintended consequences of culture interventions: a study of unexpected outcomes', *British Journal of Management*, vol.13, no.1, pp.31–49.

Harrison, R. (1972), 'How to describe your organization', *Harvard Business Review*, vol.50, May–June, pp.110–28.

Henderson, B. (1989), 'The origin of strategy', *Harvard Business Review*, November–December, pp.139–43.

Hillebrandt, P. M. (1984), *Analysis of the British Construction Industry*, MacMillan Publishers, London.

Hobhouse, H. (1971), *Thomas Cubitt, Master Builder*, Macmillan, London.

Hoopes, D. G., Madsen, T. L. and Walker, G. (2003), 'Special Edition of Strategic Management Journal on Resource Based Value', *Strategic Management Journal*, 24, October.

Huff, A. S., Floyd, S. W., Sherman, H. D. and Terjesen, S. (2009), *Strategic Management , Logic and Action*, John Wiley and Sons, New York.

Inkpen, A. C. (2001), 'Strategic alliances', in Hitt, M. A., Freeman, R. E. and Harrison, J. S. (eds), *Handbook of Strategic Management*, Oxford University Press, Oxford.

Jaques, E. (1952), *The Changing Culture of a Factory*, Dryden Press, New York.

Johnson, G. (1992), 'Managing Strategic Change: culture and action', *Long Range Planning*, 25, pp.28–36.

Johnson, G., Scholes, K. and Whittington, R. (2005), *Exploring Corporate Strategy, Text and Cases* (seventh edition), Prentice Hall Education, Harlow, Essex.

Judd, V. C. (1987), 'Differentiate with the 5th P: People', *Industrial Marketing Management*, no.16, pp. 241–7.

Kanter, R. M. (1989), *When Giants Learn to Dance: Mastering the Challenges of Strategy, Management, and Careers in the 1990s*, Unwin, London.

Kanter, R. M., Stein, B. and Jick, T. (1992), *The Challenge of Organizational Change: How Companies Experience it and Leaders Guide it*, The Free Press, New York.

Kaplan, D. and Norton, R. (1996), *The Balanced Scorecard*, Harvard Business School Press, Boston, MA.

Kay, J. (1993), *Foundations of Corporate Success*, Oxford University Press, Oxford.

King, R. K. (2004), 'Enhancing SWOT Analysis Using TRIZ and the Bipolar Conflict Graph: A Case Study on the Microsoft Corporation', *Proceedings of TRIZCON2004, 6th Annual Conference*, Altshuller Institute, Worcester, Massachusetts.

Kingsford, P. (1973), *Builders and Building Workers*, Edward Arnold, London.

Koskela, L. (2000), *An exploration towards a production theory and its application to construction*, VVT Technical Research Centre of Finland.

Koskela, L. and Howell, G. (2002), 'The underlying theory of project management is obsolete', *Proceedings of the PMI Research Conference*, pp.293–302.

Kotter, J. P. (1996), *Leading Change*, Harvard Business School Press, Boston, MA.

Kroeber, A. L. and Kluckhohn, C. (1952), *Culture: A Critical Review of Concepts and Definitions*, Vintage Books, New York.

Langford, D. and Male, S. (2001), *Strategic Management in Construction*, Blackwell Science, Oxford.

Lascelles, D. M. and Dale, B. G. (1993), *The Road to Quality*, IFS Limited, Bedford.

Latham, M. (1994), *Constructing the Team*, HMSO, London.

Lawrence, F. (2004), *Not on the Label: What Really Goes into the Food on Your Plate*, Penguin, London.

Leonard-Barton, D. (1995), *Wellsprings of Knowledge*, Harvard Business School Press, Boston, MA.

Lenz, R. and Lyles, M. (1985), 'Is your planning becoming too rational?' *Long Range Planning*, August.

Levitt, T. (1960), 'Marketing myopia', *Harvard Business Review*, July–August, pp.45–56.

Levitt, T. (1983), *The Marketing Imagination*, The Free Press, New York.

Lewin, K. (1946), 'Action research and minority problems' in *Resolving Social Conflict* (published in 1948), edited by Lewin, G. W. and Allport, G. W., Harper and Row, London.

Lynch, R. (2006), *Corporate Strategy* (fourth edition), Prentice Hall Education, Harlow, Essex.

Madelin, R. and Partridge, D. (2007) 'Summer Review Day 2007', The Regency Hotel, London, 18 July 2007.

Mansfield, E. (1992), 'The diffusion of industrial innovations – a citation-classic commentary on technical change and the rate of imitation', *Arts and Humanities*, May, 25 (11), pp.14–20.

March, J. G. and Simon, H. A. (1958), *Organizations*, Wiley, New York.

Mbachu, J. I. (2002), 'Modelling Client Needs and Satisfaction in the Built Environment', Unpublished PhD dissertation, University of Port Elizabeth, SA.

McCabe, S. (1998), *Quality Improvement Techniques in Construction*, Addison Wesley Longman, Harlow, Essex.

McCabe, S. (2001), *Benchmarking in Construction*, Blackwell Science, Oxford.

McCabe, S. (2006), 'Valuing people in construction: a historical perspective', in *Construction Information Quarterly*, vol.8, no.2., CIOB, Ascot, pp.63–9.

McCabe, S. (2007), 'Respect for people – the dawn of a new era or mere rhetoric', (eds) Dainty, A., Green, S. and Bagilhole, B., *People and Culture in Construction*, A reader, Taylor and Francis, London, pp.300–15.

McGeorge, D. and Palmer, A. (1997), *Construction Management – new directions*, Blackwell Science, Oxford.

Miles, R. E. and Snow, C. C. (1978), *Organisation Strategy: Structure and Process*, McGraw-Hill, New York.

Miller, E. (1967), *Systems of Organisation*, Tavistock, London.

Millington, A. F. (2000), *Property Development*, Estates Gazette.

Mintzberg, H. (1978), 'Patterns in strategy formation', *Management Science*, 24 (9), pp.934–48.

Mintzberg, H. (1979), *The Structuring of Organisations*, Prentice Hall, New York.

Mintzberg, H. (1987), 'Crafting strategy', *Harvard Business Review*, vol.65, no.4, pp.66–75.

Mintzberg, H. (1994), 'The fall and rise of strategic planning', *Harvard Business Review*, January–February, pp.107–14.

Murray, M. and Langford, D. (eds) (2003), *Construction Reports 1944–98*, Blackwell Science, Oxford.

Nadler, D. A. (1988), 'Concepts for the management of organizational change', in *Readings in the Management of Innovation*, edited by Tushman, M. L. and Moore, W. L., Ballinger, New York.

Nadler, D. A. (1993), 'Concepts for the management of strategic change', in *Managing Change* (second edition), edited by Mabey, C. and Mayon-White, B., The Open University/Paul Chapman Publishing, London.

Nelson, R. R. and Winter, S. G. (1982), *An Evolutionary Theory of Economic Change*, Belknap Press, Cambridge, MA.

Newcombe, R. (1994), 'Procurement path: a proper paradigm', In *East Meets West: Proceedings of CIB W92 Procurement Systems Symposium*, edited by Rowlinson, S. M., University of Hong Kong, pp.245–50.

Nonaka, I. (1990), 'Redundant overlapping organizations: a Japanese approach to managing the innovation process', *California Management Review*, Spring, p.27.

Nonaka, I. (1991), 'The knowledge-creating company', *Harvard Business Review*, November–December.

Nonaka, I. and Takeuchi, H. (1995), *The Knowledge-Creating Company*, Oxford University Press, Oxford.

Oakland, J. S. (1999), *Total Organizational Excellence, achieving world-class performance*, Butterworth-Heinemann, Oxford.

Office for National Statistics (2007), UK Standard Industrial Classification of Economic Activities 2007, Classifications and Harmonisation Unit, Fareham.

Ohmae, K. (1983), *The Mind of the Strategist*, Penguin, Harmondsworth.

Parasuraman, A., Zeithaml, V. A. and Berry, L. L. (1988), 'SERVQUAL: A multiple-item scale for measuring consumer perceptions of service quality', in *Journal of Retailing*, vol.64, no.1, Spring, pp.12–40.

Pascale, R. T. and Athos, A. G. (1981), *The Art of Japanese Management*, Simon and Schuster, New York.

Penrose, E. (1959), *The Theory of the Growth of the Firm*, Basil Blackwell, Oxford.

Peters, T. J. (1993), *The Tom Peters Seminar: Crazy Times Call for Crazy Organizations*, Vintage Books, London.

Peters, T. J. and Waterman, R. H. (1982), *In Search of Excellence, Lessons from America's Best-Run Companies*, Harper Collins, New York.

Pettigrew, A. M. (1973), *The Politics of Organizational Decision-Making*, Tavistock, London.

Pettigrew, A. M. (1987), 'Context and action in the transformation of the firm', *Journal of Management Science*, 24(6), pp.649–70.

Pettigrew, A. M. and Whipp, R. (1991), *Managing Change for Competitive Success*, Blackwell Business, Oxford.

Pettigrew, A. M. and Whipp, R. (1993a), *Managing for Competitive Success*, Blackwell Business, Oxford.

Pettigrew, A. M. and Whipp, R. (1993b), 'Understanding the environment', in *Managing Change* (second edition), edited by Mabey, C. and Mayon-White, B., The Open University/Paul Chapman Publishing, London.

Phelps Brown, E. (1968), *Report of the Committee of Inquiry into Certain Matters concerning Labour in Building and Civil Engineering*, Cmnd 3714, HMSO, London.

Porter, M. (1980), *Competitive Strategy*, The Free Press, New York.

Porter, M. (1985), *Competitive Advantage*, The Free Press, New York.

Porter, L. and Tanner, S. (1996), *Assessing Business Excellence*, Butterworth-Heinemann, Jordan Hill, Oxford.

Postgate, R. (1923), *The Builder's History*, Labour Publishing, London.

Price, R. (1980), *Masters, Unions and Men*, Cambridge University Press, Cambridge.

Pursell, C. (1994), *White Heat – People and Technology*, BBC Books, London.

Pursell, J. (1991), 'The impact of corporate strategy on human resource management', in Storey, J. (ed.), *New Perspectives on Personnel Management*, Routledge, London.

Quinn, J. B. (1980), *Strategies for Change: Logical Incrementalism*, Irwin, Burr Ridge, MN.

Quinn, J. B. (1985), 'Managing innovation: controlled chaos', *Harvard Business Review*, May–June, p.73.

Quinn, J. B. (1991), *The Strategy Process*, Prentice Hall, New York.

Ricardo, D. (1817), *Principles of Political Economy and Taxation*, J. Murray, London.

Rosen, R. (1995), *Strategic Management: an introduction*, Pitman Publishing, London.

Rowlinson, S. M. (1999), 'The selection criteria', in *Procurement Systems: A Guide to Best Practice in Construction*, Rowlinson, S. M. and McDermott, P. (eds), Spon, London, pp.276–9.

Rumelt, R. (1980), 'The evaluation of business strategy', in *Business Policy and Strategic Management* (edited by Glueck, W. F.), McGraw-Hill, New York.

Sadler, P. (1995), *Managing Change*, Kogan Page, London.

Sako, M. (1992), *Price, Quality and Trust: Inter-firm Relations in Britain and Japan*, Cambridge Studies in Management, Cambridge.

Salway, P. (1981), *Roman Britain (Oxford History of Britain)*, Oxford University Press, Oxford.

Schein, E. (1985), *Organisational Culture and Leadership: A Dynamic View*, Jossey Bass, San Francisco, California.

Schein, E. (1990), *Organisational Psychology* (second edition), Prentice Hall, New York.

Schein, E. (1997), *Organisational Culture and Leadership* (second edition), Jossey Bass, San Francisco, California.

Schuler, R. S. and Jackson, S. E. (1987), 'Linking competitive strategies with human resource management practices', *Academy of Management Executive*, no.3, August.

Schumpeter, J. (1934), *The Theory of Economic Development*, Harvard University Press, Harvard, MA.

Schumpeter, J. (1942), *Capitalism, Socialism and Democracy*, Harper and Row, New York.

Semler, R. (1993), *Maverick: The Success Story Behind the World's Most Unusual Workplace*, Century Books, London.

Senge, P. M. (1990), 'The leader's new work: building learning organisations', *Sloan Management Review*, Fall, pp.7–22.

Senge, P. M., Kleiner, A., Roberts, C., Ross, R. B. and Smith, B. J. (1994), *The Fifth Discipline Fieldbook: strategies and tools for building a learning organization*, Nicholas Brealey, London.

Senior, B. (2002), *Organisational Change* (second edition), Pitman, London.

Sims, D., Fineman, S. and Gabriel, Y. (1993), *Organizing and Organizations, An Introduction*, Sage, London.

Sims, A. (2007), *Tescopoly: How One Shop Came Out on Top and Why It Matters*, Constable, London.

Slatter, S. (1984), *Corporate Recovery: Successful Turnaround Strategies and Their Implementation*, Penguin, Harmondsworth.

Sloan, A. P. (1963), *My Years with General Motors*, Sedgewick and Jackson, London.

Smith, A. (1776), *The Wealth of Nations, An Inquiry into the Nature and Causes of the Wealth of Nations*, W. Strahan and T. Cadell, London.

Sommerville, J. and McCarney, M. (2003), 'Strategic objectives of firms with a UK construction paradigm: the impact on the micro-enterprise objectives', *Proceedings of The RICS Foundation Construction and Building Research Conference* (edited by David Proverbs) COBRA (RICS) in conjunction with School of Engineering and the Built Environment, The University of Wolverhampton, 1s^t and 2n^d September.

Spear, S. (2004), 'Learning to lead at Toyota', *Harvard Business Review*, May, pp.1–9.

Stacey, R. D. (1992), *Managing Chaos: Dynamic business strategies in an unpredictable world*, Kogan Page, London.

Stacey, R. D. (1993), *Strategic Management and Organisational Dynamics*, Pearson Education, London.

Strategic Forum for Construction (2002), *Accelerating Change*, Construction Industry Council, London.

Strebel, P. (1992), *Breakpoints*, Harvard Business School Press, Boston, MA.

Sun Tzu (2007), *The Art of War*, (translated by Lionel Giles), BN Publishing, California, USA.

Taylor, F. W. (1911), *The Principles of Scientific Management*, Harper, New York.

Teece, D. J., Pisano, G. and Shuen, A. (1997), 'Dynamic capabilities and strategic management', *Strategic Management Journal*, 18(7), pp.509–33.

Thompson, J. L. (2001), *Strategic Management* (fourth edition), Thompson Learning, London.

Thompson, A. A., Strickland, A. J. and Gamble, J. E. (2008), *Crafting and Executing Strategy, The Quest for Competitive Advantage, Concepts and Cases* (sixteenth edition), McGraw-Hill, New York.

Thompson, J. L. and Martin, F. (2005), *Strategic Management – Awareness and Change* (fifth edition), Thompson Learning, London.

Tichy, N. (1983), *Managing Strategic Change*, Wiley, New York.

Tidd, J., Bessant, J. and Pavitt, K. (2001), *Managing Innovation: Integrating Technological Market and Technological Change* (second edition), Wiley, New York.

Tidd, J., Bessant, J. and Pavitt, K. (2005), *Managing Innovation: Integrating Technological Market and Technological Change* (third edition), Wiley, New York.

Tiles, S. (1963), 'How to evaluate business strategy', *Harvard Business Review*, July–August, pp.111–22.

Torrington, D. P. and Chapman, J. B. (1979), *Personnel Management*, Prentice Hall, Hemel Hempstead.

Torrington, D. P., Hall, L. A. and Taylor, S. (2002), *Human Resource Management* (fifth edition), Pearson Education, Harlow.

Torrington, D. P., Hall, L. A. and Taylor, S. (2005), *Human Resource Management* (sixth edition), Pearson Education, Harlow.

Turner, S. (2002), *Tools for Success: A Manager's Guide*, McGraw-Hill, London.

Tyson, S. (1995), *Human Resource Strategy*, Pitman Publishing, London.

Utterback, J. M. (1996), *Mastering the Dynamics of Innovation*, Harvard Business School Press, Boston, MA.

Van de Vliet, A. (1998), 'Back from the brink', *Management Today*, January.

Vroom, V. H. and Yetton, P. W. (1973), *Leadership and Decision-Making*, University of Pittsburg Press, Pittsburg.

Weick, K. E. (1979), *The Social Psychology of Organising* (second edition), Addison-Wesley, Reading, MA.

Weick, K. E. (2000), 'Emergent change as a universal in organisations', in *Breaking the Code of Change* (edited by Beer, M. and Nohira, N.), Harvard Business School Press, Boston, MA.

Weitzel, W. and Johnson, E. (1989), 'Decline in organizations – a literature extension and integration', *Administrative Science Quarterly*, 34(1), March.

Wernerfelt, B. (1984), 'A resource-based view of the firm', *Strategic Management Journal*, 5(2), pp.171–80.

Wheelan, T. and Hunger, D. (1992), *Strategic Management and Business Policy*, Addison-Wesley, Reading, MA.

Whipp, P. (1992), 'Human resource management, competition and strategy: some productive tensions', in Blyton, P. and Turnbull, P. (eds), *Reassessing Human Resource Management*, Sage, California.

Whittington, R. (1993), *What is Strategy – and does it Matter?* Routledge, London.

Wickham, P. A. (2000), *Financial Times Corporate Strategy Casebook*, Pearson Education, Harlow, Essex.

Williamson, O. E. (1991), 'Strategizing, economizing and economic organization', *Strategic Management Journal*, 12: pp.75–94.

Wilson, I. (1978), 'Scenarios' in *Handbook of Future Research* (edited by Fowles, J.), Greenwood Press, California, pp.225–48.

Wilson, D. C. (1992), *A Strategy of Change*, Routledge, London.

Wolmar, C. (2007), *Fire and Steam: A New History of the Railways in Britain*, Atlantic Books, London.

Womack, J. P. and Jones, D. T. (1996), *Lean Thinking – banish waste and create wealth in your corporation*, Simon and Schuster, New York.

Woodward, J. (1965), *Industrial Organization: Theory and Practice*, Oxford University Press, Oxford.

Further reading

'Live as if you were to die tomorrow. Learn as if you were to live forever',
Mahatma Gandhi (1869–1948), Indian philosopher and advocate of non-
violent protest

The key to understanding strategic management is to gain as great an appreciation as possible of
the factors that influence theoretical development and its application by organisations which, of
course, applies to those operating in construction as much as in any other sector. The list of texts
below is intended to provide material that will assist in achieving this objective; the author has
certainly found them informative and many have been inspirational. Greater appreciation of the
concepts described in the preceding chapters in this book will be assisted by developing a deeper
understanding of context and the provenance of particular issues. Ultimately, there is no end to
what can be read and it is fully accepted that there are many other valuable texts that do not
appear in the list. As a student of construction, you should not assume that any one book (including this one) contains everything you need to know; understanding the world around us is a
never-ending process.

The themes use to categorise the list:

- General strategic texts;
- Texts describing aspects of construction, its development and its importance to contemporary
  society;
- Contemporary analysis of the economic environment;
- Selected construction management research papers.

General strategic texts

Andrews, K. (1998), 'The concept of corporate strategy', in *The Strategy Process*, edited by
Mintzberg, H., Quinn, J. B. and Ghoshal, S. (eds), Prentice Hall, Harlow, Essex.
Barney, J. B. and Hesterley, W. S. (2006), *Strategic Management and Competitive Advantage*,
Pearson Prentice Hall, New Jersey.
Clarke-Hill, C. and Glaister, K. (1995), *Cases in Strategic Management*, Pitman Publishing,
London.
Coulter, M. (2002), *Strategic Management in Action* (second edition), Prentice Hall, New Jersey.
David, F. (2009), *Strategic Management, Concepts*, Prentice Hall International, London.
De Wit, B. and Meyer, R. (1994), *Strategy, Process, Content, Context*, West Publishing, St Paul,
Minneapolis.
Grant, R. M. (2008), *Contemporary Strategic Analysis*, John Wiley and Sons, New Jersey.

Huff, A. S., Floyd, S. W., Sherman, H. D. and Terjesen, S. (2009), *Strategic Management, Logic and Action*, John Wiley and Sons, New Jersey.

Johnson, G., Langley, A., Melin, L. and Whittington, R. (2007), *Strategy as Practice, Research Directions and Resources*, Cambridge University Press, Cambridge.

Miller, A. and Dess, G. G. (1996), *Strategic Management*, McGraw-Hill, New York.

Mintzberg, H., Lampel, J., Quinn, J. B. and Ghoshal, S. (2003), *The Strategy Process, Concepts, Contexts, Cases*, Pearson Eductaion, Harlow, Essex.

Rosen, R. (1995), *Strategic Management, An Introduction*, Pitman Publishing, London.

Wheelan, T. L. and Hunger, J. D. (2002) *Cases in Strategic Management and Business Policy*, Prentice Hall, New Jersey.

White, C. (2004) *Strategic Management*, Palgrave Macmillan, Hampshire.

Williamson, D., Jenkins, W., Cooke, P. and Moreton, K. M. (2004), *Strategic Management and Business Analysis*, Elsevier Butterworth-Heinemann, Oxford.

Since the writing of this book, new versions of two books, which have featured heavily, have been published:

Johnson, G., Scholes, K. and Whittington, R. (2008), *Exploring Corporate Strategy, Text and Cases* (eighth edition), Prentice Hall Education, Harlow, Essex.

Lynch, R. (2009), *Strategic Management* (fifth edition, previously *Corporate Strategy*), Prentice Hall Education, Harlow, Essex.

Notably, there is now a concise edition of the first of these books:

Johnson, G., Scholes, K. and Whittington, R. (2009), *Fundamentals of Strategy*, Prentice Hall Education, Harlow, Essex.

Texts describing aspects of construction, its development and its importance to contemporary society:

Buckley, P. J. and Enderwick, P. (1989), 'Manpower management', in *The Management of Construction Firms, Aspects of Theory* (Hillebrandt, P. M. and Cannon, J. eds), pp.108–127, Macmillan Press, Hampshire.

Charlesworth, G. (1984), *A History of British Motorways*, Thomas Telford, London.

CIOB (Chartered Institute of Building) (1995), *Time for Real Improvement: Learning from Best Practice in Japanese Construction R&D*, Report of the DTI Overseas Science and Technology Expert Mission to Japan, December 1994, CIOB Publications, Ascot.

Coleman, T. (1965), *The Railway Navvies, A History of the Men Who Made the Railways*, Hutchinson, London.

Cooney, E. (1955), 'The origins of the Victorian master builders', *Economic History Review*, VIII, pp.167–76.

Cowley, U. (2001), *The Men Who Built Britain, A History of the Irish Navvy*, Wolfhound Press, Dublin.

Halliday, S. (1999), *The Great Stink of London, Sir Joseph Bazalgette and the Cleansing of the Victorian Metropolis*, Sutton Publishing, Gloucestershire.

Kingsford, P. W. (1973), *Building and Building Workers*, Edward Arnold, London.

Morton, R. and Ross, A. (2007), *Construction UK: Introduction to the Industry*, Wiley-Blackwell, Oxford.

Murray, M. and Langford, D. (eds) (2003), *Construction Reports 1944–98*, Blackwell Science, Oxford.

Pinnington, A. and Edwards, T. (2000), *Introduction to Human Resource Management*, Oxford University Press, Oxford.

Postgate, R. (1923), *The Builder's History*, Labour Publishing, London.
Powell, C. (1996), *The British Building Industry Since 1800: An Economic History*, Routledge, London.
Respect for People Working Group, (2000), *A Commitment to People, Our Biggest Asset (interim report)*, Rethinking Construction Limited, London.
Respect for People Working Group, (2002), *Respect for People – A Framework for Action (final report)*, Rethinking Construction Limited, London.
Tomlinson, R. (2003), *Ricky*, Time Warner Books, London.[*]
Warren, D. (2007), *The Key to my Cell*, Living History Library, Liverpool.[**]
Wellings, F. (2006), *British Housebuilders, History and Analysis*, Blackwell, Oxford.
Wolmar, C. (2004), *The Subterranean Railway, How London Underground was Built and How it Changed the City Forever*, Atlantic Books, London.
Wolmar, C. (2007), *Fire and Steam, A New History of the Railways in Britain*, Atlantic Books, London.
Wood, L. (1979), *A Union to Build, The Story of UCATT*, Lawrence and Wishart, London.

Contemporary analysis of the economic environment:

Blythman, J. (2004), *Shopped: The Shocking Power of British Supermarkets*, Perennial Publishers, London.
Cooper, G. (2008), *The Origins of Financial Crisis: Central Banks, Credit Bubbles and the Efficient Market Fallacy*, Harriman House Publishing, London.
Ferguson, N. (2008), *The Ascent of Money*, Allen Lane, London.
Foster, J. B. and Magdoff, F. (2009), *The Great Financial Crisis: Causes and Consequences*, Monthly Review Press, New York.
Harrison, F. (2007), *Boom Bust, House Prices, Banking and the Depression of 2010*, Shepheard-Walwyn Publishers, London.
Hiraoka, L. S. (2000), *Global Alliances in the Motor Vehicle Industry*, Westport, CT, USA.
Klein, N. (2001), *No Logo*, Flamingo, London.
Krugman, P. (2008), *The Return of Depression Economics and the Crisis of 2008*, Allen Lane, London.
Lawrence, F. (2004), *Not on the Label: What Really Goes into the Food on Your Plate*, Penguin, London.
Marr, A. (2007), *A History of Modern Britain*, Pan Books, London.
Peston, R. (2008), *Who Runs Britain, and Who's to Blame for the Economic Mess We're in?* Hodder and Stoughton, London.
Reid, A. J. (2004), *United We Stand, A History of Britain's Trade Unions*, Penguin Books, London.
Sampson, A. (2005), *Who Runs this Place? The Anatomy of Britain in the 21st Century*, John Murray Publishers, London.
Taleb, N. N. (2008), *The Black Swan, The Impact of the Highly Improbable*, Penguin, London.

Selected construction management research papers:

Barrett, P. (2007), 'Revaluing Construction: a holistic model', *Building Research and Information*, 35(3), pp.268–86.

---

[*]The autobiography of the actor, Ricky Tomlinson, describes his involvement in the 1972 building strike that led to, he argues, his (and Dessie Warren's – see dedication) wrongful conviction and subsequent imprisonment for conspiracy to cause a riot in Shewsbury.

[**]The autobiography of Desmond (Dessie) Warren who was jailed with Ricky Tomlinson for his alleged part in the 1972 building strike. This is a re-publication of the book originally published in 1982.

Bergstr m, M. and Stehn, L. (2005), 'Benefits and disadvantages of ERP in industrialised timber frame housing in Sweden', *Construction Management and Economics*, (October) 23, pp.831–8.

Best, R. and Langston, C. (2006), Evaluation of construction contractor performance: a critical analysis of some recent research, *Construction Management and Economics*, (April) 24, pp.439–45.

Bresnen, M. and Marshall, N. (2000), 'Building partnerships: case studies of client–contractor collaboration in the UK construction industry', *Construction Management and Economics*, 18, pp.819–32.

Bresnen, M. and Marshall, N. (2000), 'Motivation, commitment and the use of incentives in partnerships and alliances', *Construction Management and Economics*, 18, pp.587–98.

Bresnen, M. and Marshall, N. (2000), 'Partnering in construction: a critical review of issues, problems and dilemmas', *Construction Management and Economics*, 18, pp. 229–37.

Briscoe, G. H., Dainty, A. R. J., Millet, S. J. and Neale, R. H. (2004), 'Client-led strategies for construction supply chain improvement', *Construction Management and Economics*, (February) 22, pp.193–201.

Bröchner, J., Josephson, P. E. and Kadefors, A. (2002), 'Swedish construction culture, quality management and collaborative practice', *Building Research & Information*, 30(6), pp.392–400.

Drucker, J., White, G., Hegewisch, A. and Mayne, L. (1996), 'Between hard and soft HRM: human resource management in the construction industry', *Construction Management and Economics*, 14, pp.405–16.

Flynn, B. B., Schroeder, R. G. and Flynn, E. J. (1999), 'World class manufacturing: an investigation of Hayes and Wheelwright's foundation', *Journal of Operations Management*, 17, pp.249–69.

Fraser, C. and Zarkada-Fraser, A. (2001), 'The philosophy, structure and objectives of research and development in Japan', *Construction Management and Economics*, 19, pp.831–40.

Gann, D. M. (1996), 'Construction as a manufacturing process? Similarities and differences between industrialized housing and car production in Japan', *Construction Management and Economics*, 14, pp.437–50.

Garnett, N. and Pickrell, S. (2000), 'Benchmarking for construction: theory and practice', *Construction Management and Economics*, 18, pp.55–63.

Green, S. D., Larsen, G. D. and Kao, C. C. (2008), 'Competitive strategy revisited: contested concepts and dynamic capabilities', *Construction Management and Economics*, (January) 26, pp.63–78.

Iwashita, S. I. (2001), 'Custom made housing in Japan and the growth of the super subcontractor', *Construction Management and Economics*, 19, pp.295–300.

Kale, S. and Arditi, D. (2006),'Diffusion of ISO 9000 certification in the precast concrete industry', *Construction Management and Economics*, (May) 24, pp.485–95.

Landin, A. and Nilsson, C. H. (2001), 'Do quality systems really make a difference?' *Building Research & Information*, 29 (1), pp.12–20.

Mason, J. M. (2007), 'The views and experiences of specialist contractors on partnering in the UK', *Construction Management and Economics*, (May) 25, pp.519–27.

Matsumura, S. (2001), 'Perspectives on component-based design in Japanese construction', *Construction Management and Economics*, 19, pp.317–9.

McCabe, S. (2006). 'Valuing people in construction: a historical perspective', in *Construction Information Quarterly*, vol.8, no.2., CIOB, Ascot, pp.63–9.

McCabe, S. (2007), 'Respect for people – the dawn of a new era or mere rhetoric?' in *People and Culture in Construction – A reader*, edited by A. Dainty, S. Green, and B. Bagilhole, E. and F.N. Spon, London, pp.300–15.

McCabe, S., Rooke, J., Seymour, D. and Brown, P. (1998), 'Quality managers, authority and leadership', *Construction Management and Economics*, 16, pp.447–57.

Neto, J. de P. B. (2002), 'The relationship between strategy and lean construction', in *In the Proceedings of the Tenth International Group for Lean Construction (IGLC)*, August, Gramado, Brazil.

Neves, J. C. and Bugalho, A. (2008), 'Coordination and control in emerging international construction firms', *Construction Management and Economics*, (January), 26, pp.3–13.

New, C. (1991), 'World class manufacturing versus strategic trade-offs', *Sixth International Conference of the Operations Management Association of the UK*, University of Aston, June.

Ofori, G. (2003), 'Frameworks for analysing international construction', *Construction Management and Economics*, (June) 21, pp.379–91.

Peansupap, V. and Walker, D. H. (2006), 'Innovation diffusion at the implementation stage of a construction project: a case study of information communication technology', *Construction Management and Economics*, (March) 24, pp.321–32.

Raiden, A. B., Dainty, A. R. J. and Neale, R. H. (2006), 'Balancing employee needs, project requirements and organisational priorities in team deployment', *Construction Management and Economics*, (August) 24, pp.883–95.

Reichstien, T., Salter, A. J. and Gann, D. M. (2005), 'Last among equals: a comparison of innovation in construction, services and manufacturing in the UK', *Construction Management and Economics*, (July) 23, pp.631–44.

Rooke, J., Seymour, D. and Fellows, R. (2003), 'The claims culture: a taxonomy of attitudes in the industry', *Construction Management and Economics*, 21, pp.167–74.

Salter, A. and Torbett, R. (2003), 'Innovation and performance in engineering design', *Construction Management and Economics*, (September) 21, pp.573–80.

Seaden, G., Guolla, M., Doutriaux, J. and Nash, J. (2003), 'Strategic decisions and innovation in construction firms', *Construction Management and Economics*, (September) 21, pp.603–12.

Stewart, R. A. and Spenser, C. A. (2006), 'Six-sigma as a strategy for process improvement on construction projects: a case study', *Construction Management and Economics*, (April) 24, pp.339–48.

Tombesi, P. (2006), 'Good thinking and poor value: on the socialization of knowledge in construction', *Building Research and Information*, 34(3), pp.272–86.

Whitla, P., Walters, P. and Davies, H. (2006), 'The use of global strategies by British construction firms', *Construction Management and Economics*, (September) 24, pp.945–54.

Winch, G. and Carr, B. (2001), 'Benchmarking on-site productivity in France and the UK: a CALIBRE approach', *Construction Management and Economics*, 19, pp.577–90.

Wood, G. D. and Ellis, R. C. T. (2005), 'Main contractor experiences of partnering relationships on UK construction projects', *Construction Management and Economics*, (March) 23, pp.317–25.

Yashiro, T. (2001), 'A Japanese perspective on the decline of robust technologies and changing technological paradigms in housing construction: issues for construction management research', *Construction Management and Economics*, (2001) 19, pp.301–6.

Glossary

added value the amount that a product or service is increased by an organisation between its input and output

architecture the way an organisation is structured so as to facilitate relationships and networks that support its strategy

backward integration the process of attempting to gain control (or influence) of those who supply goods and/or services

benchmarking a technique by which organisations identify areas for improvement and compare their practices with those adopted by proven performers who have demonstrated their excellence by, for example, the use of the EFQM Excellence Model

branding the ability to generate a strong perception of the attributes of a product and/or service in the mind of customers/consumers. For instance, some brands are strongly identified as being associated with quality

break-even the point at which the revenue received is equal to the costs of production

business-level strategy the way that an organisation competes in a particular market (or environment particularly if the intention is not to make profit)

business model the particular approach that an organisation uses to attain its strategic goals (objectives)

business process re-engineering an approach to organisational design intended to reconsider the way that processes (tasks) are carried out to achieve operational objectives which, in turn, will lead to attainment of strategic goals. It is often associated with reorganisation intended to result in greater efficiency

'cash cow' a product or service with high relative market share (compared to a range or portfolio of other goods or services)

change agent a person, who may or may not be a manager, who actively supports proposed organisational or strategic change and will facilitate its implementation. This may require them to convince those who are fearful or resistant

change options matrix a method of linking the way that people are deployed through HRM (see below) to achieve strategic change in three areas: tasks carried out; culture (see below); and the political environment that exists in the organisation

changeability of the environment the amount to which the environment (see below) is likely to change. This will require testing of assumptions and garnering of data on the likelihood of change occurring in, for example, customer behaviour (which could be influenced by trends in demography)

competences the skills and/or abilities which are possessed by an organisation and which it can deploy in order to carry out tasks in pursuit of strategic objectives

competitive advantage the effective 'head start' that an organisation has over its competitors which means that it enjoys the perception that it offers something better in terms of value or features

305

competitor profiling the analysis of what other competitors do and how they achieve their strategy; most particularly if they are successful and have advantage; the use of benchmarking (see above) may be a useful tool

complementor if an organisation provides a product or service that adds value to another product or service (which may be provided by another organisation), it is seen to be complementary. For example, a housebuilder might use only certain types of fittings (in kitchens or bathrooms) which make its product more attractive to potential purchasers

consumers those who are provided with a product or service but may not have paid directly for the privilege. For instance, those who 'consume' the health service may not have paid tax or national insurance

cooperation the willingness of organisations to join together to develop increased ability to provide products or services. This may be through what are known as 'alliances' or 'partnerships' (see cooperative strategy below)

cooperative strategy in which organisations are willing to develop a shared approach to their desire to achieve goals. There should be a mutual benefit for all concerned

core areas of strategic management the key areas in which an organisation is able to compete

core competences the skills and/or attributes that are central to the organisation's ability to deliver strategy and maintain competitive advantage. In effect, these may be based on the organisational structure, its possession of technology (especially if innovative and difficult to imitate) and the knowledge that people have about the tasks they carry out

core resources the essential parts of the organisation that are used to deliver strategy: equipment, materials, people, knowledge

corporate-level strategy the overall purpose that the organisation seeks to pursue

corporate social responsibility the various ways in which an organisation operates to ensure that it conducts its affairs with emphasis on ethical behaviour and with consideration of its stakeholders and the environment

critical success factors (CSFs) the aspects of performance that are considered to be essential to the way in which customers derive benefit (added value). Therefore, CSFs should be explicit to everyone involved in day-to-day tasks that will lead to delivery of strategic objectives

cultural orientation the beliefs that are routinely held by the members of the organisation. This can be strongly influenced by those in leadership

culture the norms, beliefs and symbols that provide a way of identifying the essence of the organisation, frequently referred to as 'organisational culture'

customers those who purchase the results of production (can be both services and products). There is a differentiation between what may be considered the 'end customer' (which we all tend to be) or 'internal customers' who purchase partially completed goods, services or components that will be used in the assembly of the final product

customer–competitor matrix used to consider the needs of potential customers and analyses how other organisations deliver value intended to attract their custom

customer-driven strategy how an organisation ensures that all operations are carried out in a way that deliberately focuses on the needs of the customer and, most particularly, gives added value

customer profiling the analysis to identify potential customers and their specific needs

differentiation the way that the products or services of an organisation are produced or delivered so as to emphasise their uniqueness when compared to those offered by competitors. Differentiation may be based on price (being cheaper) or by having additional features built in or available. Higher levels of customer service may be the way that an organisation differentiates itself

diversification strategy when an organisation moves away from 'core business' into new markets to develop a wider portfolio of goods and/or services and, it is believed, spreads the risk by avoiding dependence on only one 'line'

division a part of an organisation (sometimes referred to as a 'department'); in a large organisation there are many divisions

'dogs' the products or services that are considered to have low market share and very limited potential for development

double-loop learning the procedure by which the organisational members are encouraged to question the value of standard, accepted methods of carrying out tasks (people who are taught to obey procedures engage in 'single-loop' learning). The intention is that people who do this will find ways of improving processes and raising efficiency

EFQM Excellence the model which contains nine criteria designed to assess an organisation's commitment to the achievement of improvement leading to the eventual accolade of being considered as 'world class' (see below)

economic rent the amount that a resource earns over the minimum required to ensure that it continues in its present use

economies of scale the reduction of unit cost resulting from production at higher levels of output

emergent change in which strategy is altered in accordance with the way that the environment shifts

emergent strategic management in which the overall strategy is developed as the circumstances alter; there is no predetermined plan and resources are utilised in as flexible a way as possible so as to remain dynamic

empowerment the devolvement of authority to take decisions to the people who are closer to customers and who can be trusted to do what is best

environment that which creates the particular circumstances in which the organisation must operate; a market which is based on society, economy, technology and other key influences (see also PESTEL below)

'fit' the way that all that is carried out in the organisation is to a standard that ensures consistency, coherence and congruence

focus strategy in which the organisation develops and implements an approach to achieving its objectives that is based on targeting one segment of a market or what is known as a 'niche'

formal organisational structure the way that configuration of departments and people is achieved which is based on the belief that tasks, responsibilities and relationships occur in a 'hierarchy'

forward integration in which an organisation seeks greater control over its outputs by input or acquisition over distribution or retailing

functional organisational structure in which all activities are based on key functions such as production, HR, R and D, and marketing; there will usually be little integration

game-based theories of strategy where the organisation bases its decisions on testing of games (simulations based upon assumed data and circumstances) to ascertain likely outcomes

generic strategies are threefold: cost leadership, differentiation and niche (focus)

hierarchy of resources the way that resources are used at different levels within the organisation (strategic at the top and operational at the bottom)

horizontal integration where an organisation seeks to acquire or develop greater control over others whose activities are complementary (this could include the takeover of competitors)

human resource audit requires the organisation to examine its approach to HRM (see below), and any changes that will increase the likelihood of success are implemented

HRM (Human Resource Management) the function that is dedicated to utilising the 'human element' in a way that ensures greatest effectiveness and efficiency. As such, the organisation will develop its strategy based on people and incorporating increased use of training and education to facilitate improvement

informal organisational structures the way that resources(especially people) are configured on the basis of carrying out tasks which recognises informality and flexibility

innovation the generation and exploitation of new ideas and technology which will be used to develop either new products and/or services or to enhance the features of those that exist

innovation and knowledge-based theories in which the organisation develops an approach to strategy that emphasises a culture of continuous enquiry and learning that ensures innovation

innovative capability the ability of people within an organisation to constantly create new ideas

Japanese approach to management in which the emphasis is placed on continuous improvement of all processes and in which every person is encouraged to be actively involved. This was the basis of success that was demonstrated by the producers of cars and electronics who implemented the advice of quality gurus, Dr Deming and Dr Juran, in the aftermath of the Second World War

kaizen the Japanese word that means continuous improvement of all processes that is intended to result in superior levels of customer satisfaction and, as a consequence, will be a strategic weapon that will give competitive advantage (see Japanese approach to management – above, and TQM – below)

knowledge the accumulated mixture of experience, insights and understanding of processes that an organisation has achieved and which gives it particular ability (capability) to implement particular strategy

knowledge management the way that an organisation is able to muster knowledge and ensure that it is applied in a way that maximises the likelihood of achieving success and competitive advantage

learning organization where the emphasis is explicitly on the desire to constantly search for new opportunities, innovation and potential for carrying out processes and use of technology differently. There is a recognition that constant experimentation is desirable (even though mistakes and errors will result)

life cycle in which there is an assumed evolution of a product or service through various stages of introduction, growth, maturity and decline. An organisation should have a range of strategies to deal with each stage

low-cost strategy in which the emphasis is on developing a strategy which delivers its products and/or services at a lower cost than all other competitors

market development where existing markets are increased by enhanced strategy based on appropriate logic

market options matrix which explores the products and/or services that an organisation provides and considers its options for development or withdrawal

market penetration in which market share is increased by a successful strategy

market segmentation in which the overall market is divided into groups of potential customers to ensure that strategy is more appropriately focused

merger an agreement between two organisations to combine into one and to develop an agreed joint strategy

mission the broad overall direction that the organisation wishes to pursue and which contains the major objectives

mission statement in which the mission is articulated to outsiders such as key stakeholders (such as suppliers), potential investors and customers

network-based strategy which considers ways in which cooperation between organisations or industries can lead to synergy and greater chances of strategic success

niche market the focus on a particular group of customers or consumers in such a way as to develop strategy that is appropriate to their needs

'no frills' a strategy that is intended to give the customer no more than they really want at lowest possible cost

objectives the particular goals that an organisation will pursue in its strategy and which are usually contained in the mission

operational strategy the organisation's development of an approach that ensures all resources used in carrying out day-to-day tasks are directed towards pursuance of the overall business and corporate-level strategy

organization the social entity that collectively provides the particular combination of ability and expertise that will enable its strategy to be implemented. As such, the importance of the human input is key to achievement of success

organisational capability the skills, ability, resources and particular management that an organisation possesses and which allows it to do the things that make its strategy successful (or not!)

paradigm the set of assumptions and beliefs that provide the accepted model of thinking (mental model) in an organisation

PESTEL the acronym of Political, Economic, Social, Technological, Environmental and Legal, which should all be considered by any organisation in terms of the environment in which it operates (competes)

portfolio matrix which explores the range of products and/or services that an organisation provides and which are considered against two criteria: relative market share and market growth

power the magnitude of authority that the organisational leaders have and which is used in making decisions, developing strategy and its subsequent implementation

prescriptive change the method of creating change based on definite intentions and deliberate plans

prescriptive strategic management the intention to implement strategy based on a set of assumptions that result in the belief that a deliberate plan is appropriate (with requisite resources dedicated)

'problem child' the products and/or services that have low market share but which are in a market with potential for high growth

processes the way that inputs are turned into outputs in an organisation

producer those who procure inputs (materials, labour, machinery, plant or equipment) and create outputs

profitability the measure of profits divided by the capital employed in a business. It is important to know whether profitability is gross (before the deduction of expense, tax and any dividend payable) or net after which all expenses have been deducted. Reference is also made to the 'ROCE' ration (see below)

reputation by which an organisation develops a perception amongst customers (existing and potential) of its ability to deliver products or services with particular features or to certain levels of quality

resource allocation the dedicating of resources to achieve a particular strategy

resource-based view which emphasises the importance of the resources that an organisation has in achieving competitive advantage. It tends to be associated with prescriptive strategy (see above)

retained profit the money that a business retains after all expenses have been deducted, tax paid and dividends paid on shares

return on capital employed (ROCE) a key financial ratio that measures the amount earned compared to investment

Ricardian rent the economic rent that comes from the use of a resource that allows the organisation to develop competitive advantage and attain high returns

scenarios models of the future that allow the organisation to test assumptions and monitor the effects. These can utilise simulations based on, for instance, computer models of market conditions

Schumpeterian rent the return which can be achieved by an organisation because it is able to develop an innovative or novel solution for which customers are willing to pay a premium to obtain

single-loop learning in which people are taught how to comply with organisational instructions or procedures (usually by some process of checking, normally referred to as 'auditing'); people are not encouraged to ask questions or to seek alternative ways of carrying out tasks

stakeholder those who have a vested interest in the organisation (suppliers and employees) and who may want to influence strategy to operate in their best interests (or certainly not put them at risk)

'star' those products or services that have high relative market share and are in a high-growth market

strategic alliances which are cooperative and based on the desire of two or more organisations to combine their strategies and resources for mutual benefit

strategic business unit the parts of the organisation that are responsible for implementation of a particular strategy (usually within a large entity)

strategic capability the ability of the organisation to deliver strategy because of its possession of resources and skills, expertise or ability

strategic change in which the organisation intends to alter its strategy to pursue objectives (contained in the mission) by moving that which currently exists. This can be achieved by either prescriptive or emergent means

strategic leadership the ability of those within the organisation to influence the strategy by their inspiration, beliefs and personal commitment

strategic management the process by which the custodians of an organisation identify what its purpose is, the environmental conditions that exist, the plans that are available and the decisions that must be taken about which strategy is most appropriate

strategic planning the method by which formal methods are used to decide on resources and how they should be implemented to achieve strategic objectives

strategic thinking the cognitive processes that are conducted by those who will carry out strategic management (see above)

suppliers those who supply goods or services to an organisation that will use them to engage in value-adding

sustainable competitive advantage the competitive advantage that can be maintained over a long period of time, probably because competitors cannot easily imitate

SWOT the acronym of Strengths, Weaknesses, Opportunities and Threats

synergy the belief that combining two parts will result in a sum that is greater; usually applied to groups of people (or organisations) who come together

TQM (Total Quality Management) the assumptions and paradigm based on the belief that continuous improvement is possible and which will result in higher levels of customer satisfaction. It also involves the desire to continuously measure processes, engage people throughout the value chain (see below) and emphasise their importance. The total is essential as TQM applies to every part of the process from beginning (including obtaining raw materials) through to eventual distribution and delivery to the end customer

value chain which identifies where value is created throughout all of the processes that are utilised within an organisation to create output. Careful analysis of the value chain that exists will enable an organisation to seek improvement

value network (system) the greater set of inter-relationships and organisational links that exist within (and without) an industry. By developing these links and reducing waste and increasing efficiency, every organisation will be assumed to be able to improve and the industry as a whole will be collectively enhanced in terms of ability and achievement

vertical integration this can occur if the organisation wishes to exert control over its suppliers or, alternatively, the distributors or retailers of its goods or services. As such, the former is 'backward' and the latter 'forward'

vision the beliefs that usually emanate from an organisational leader and create the sense of belief and aspiration about future strategy

world class the state at which an organisation is perceived to be by demonstrating its ability to perform at an extremely high level and, most especially, by delivering customer service and satisfaction (although some argue that this word is too passive and, instead, that delight should be used)

Index